# New Materials, Processing and Manufacturability

**Scrivener Publishing**
100 Cummings Center, Suite 541J
Beverly, MA 01915-6106

*Publishers at Scrivener*
Martin Scrivener (martin@scrivenerpublishing.com)
Phillip Carmical (pcarmical@scrivenerpublishing.com)

# New Materials, Processing and Manufacturability

## Fabrication and Processing of Advanced Materials


Edited by

**R. Thanigaivelan**

*AKT Memorial College of Engineering and Technology, Kallakurichi, Tamilnadu, India*

**Pradeep Kumar Krishnan**

*Dept. of Mechanical and Industrial Engineering, National University of Science and Technology, Muscat, Sultanate of Oman*

**Kamalakanta Muduli**

*Dept. of Mechanical Engineering, Papua New Guinea University of Technology, Lae, Papua New Guinea*

and

**Santosh Kumar Tamang**

*Dept. of Mechanical Engineering, North Eastern Regional Institute of Science & Technology (Deemed to be University), Nirjuli, Arunachal Pradesh, India*



Scrivener Publishing

WILEY

This edition first published 2024 by John Wiley & Sons, Inc., 111 River Street, Hoboken, NJ 07030, USA and Scrivener Publishing LLC, 100 Cummings Center, Suite 541J, Beverly, MA 01915, USA

For more information about Scrivener publications please visit www.scrivenerpublishing.com.



**Wiley Global Headquarters**
111 River Street, Hoboken, NJ 07030, USA

For details of our global editorial offices, customer services, and more information about Wiley products visit us at www.wiley.com.



***Library of Congress Cataloging-in-Publication Data***

ISBN 978-1-394-21254-5

Front cover images supplied by Wikimedia Commons
Cover design by Russell Richardson

Set in size of 11pt and Minion Pro by Manila Typesetting Company, Makati, Philippines

Printed in the USA

10 9 8 7 6 5 4 3 2 1

# Contents

# Preface

The book focuses on the wide areas of manufacturing, including cutting-edge material processing technologies, custom-made materials, metallic and non-metallic materials, new engineering experiments, contemporary machining, joining, surface modification, and process optimization techniques.

Chapter 1 concentrates on the variety of aluminum grades and their applications. Chapter 2 provides a detailed overview of cold spray additive manufacturing, including its history, principles, applications, and recent advancements. Chapter 3 discusses the Gas Metal Arc Welding (GMAW) that is utilized to clad low-carbon steel plates (EN30). The cladding parameters considered were voltage (V), wire feed rate (WFR), travel speed (S), and gas flow rate (G). The resulting welds were evaluated by measuring bead profile parameters, including bead width, bead height, depth of penetration, and dilution.

Chapter 4 presents an innovative approach to the design and development of a brake pedal using topology optimization techniques. Chapter 5 provides a comprehensive analysis of influencing factors in resistance-spot welding weld quality and the role of modeling and optimization in improving welding quality. Chapter 6 discusses the shot peening influence on the corrosive behavior of SiC Particle Reinforced Aluminum Composite.

Chapter 7 highlights the importance of factors including material selection, design considerations, and welding quality in affecting the component strength of 3D printed components. This chapter also explores the difficulties and factors to consider while examining welded joints. Chapter 8 reviews the 3D-milling operations using ABAQUS. Chapter 9 delves into the rare ferroelectric material: Fresnoite ($Ba_2TiSi_2O_8$).

Chapter 10 surveys various technologies used to analyze the characteristics of the weld pool based on visual sensing technologies. Also, this chapter quantifies the research of the latest implements in advanced GTAW and its monitoring techniques to academicians and industrialists. Chapter 11 simulates the Claus Sulfur Recovery Unit (SRU) using Aspen

Plus simulation software. The accuracy of the model was justified by comparison between the simulation results and the industrial data of South Pars Refinery Iran and the industrial data of Tata Steel Ltd. India. Chapter 12 discusses the most up-to-date findings from studies conducted on the ultrasonic-assisted stir-casting method for creating metal matrix nanocomposites (MMNC) using a magnesium material.

Chapter 13 focuses on joining two dissimilar magnesium alloys, namely AZ31 and AZ61, via experimental approaches. Chapter 14 explores various welding processes, their applications, and resulting properties. Chapter 15 goes further into Advanced 3D Printing for Industrial Components: Welded Joint Analysis and Strength Assessment. Chapter 16 discusses advancements in Electrochemical Surface Coatings: Innovations, Applications, and Future Prospects. Chapter 17 delves into an exploration of four unconventional machining techniques: Electro-Discharge Machining (EDM), Electrochemical Machining (ECM), Abrasive Jet Machining (AJM), and Ultrasonic Machining (USM).

Chapter 18 discusses various steps involved in surface coating using pulsed power systems and performance evaluation criteria for such coatings. Different coating materials along with coating methods such as balanced and unbalanced sputtering processes are discussed. Chapter 19 looks at the fabrication of a natural fiber-reinforced composite and tests its water absorption percentage. Chapter 20 introduces a process parameter optimization approach that utilizes the non-dominated sorting genetic algorithm II (NSGA-II) in laser beam welding. The key process parameters considered in this study are power (P), welding speed (WS), and focal position (FP). The application of NSGA-II enables the search for multi-objective Pareto optimal solutions, thereby achieving the desired outcomes.

We are deeply grateful to all the contributors to this book and highly appreciate the dedicated support and valuable assistance rendered by Martin Scrivener and the Scrivener Publishing team during publication.

**The Editors**
April 2024

1

# Aluminum and Its Different Graded Alloys

**G. Avinash, V. V. N. Sarath* and A. Yeswanth**

*Department of Mechanical Engineering, Pragati Engineering College, Surampalem, Peddapuram, India*

***Abstract***

Mixtures of chemical elements called alloys must contain at least one metal. The properties of the metal, such as conductivity, flexibility, opacity, and luster, will all be retained in the final alloy product, unlike metal-based chemical compounds. However, alloys may also have additional properties not found in pure metals, such as increased strength or hardness. One well-known alloy among all was aluminum; based on its availability and its individual properties, usage of this material was increased. It was highly occupied in aerospace, followed by the automobile and industrial sectors. These alloys are categorized into different series grades based on their elements, which play a prominent role in their utilization due to their properties. This paper mainly focused on different grades of alloys, and the work performed by various researchers was presented. Among these, an intense study was conducted on a few of them, creating future scope for the others. It summarizes the following: parameters and powder materials, welding on alloys, recrystallization evolution, microstructure and texture analysis, extrusion profiles and wear study, production steps, precipitation mechanisms, heat treatment evolution, etc. Some of the applications were also presented for a better understanding of alloys.

***Keywords*:** Aluminum alloys, heat treatment process, recrystallization evolution, micro structural and surface texture analysis, powder form materials

## 1.1 Introduction

An aluminum alloy is an alloy with aluminum (Al) as the primary metal. Copper, magnesium, manganese, silicon, tin, nickel, and zinc are common

**Corresponding author*: Nagasarath345@gmail.com

R. Thanigaivelan, Pradeep Kumar Krishnan, Kamalakanta Muduli and Santosh Kumar Tamang (eds.) New Materials, Processing and Manufacturability: Fabrication and Processing of Advanced Materials, (1–20) 

alloying elements as presented in Figure 1.1. Casting and wrought alloys were categorized into heat-treatable and non-heat-treatable. Most of the aluminum is used to make wrought items such as rolled plates, foils, and extrusions.

Due to its superior performance and acceptable qualities over currently available aluminum or its alloys, a metal matrix composite like aluminum is preferable for industrial applications. The utilization of these metallic matrix materials was higher due to the various aspects that may influence the further addition of ceramic materials. Since the invention of metal-skinned aircraft, aluminum alloys played a significant role in aerospace manufacturing. In addition to being far less explosive than other alloys with a very high percentage of magnesium, aluminum-magnesium alloys are also lighter than other aluminum alloys [1].

Similarly, these materials have an excellent strength-to-weight ratio, which is one of the reasons it has been so well-liked in sectors like aerospace. Apart from this, corrosion resistance is one of aluminum's other vital advantages. Aluminum alloy offers excellent resistance to air corrosion in various environmental conditions.

Different identification standards are used for wrought and cast aluminum alloys as shown in Figure 1.2. A four-digit number specifies the alloying elements used to identify wrought aluminum.

A four- to five-digit number with a decimal point is used for cast aluminum alloys. Table 1.1 defines the digit following the decimal point denotes the form, while the digit in the hundreds place denotes the alloying components (cast shape or ingot).

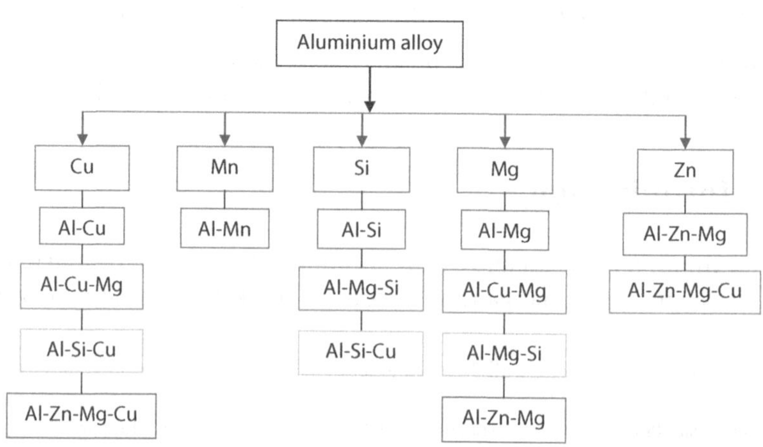


**Figure 1.1** Aluminum alloy with various elements.

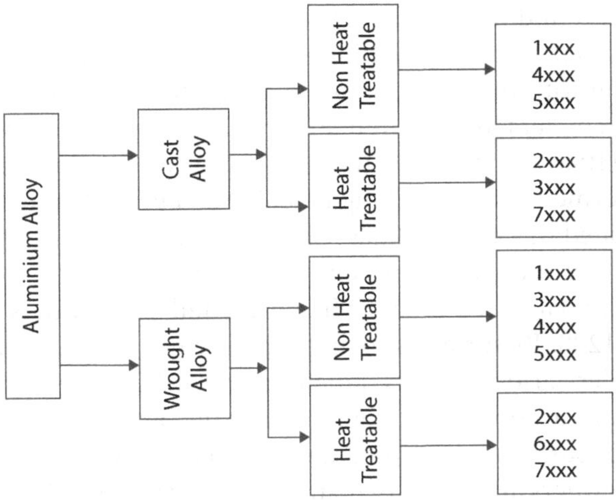


**Figure 1.2** Aluminum alloys with different graded series.

**Table 1.1** Four-digit series of aluminum alloys with main elements.

| Series no. | Series | Elements |
|---|---|---|
| 1 | 1XXX | Pure aluminum |
| 2 | 2XXX | Copper |
| 3 | 3XXX | Manganese |
| 4 | 4XXX | Silicon |
| 5 | 5XXX | Magnesium |
| 6 | 6XXX | Magnesium and silicon |
| 7 | 7XXX | Zinc |
| 8 | 8XXX | Others |

### 1.1.1 AA1XXX Series

Material shape and type significantly influence the tensile properties; a comparison was made between the AA1100 and SS400, which states that width might also impact the material's properties [2]. While under different strain rates, samples were tested and compared with the simulation analysis, which finally showed a satisfactory result [3]. The AA1100 aluminum alloy and mild steel materials were studied under the different welded, various parameters conditions and defined its mechanical properties [4]. Along with this, for maximum strength and minimum corrosion rate, tool parameters and friction stir welding (FSW) process were optimized by the RSM technique on this AA1100 [5, 6]. Friction stir spot welding was performed using a drilling machine at a variable rotational speed, it was concluded that the thickness of 3-mm sheets has a greater potential for welding, and the maximum tensile value was recorded at the highest value of rotational speed [7, 8]. Mixing Ni and Sic powders in different ratios to AA1200. Its surface alloying is performed, and microstructure is observed under various aspects. Finally, it states that hardness occurred in the alloyed zone around two to four times in the heated zone [9]. Similarly, the aluminum surface was studied under different compositions using Ni, Ti, and Sic powders and states that among various compositions, the highest percentage of Ni with Ti and Sic receives maximum hardness, which is 13 times higher than pure aluminum [10].

### 1.1.2 AA2XXX Series

Tensile and notch strengths were determined in the range of 78° to −423°F for alloys of 2014, 2024, 2219, and 6061 in various conditions and observed toughness and flexibility [11]. A comparative study was made between the three different AA2000 series in the aspect of uniaxial fatigue loading and its responses; finally, AA 2624-T351 alloy is the most advisable material [12, 13].

Similarly, with NaCl on AA2024-T351 aluminum alloy, its intergranular corrosion and crystallographic pitting were explored [14]. The essential goal of this article is to discover the improvement and utilization of aluminum alloys in aerospace industries. In this contrast, the upgrades in mechanical properties of the Al-Cu, Al-Zn, and Al-Li alloys are mentioned [15]. Corrosion, hardness, and microstructural change brought on by SiC and Zr additions to a 2xxx series of Al-Cu alloy was investigated [16, 17].

### 1.1.3 AA3XXX Series

This paper examines hot-tearing sensitivity on AA3000 with different alloy contents utilizing restricted rod casting molds. The results validated that Cu and Fe product affects hot-tearing resistance [18]. On AA3003, the recrystallized evolution, grains, and kinetics were also carried out on cold-rolled continuous cast of aluminum alloys. Similarly, microstructure and texture analysis were also performed under the different thicknesses of 3003 alloys. Along with this to study its mechanical behavior in samples with different pores, FE analysis was conducted on it [19–22].

On AA3004 alloy, tensile tests were performed in the extruded alloy with equal-channel angular pressing (ECAP) under different temperatures. Hot and cold rolled samples received good mechanical values at 573 K. During the study on this alloy, intermetallic particle control is necessary during annealing and strip casting processes, which may influence cube texture after recrystallization [23, 24]. Apart from this, the impact of temperature and stress on texture improvement in aluminum alloy AA 3004 has been studied in a sequence of multi-pass rolling experiments; finally, the texture is observed without appreciable results except for an interactive effect [25]. Similarly, a commercial grade 3004 alloy that has been solution-treated at two different temperatures (580°C and 400°C) has had the impact of warm/hot-rolling temperature evaluated on its microstructure and texture [26].

### 1.1.4 AA4XXX Series

The morphology study on the 4XXX series, i.e., AlSi5Cu2Mg alloy, revealed their chemical composition, shape, and three intermetallic phases based on parameters [27]. Furthermore, the hot compression–induced deformation behavior of aluminum alloy 4032 was investigated and concluded that flow stress was highly influenced by temperature and strain rate [28]. The aggregate of silicon and nickel in an aluminum matrix affects exquisite wear resistance. Similarly, rice husk ash (RHA) and coal ash were also used in Al 4032 and studied its wear resistance [29–31].

In 4032 series, A silicon hollow glass microspheres (SIHGM) on a metal matrix composites are presented and its influence on mechanical and tribological properties was studied, along with this, the wire's microstructure and modification performance on 4032 aluminum alloy may be conducted, from which it was discovered that preparation procedures had a substantial impact on the microstructures of Al-10Sr, which in turn affected its performance [32, 33]. The hot extrusion procedure was conducted on Al-Si chips in several sizes and morphologies. Later on, anisotropy and microstructures may also be studied. Along with fractography, evaluation was done to examine the character of extruded profiles [34].

### 1.1.5 AA5XXX Series

The effect of friction joining and post-weld heat treatment was studied from the tensile strength and metallurgical characteristics of dissimilar

Ti-5083 friction-welded joints [35]. A 5083 alloy's grain size can be effectively refined by rapid solidification and plastic consolidation by hot extrusion [36]. The study was on different clad sheets fabricated under 50 and 75% rolling reduction by the cold roll bonding process. Similarly, a simple model explains the effects of the rolling reduction and area fraction of the deposit of Al alloy [37]. In a 5XXX series, alloy surface quality is affected due to Luders lines, while the traditional methods demonstrate that a reduced Luders elongation occurs when the grain size may increase. This research reveals that by including Zn/Cu elements into this alloy and preaging them, Luders elongation can be suppressed at a reduced grain size [38].

Similarly, Luders elongation on AA5182 alloy was also investigated concerning their strain and grain size [39]. Thus, they showed that nondestructive testing strategies could help assess how Mg concentration affects the corrosion and mechanical characteristics of chill-rolled Al-Mg alloys. Furthermore, the conductivity dimension had a greater level of sensitivity for identifying the precipitation of the reinforcing stage compared to the ultrasonic screening [40].

### 1.1.6 AA6XXX Series

An extent in the fatigue specimen's stress attention or the presence of a weldment substantially decreases the effect on corrosion pre-exposure [41]. Eight different-sized quartz particles with suggested sizes varying between 37.5 and 655 m were used to study the erosion wear of AA6063 alloy by a slurry pot tester [42]. This review paper aims to compile the most recent information on alloy design, industrial processing, and property development, and future applications for AA6XXX alloys were presented. Along with the sources currently available, production steps and precipitation strengthening mechanism of aluminum alloys are also defined; usually, these 6XXX series are highly employed in the aerospace and defense sectors in various shapes, including sheets and rods [43, 44]. By changing different filler rods, welding was performed on 6061 aluminum alloy by using tungsten inert gas and metal inert gas welding processes. For a better assessment of welded joints, a microstructure study was also conducted [45].

### 1.1.7 AA7XXX Series

To create ingots using 7050 alloys, the low-frequency electromagnetic casting (LFEC) process was employed, and its impact on the ingot's as-cast microstructure was investigated [46]. Furthermore, the effect on microstructure and properties was also studied after the age-hardening precipitation of the double peak process [47]. A detailed study on Al 7150 by exfoliation behaviors with different tempers was presented; similarly, electrochemical impedance spectroscopy (EIS) was used to examine the pitting corrosion behavior in a solution of sodium chloride and hydrochloric acid [48, 49]. The electrochemical conduct of the octadecyltrimethoxysilane (ODTMS) and PTMSSAM-covered AA7075 alloy has proven the corrosion safety of the self-assembled monolayers (SAMs) [50]. This review paper describes the improvement in utilization of Al 2000 series and 7000 series, along with Al-Li, which may also be preferable in lightweight applications [51]. Similarly, a detailed evolution of heat treatment of 7XXX series alloy, reported by various researchers, was presented in this paper [52].

### 1.1.8 AA8XXX Series

The production-feasible diffusion bonding (DB) condition has been successfully used on A1 alloy 8090 [53]. Furthermore, the alloy was used and stated that variation in texture intensity of a matrix material depends on the particle's volume fraction [54]. This study conducted welding tests on 8090 Al-Li plates using a 5% Mg filler metal and an electric arc, along with a new heat treatment technique called directional quenching that is used to enhance mechanical properties [55, 56]. This study shows the microstructure and formability of sheets manufactured with the alloy AA8011 under uniaxial and biaxial deformation modes; similarly, by using this alloy as a primary manufacturing material, heat exchanger fins were made and studied in this paper [57, 58]. The impact of peanut shell ash on AA8079 under different percentages was observed by studying its mechanical properties and a low silicon content impact was also observed along with its microstructure study [59, 60].

## 1.2 Other Aluminum Materials

Pure aluminum powder material with a specified mesh size mixed with different alloys was used to study its mechanical properties, which defines the impact in strength observed due to the rise of interfacial bonding in elements [61]. Lightweight aluminum steel matrix compound is produced by including 10 as well as 20 wt % of wet grinder stone dust, a type of industrial waste made while refining normally takes place in quarry rocks [62].

**Table 1.2** Detailed sources of different aluminum series.

| Aluminum series | Advantages | Disadvantages | Challenges and limitations |
|---|---|---|---|
| AA1XXX | • Excellent corrosion resistance | • Low strength compared to other series | • Limited availability of alloys within the series |
| | • High thermal and electrical conductivity | • Limited to non-structural applications | • Limited fatigue resistance |
| | • Good formability | | • Limited wear resistance |
| AA2XXX | • High strength-to-weight ratio | • Poor corrosion resistance | • Limited formability |
| | • Good fatigue resistance | • Limited to lower temperature applications | • Limited availability of alloys within the series |
| | • Good machinability | | • Limited resistance to stress corrosion cracking |
| AA3XXX | • Good corrosion resistance | • Limited strength compared to other series | • Limited high-temperature properties |
| | • Good formability | • Limited to non-structural applications | • Limited availability of alloys within the series |
| | • Good weldability | | • Limited fatigue resistance |

(*Continued*)

**Table 1.2** Detailed sources of different aluminum series. (*Continued*)

| Aluminum series | Advantages | Disadvantages | Challenges and limitations |
|---|---|---|---|
| AA4XXX | • Good castability | • Limited to lower temperature applications | • Limited availability of alloys within the series |
| | • Good machinability | • Limited strength compared to other series | • Limited weldability |
| | • Good thermal conductivity | | |
| AA5XXX | • Excellent corrosion resistance | • Excellent corrosion resistance | • Limited formability |
| | • Good weldability | • Good weldability | • Limited resistance to stress corrosion cracking |
| | • Good strength-to-weight ratio | • Good strength-to-weight ratio | |
| AA6XXX | • Good strength-to-weight ratio | • Poor corrosion resistance compared to other series | • Limited resistance to stress corrosion cracking |
| | • Good formability | • Limited to lower temperature applications | • Limited wear resistance |
| | • Good machinability | | |

(*Continued*)

**Table 1.2** Detailed sources of different aluminum series. (*Continued*)

| Aluminum series | Advantages | Disadvantages | Challenges and limitations |
|---|---|---|---|
| AA7XXX | • High strength and toughness | • Poor corrosion resistance compared to other series | • Limited weldability |
| | • Good fatigue resistance | • Limited to lower temperature applications | • Limited formability |
| | • Good machinability | | |
| AA8XXX | • Excellent strength-to-weight ratio | • Limited availability and higher cost compared to other series | • Limited fatigue resistance |
| | • Good fatigue resistance | • Limited to high-performance applications | • Limited resistance to stress corrosion cracking |
| | • Good weldability | | |

On different aluminum alloys welded by the FSW process, variations of its tensile strength were presented based on parameters [63].

A close inspection was done of the developed magnesium-based product's specific strength and the two different aluminum alloys in structural and non-structural applications. This paper discusses how the various parameters significantly influence the mechanical properties of aluminum material [64]. By evaluating the experimental, numerical, and logical work with architectural lightweight aluminum alloys, this paper goals to completely examine the area of research. It also sums up examinations of the mechanical attributes of alloys under monotonic, cyclic, and thermal loading [65]. This paper proposes some of the design and developing methods

**Table 1.3** Properties and aplications of alloys.

| Aluminum grade | 1XXX | 2XXX | 3XXX | 4XXX | 5XXX | 6XXX | 7XXX | 8XXX |
|---|---|---|---|---|---|---|---|---|
| Main alloying elements | Unalloyed (99% Al) | Copper | Silicon, magnesium, copper | Silicon | Magnesium | Magnesium, silicon | Zinc | Other (lithium, iron, tin) |
| Heat-treatable | No | Yes | Yes | Yes | No | Yes | Yes | – |
| Strength | Poor | Excellent | Good | Good | Very good | Very good | Excellent | – |
| Corrosion resistance | Excellent | Average | Very good | Average | Excellent | Good | Excellent | – |
| Workability/ formability | Excellent | Average | Excellent | Excellent | Excellent | Very good | Average | – |
| Common applications | Electrical wires, chemical tanks, pipe systems, aluminum foil | Aerospace, military vehicles, rocket parts | Roofing and guttering, street signs, cooking utensils | In the automotive industry, welding wire | Marine applications, welding wire | Architectural, marine, welding wire | Aerospace, armored vehicles, sporting equipment | Experimental in design, created for specific purposes within specialist industries like aerospace. |

for lightweight alloys, along with this, the latest innovations in the casting and forming process by these alloys were also summaries [66]. Table 1.2 shows different grades of aluminum materials with their advantages, disadvantages, and challenges.

## 1.3 Applications

The corrosion behavior of the wrought aluminum alloy under various cooking conditions was investigated experimentally [67]. Table 1.3 shows the properties and applications of various aluminum series [68]. In this paper, defect diagnosis systems for the aluminum process are reviewed, and suggestions for their improvement are covered [69]. It defines that the creation of free ear material might be acknowledged by controlling the microstructural development during the entire manufacturing plan and the place of the recrystallizations during the hot-rolling and cold-moving successions [70]. Here are the current outcomes of a collection of laboratory experiments, sliding saline ice on aluminum and controlling for roughness and temperature. It locates that the friction of saline ice on aluminum $\mu_{\text{ice-al}} = 0.1$ typically; however this cost varies with sliding conditions [71]. The impact of heat-treatable and non-heat-treatable aluminum alloys under various parameters was thoroughly investigated [72, 73].

## 1.4 Conclusion

Among all alloys, aluminum was highly noticed in the aerospace field and other areas. This paper mainly focuses on the various works presented on different grades of alloys and their applications by researchers, from which it concludes that

- ❖ Apart from the 1XXX and 2XXX aluminum series, its properties, microstructure welding, and various parameters influencing these materials were studied. Similarly, the study of properties was continued by the addition of different compositions of powders into it.
- ❖ Meanwhile, from the 3XXX series, hot-tearing susceptibility, recrystallization evolution, corrosion reduction, and surface texture analysis were carried out.
- ❖ The 4XXX, 5XXX, and 6XXX series presented various elements and their impact on morphology, wear resistance,

grain size determination, Mg effects on corrosion, and erosion wear along with production steps and precipitation strengthening mechanism on the 6XXX series was shown.

- ❖ Age hardening, exfoliation, and corrosion behavior conditions under chemical solutions, with a set of Al-Li alloys, and other lightweight applications were presented in the case of the 7XXX series.
- ❖ Diffusion bonding, texture intensity, welding test, heat treatment, and powder form of material on 8XXX series were studied. Simultaneously, other works reported on different grades of alloys were also presented in this paper, including various powder form–treated aluminum material properties, specific strength of Mg-based products, and mechanical characteristics of alloys.
- ❖ Similarly, different grades of alloys, their properties, and their main elements and applications were also presented. Finally, most of them utilized the 1XXX, 2XXX, 3XXX, 7XXX, and 8XXX series of alloys from the above. Less work was reported in the 6XXX grade, and the next was on the 4XXX and 5XXX series.

## 1.5 Future Scope

- ❖ The works presented on different graded alloys may be considered a basic idea for the work out on remaining alloys.
- ❖ Similarly, the powder form of the material may also be considered for addition to the alloys to enhance its properties.
- ❖ Mainly corrosion reduction, coating on alloys, powder form and its particle size impact, and alloys with the addition of different graded sets of elements and composition may be considered for further studies. These may increase the scope for expanding aluminum alloys into various fields.

## References

1. Hombergsmeier, E., Magnesium for aerospace applications. In: *2nd international conference and exhibition "Magnesium—Broad Horizons"*, p. 1–13, 2007.

2. Sulamet-Ariobimo, R.D., Soedarsono, J.W., Sukarnoto, T., Rustandi, A., Mujalis, Y., Prayitno, D., Tensile properties analysis of AA1100 aluminium and SS400 steel using different JIS tensile standard specimen. *J. Appl. Res. Technol.*, 14, 2, 148–153, Apr. 2016, doi: 10.1016/j.jart.2016.03.006.
3. Sahu, S., Pada Mondal, D., Goel, M.D., Ansari, M.Z., Finite element analysis of AA1100 elasto-plastic behaviour using Johnson-Cook model. *Mater. Today Proc.*, 5, 2, 5349–5353, 2018, doi: 10.1016/j. matpr.2017.12.120.
4. Khalfallah, F., Boumerzoug, Z., Rajakumar, S., Raouache, E., Optimization by RSM on rotary friction welding of AA1100 aluminum alloy and mild steel. *Int. Rev. Appl. Sci. Eng.*, 11, 1, 34–42, Apr. 2020, doi: 10.1556/1848.2020.00005.
5. Pandipati, S., Sreeramulu, D., Potnuru, S., Friction Stir Lap Welded Joints for Aluminum Alloys: An Evaluation. *Int. J. Mech. Eng.*, 7, 2, 466–472, Feb. 2022.
6. Rajakumar, S. and Balasubramanian, V., Multi-Response Optimization of Friction-Stir-Welded AA1100 Aluminum Alloy Joints. *J. Mater. Eng. Perform.*, 21, 6, 809–822, Jun. 2012, doi: 10.1007/s11665-011-9979-z.
7. Yadav, V.D. and Bhatwadekar, S.G., Friction Stir Welding of Dissimilar Aluminium alloys AA1100 to AA6101-T6. *International Journal of Research in Aeronautical and Mechanical Engineering (IJRAME)*, 3, 1–6, 2015.
8. Jassim, A.K. and Al-Subar, R.K., Studying the Possibility to Weld AA1100 Aluminum Alloy by Friction Stir Spot Welding. *International Journal of Chemical, Molecular, Nuclear, Materials and Metallurgical Engineering*, 11, 9 2017, doi: 10.5281/zenodo.1132443.
9. Mabhali, L.A.B., Pityana, S.L., Sacks, N., Laser Surface Alloying of Aluminum (AA1200) with Ni and SiC Powders. *Mater. Manuf. Process.*, 25, 12, 1397–1403, Dec. 2010, doi: 10.1080/10426914.2010.498073.
10. Mabhali, L.A.B., Pityana, S.L., Sacks, N., Laser Surface Alloying of Aluminium AA1200. *Mol. Cryst. Liq. Cryst.*, 555, 1, 138–148, Apr. 2012, doi: 10.1080/15421406.2012.635095.
11. Christian, J.L. and Watson, J.F., Mechanical Properties of Several 2000 and 6000 Series Aluminum Alloys, in: *Advances in Cryogenic Engineering*, K.D. Timmerhaus (Ed.), pp. 63–76, Springer US, Boston, MA, 1965, doi: 10.1007/978-1-4684-3108-7_8.
12. Owolabi, G.M., Thom, M., Ajide, O.O., Kumar, N., Odeshi, A.G., Warner, G., Tensile Properties and Fractography of Three AA 2000 Series Aluminum Alloys Used for Aerospace Applications. *Trans. Indian Inst. Met.*, 72, 10, 2623–2630, Oct. 2019, doi: 10.1007/s12666-019-01731-y.
13. Owolabi, G.M. *et al.*, Fatigue Responses of Three AA 2000 Series Aluminum Alloys. *J. Mater. Sci. Chem. Eng.*, 07, 03, 32–48, 2019, doi: 10.4236/msce.2019.73003.
14. Zhang, X., Zhou, X., Hashimoto, T., Liu, B., Localized corrosion in AA2024-T351 aluminium alloy: Transition from intergranular corrosion to crystallographic pitting. *Mater. Charact.*, 130, 230–236, Aug. 2017, doi: 10.1016/j.matchar.2017.06.022.

15. Rajan, R., Kah, P., Mvola, B., Martikainen, J., Trends in Aluminium Alloy Development and Their Joining Methods. *Rev. Adv. Mater. Sci.*, 44. 383–397, 2016.
16. Cooke, R.W., Hexemer, R.L., Donaldson, I.W., Bishop, D.P., Press- and-sinter processing of a PM counterpart to wrought aluminum 2618. *J. Mater. Process. Technol.*, 230, 72–79, Apr. 2016, doi: 10.1016/j. jmatprotec.2015.11.011.
17. Kairy, S.K. *et al.*, Simultaneous improvement in corrosion resistance and hardness of a model 2xxx series Al-Cu alloy with the microstructural variation caused by Sc and Zr additions. *Corros. Sci.*, 158, 108095, Sep. 2019, doi: 10.1016/j.corsci.2019.108095.
18. Razaz, G. and Carlberg, T., Hot Tearing Susceptibility of AA3000 Aluminum Alloy Containing Cu, Ti, and Zr. *Metall. Mater. Trans. A*, 50, 8, 3842–3854, Aug. 2019, doi: 10.1007/s11661-019-05290-1.
19. Zhang, J.X., Sun, H.Y., Du, P., Liu, W.C., Effect of Processing Method on the Recrystallization Behavior of Cold-Rolled AA3003 Aluminum Alloy. *J. Mater. Eng. Perform.*, 29, 7, 4286–4294, Jul. 2020, doi: 10.1007/ s11665-020-04963-y.
20. N.L.A., Wisdom, J.O., Onyinyechi, N., Corrosion Inhibition Study of Aluminum Alloy AA3003 in Alkaline Medium by Palisota Hirsute Extract. *Int. J. Eng. Res. Rev.*, 2, 4, 113–118, October - December 2014.
21. Cai, C., Ji, Z., Zhang, H., Wang, G., Study on the Microstructure and Texture of 3003 Aluminum Sheets Rolled by Laser-Textured Roll. *J. Metall.*, 2009, 1–6, May 2009, doi: 10.1155/2009/587938.
22. Al-Zubaidy, B.M., AbidAli, A.K., Athaib, N.H., Finite Element Analysis Of Porous Aluminium Aa3003 Alloy Under Compression And Bending Loading. *Eur. J. Mol. Clin. Med.*, 7, 735–746, 2020.
23. Cabibbo, M., Mehtedi, M.E., Barone, L., Prados, E.F., Ferrante, M., Mechanical properties at high temperature of an AA3004 after ECAP and cold/hot rolling. *Rev. Adv. Mater. Sci.*, 25, 183–188, 2010.
24. Singh, R., Khatirkar, R.K., Chouhan, R.N., Sapate, S.G., Development of Cube Recrystallization Texture in Strip Cast AA3004 Aluminium Alloy. *Trans. Indian Inst. Met.*, 69, 10, 1833–1841, Dec. 2016, doi: 10.1007/ s12666-016-0842-7.
25. Hollinshead, P.A. and Sheppard, T., Development of rolling textures in aluminum alloy 3004 subjected to varying hot-rolling deformation. *Metall. Trans. A*, 20, 8, 1495–1507, Aug. 1989, doi: 10.1007/BF02665506.
26. Schumann, G.O., Crosky, A., Hatherly, M., The Microstructure and Texture of Warm Rolled 3004 Aluminium, *Texture, Stress, Microstructure*, 26, Article ID 749093, 18, 1996.
27. Mrowka-Nowotnik, G., Sieniawski, J., Wierzbińska, M., Morphology prediction of intermetallics formed in 4xxx type of aluminium alloy. *J. Achiev. Mater. Manuf. Eng.*, 31, 2008.
28. Fang, G. and Zeng, P., Investigation of Thermal Formability of Aluminum Alloy 4032. *J. Mater. Eng. Perform.*, 15, 5, 535–539, Oct. 2006, doi: 10.1361/105994906X136043.

29. Hohman, D.W., Aluminum Alloy Deltalloy™ 4032: Elimination of Anodizing for Wear Resistant Applications. Presented at the *International Congress & Exposition*, p. 930708, Mar. 1993, doi: 10.4271/930708.
30. Gudimetla, A. and Lingaraju, D., Influence of RHA Reinforcements on Mechanical and Wear Behavior of Al 4032 Composites. *Emerg. Mater. Res.*, 9, 4, 1–16, 30 October 2020, doi: 10.1680/jemmr.20.00104.
31. Gudimetla, A., Modelling and optimization of wear parameters of Al 4032 reinforced with coal ash using Taguchi and RSM approach. *Compos. Theory Pract.*, 21, 1–2, 3–11, 2021.
32. Liao, C., Chen, J., Li, Y., Chen, H., Pan, C., Modification performance on 4032 Al alloy by using Al–10Sr master alloys manufactured from different processes. *Prog. Nat. Sci. Mater. Int.*, 24, 2, 87–96, Apr. 2014, doi: 10.1016/j.pnsc.2014.03.002.
33. Avinash, G., Lingaraju, D., Prasad, S.S., Investigation of mechanical and tribological behaviour of Al 4032-SiHGM MMC. *Compos. Theory Pract.*, 20, 3–4, 142–156, 2020.
34. Tokarski, T., Mechanical Properties of Solid-State Recycled 4xxx Aluminum Alloy Chips. *J. Mater. Eng. Perform.*, 25, 8, 3252–3259, Aug. 2016, doi: 10.1007/s11665-016-2194-1.
35. Fuji, A., Kimura, M., North, T.H., Ameyama, K., Aki, M., Mechanical properties of titanium-5083 friction joints. *Mater. Sci. Technol.*, 13, 8, 673–678, 1997.
36. Tokarski, T., Thermo-Mechanical Processing of Rapidly Solidified 5083 Aluminium Alloy - Structure and Mechanical Properties. *Arch. Metall. Mater.*, 60, 1, 177–180, Apr. 2015, doi: 10.1515/amm-2015-0028.
37. Miyajima, Y., Iguchi, K., Onaka, S., Kato, M., Effects of Rolling Reduction and Strength of Composed Layers on Bond Strength of Pure Copper and Aluminium Alloy Clad Sheets Fabricated by Cold Roll Bonding. *Adv. Mater. Sci. Eng.*, 2014, 1–11, 2014, doi: 10.1155/2014/614821.
38. Geng, Y., Zhang, D., Zhang, J., Zhuang, L., On the suppression of Lüders elongation in high-strength Cu/Zn modified 5xxx series aluminum alloy. *J. Alloys Compd.*, 834, 155138, Sep. 2020, doi: 10.1016/j. jallcom.2020.155138.
39. Lloyd, D.J., Court, S.A., Gatenby, K.M., Lüders elongation in AI-Mg alloy AA5182. *Mater. Sci. Technol.*, 13, 8, 660–666, Aug. 1997, doi: 10.1179/mst.1997.13.8.660.
40. Tzeng, Y.-C., Chen, R.-Y., Lee, S.-L., Nondestructive tests on the effect of Mg content on the corrosion and mechanical properties of 5000 series aluminum alloys. *Mater. Chem. Phys.*, 259, 124202, Feb. 2021, doi: 10.1016/j.matchemphys.2020.124202.
41. Dif, R. and Ehrström, J.-C., Corrosion-Fatigue Interactions for High Strength Aluminium Alloys. *Mater. Sci. Forum*, 396–402, 1285–1290, Jul. 2002, doi: 10.4028/www.scientific.net/MSF.396-402.1285.

42. Desale, G.R., Gandhi, B.K., Jain, S.C., Particle size effects on the slurry erosion of aluminium alloy (AA 6063). *Wear*, 266, 11–12, 1066– 1071, May 2009, doi: 10.1016/j.wear.2009.01.002.
43. Mukhopadhyay, P., Alloy Designation, Processing, and Use of AA6XXX Series Aluminium Alloys. *ISRN Metall.*, 2012, 1–15, Apr. 2012, doi: 10.5402/2012/165082.
44. Baruah, M. and Borah, A., Processing and precipitation strengthening of 6xxx series aluminium alloys: A review. *Int. J. Mater. Sci.*, 1, 1, 40–48, Jan. 2020, doi: 10.22271/27078221.2020.v1.i1a.10.
45. Ramji, B.R., Bharathi, V., Swamy, N.R.P., Characterization of TIG and MIG welded Aluminium 6063 alloys. *Mater. Today Proc.*, 46, 8895–8899, 2021, doi: 10.1016/j.matpr.2021.05.356.
46. Zuo, Y.B., Zhao, Z.H., Cui, J.Z., Grain Refinement of Direct Chill Cast 7050 Aluminium Alloy with Low Frequency Electromagnetic Field. *Adv. Mater. Res.*, 402, 850–853, Nov. 2011, doi: 10.4028/www.scientific. net/AMR.402.850.
47. He, C.D., Ren, J.P., Xu, B., Effect of Two-Stage Double Peak Aging on Microstructure and Properties in 7050 Aluminum Alloy. *Appl. Mech. Mater.*, 157–158, 331–337, Feb. 2012, doi: 10.4028/www.scientific. net/AMM.157-158.331.
48. Li, J.F., Jia, Z.Q., Li, C.X., Birbilis, N., Cai, C., Exfoliation corrosion of 7150 Al alloy with various tempers and its electrochemical impedance spectroscopy in EXCO solution. *Mater. Corros.*, 60, 6, 407–414, Jun. 2009, doi: 10.1002/maco.200805102.
49. Peng, G.S., Chen, K.H., Fang, H.C., Chao, H., Chen, S.Y., EIS study on pitting corrosion of 7150 aluminum alloy in sodium chloride and hydrochloric acid solution. *Mater. Corros.*, 61, 9, 783–789, Sep. 2010, doi: 10.1002/maco.200905413.
50. Neves, R.S., Silva, D.P.B., Motheo, A.J., Corrosion Protection of AA7075 Aluminium Alloy by Trimethoxy-Silanes Self-Assembled Monolayers. *ISRN Electrochem.*, 2013, 1–9, Apr. 2013, doi: 10.1155/2013/142493.
51. Raj, R.J., Selvam, P.P., Pughalendi, M., A Review of Aluminum Alloys in Aircraft and Aerospace Industry. *J. Huazhong Univ. Sci. Technol.*, 50, 4, 11, June 2021.
52. Rometsch, P.A., Zhang, Y., Knight, S., Heat treatment of 7xxx series aluminium alloys—Some recent developments. *Trans. Nonferrous Met. Soc. China*, 24, 7, 2003–2017, Jul. 2014, doi: 10.1016/S1003-6326(14)63306-9.
53. Sunwoo, A., Diffusion bonding of aluminum alloy, 8090. *Scr. Metall. Mater.*, 31, 4, 407–412, Aug. 1994, doi: 10.1016/0956-716X(94)90009-4.
54. Bowen, A.W. and Humphreys, F.J., Texture Development in Hot-Rolled 8090 Aluminium Alloy-SiC Metal-Matrix Composites. *Textures Microstruct.*, 14, 715–720, 1991, doi: 10.1155/TSM.14-18.715.

55. Bonaccorsi, L., Costanza, G., Missori, S., Sili, A., Tata, M.E., Mechanical and metallurgical characterization of 8090 Al–Li alloy welded joints. *Metallurgist*, 56, 1–2, 75–84, May 2012, doi: 10.1007/s11015-012-9539-2.
56. Nouri, S., Sahmani, S., Hadavi, M., Mirdamadi, S., Mechanical Properties Improvement of Al-Li 8090 Alloy by Using the New Proposed Method of Directional Quenching. *Met. Mater. Int.*, 26, 8, 1134–1143, Aug. 2020, doi: 10.1007/s12540-019-00367-3.
57. Aghaie-Khafri, M., Formability of AA8011 aluminum alloy sheet in homogenized and unhomogenized conditions. *J. Mater. Sci.*, 39, 21, 6467–6472, Nov. 2004, doi: 10.1023/B:JMSC.0000044884.25589.9b.
58. Rahimi, M., Fojan, P., Gurevich, L., Afshari, A., Aluminium Alloy 8011: Surface Characteristics. *Appl. Mech. Mater.*, 719–720, 29–37, Jan. 2015, doi: 10.4028/www.scientific.net/AMM.719-720.29.
59. Refaai, M.R.A., Reddy, R.M., Venugopal, J., Rao, M.V., Vaidhegi, K., Yishak, S., Optimization on the Mechanical Properties of Aluminium 8079 Composite Materials Reinforced with PSA. *Adv. Mater. Sci. Eng.*, 2022, 1–11, Apr. 2022, doi: 10.1155/2022/6328781.
60. Zhang, L., Kang, X., Zhong, B., Effects of Si Content on Microstructure and Mechanical Properties of 8079 Aluminum Alloy. *Res. Appl. Mater. Sci.*, 2, 1, 43–47, Jun. 2020, doi: 10.33142/msra.v2i1.1978.
61. Behera, R.K., Samal, B.P., Panigrahi, S.C., Muduli, K.K., Microstructural and Mechanical Analysis of Sintered Powdered Aluminium Composites. *Adv. Mater. Sci. Eng.*, 2020, 1–7, Apr. 2020, doi: 10.1155/2020/1893475.
62. Xavier, L.F. and Suresh, P., Wear Behavior of Aluminium Metal Matrix Composite Prepared from Industrial Waste. *Sci. World J.*, 2016, 1–8, 2016, doi: 10.1155/2016/6538345.
63. Al-Sabur, R., Tensile strength prediction of aluminium alloys welded by FSW using response surface methodology – Comparative review. *Mater. Today Proc.*, 45, 4504–4510, 2021, doi: 10.1016/j.matpr.2020.12.1001.
64. Aynalem, G.F., Processing Methods and Mechanical Properties of Aluminium Matrix Composites. *Adv. Mater. Sci. Eng.*, 2020, 1–19, Sep. 2020, doi: 10.1155/2020/3765791.
65. Georgantzia, E., Gkantou, M., Kamaris, G.S., Aluminium alloys as structural material: A review of research. *Eng. Struct.*, 227, 111372, Jan. 2021, doi: 10.1016/j.engstruct.2020.111372.
66. Luo, A.A., Recent advances in light metals and manufacturing for automotive applications. *CIM J.*, 12, 3, 79–87, Jul. 2021, doi: 10.1080/19236026.2021.1947088.
67. Adeosun, S.O., E., II, Balogun, S.A., Wrought Aluminium Alloy Corrosion Propensity in Domestic Food Cooking Environment. *ISRN Corros.*, 2012, 1–6, Oct. 2012, doi: 10.5402/2012/432342.
68. Kormax, A Guide to the Grades of Aluminium: Choosing the Right Alloy. https://kormaxdesign.com/knowledge-centre/a-guide-to-the-grades-of-aluminium-choosing-the-right-alloy/.

69. Abd Majid, N.A., Taylor, M.P., Chen, J.J.J., Young, B.R., Aluminium Process Fault Detection and Diagnosis. *Adv. Mater. Sci. Eng.*, 2015, 1–11, 2015, doi: 10.1155/2015/682786.
70. Lequeu, P., Earing Control and Deformation Processing of Aluminium Alloys. *Textures Microstruct.*, 14, 1019–1024, 1991, doi: 10.1155/ TSM.14-18.1019.
71. Wallen-Russell, C. and Lishman, B., The Friction of Saline Ice on Aluminium. *Adv. Tribol.*, 2016, 1–7, 2016, doi: 10.1155/2016/1483951.
72. Hirsch, J., Aluminium Alloys for Automotive Application. *Mater. Sci. Forum*, 242, 33–50, Jan. 1997, doi: 10.4028/www.scientific.net/MSF.242.33.
73. Mythili, T. and Thanigaivelan, R., Optimization of wire EDM process parameters on Al6061/$Al_2O_3$ composite and its surface integrity studies. *Bull. Pol. Acad. Sci.: Tech. Sci.*, 68, 6, 1403–1412, 2020.

# 2

# Cold Spray Additive Manufacturing

**Jagadeesha T.[1]*, Arun Prakash J.[2], Avinash Malladi[3] and Seeniappan Kaliappan[4]**

[1]*Department of Mechanical Engineering, National Institute of Technology NIT, Calicut, Kerala, India*
[2]*Department of Aeronautical Engineering, Nehru Institute of Engineering and Technology, Coimbatore, Tamil Nadu, India*
[3]*Department of Mechatronics Engineering, Faculty of Science and Technology, IcfaiTech, Icfai Foundation for Higher Education, Hyderabad, Telangana, India*
[4]*Department of Mechatronics Engineering, KCG College of Technology, KCG Nagar, Karapakkam, Chennai, Tamil Nadu, India*

***Abstract***
Cold spray additive manufacturing is an emerging manufacturing process that uses high-pressure gas streams to accelerate metallic or composite powders and deposit them onto a substrate to form 3D structures. Unlike other additive manufacturing processes, cold spray additive manufacturing does not require high temperatures, making it suitable for a wide range of materials and applications. In this report, we provide a detailed overview of cold spray additive manufacturing, including its history, principles, applications, and recent advancements. We also review the relevant literature on this topic and discuss the key challenges and opportunities in this field. Our report concludes with a discussion of the potential future directions and applications of cold spray additive manufacturing.

***Keywords*:** Additive manufacture, cold spray, laser, high pressure deposition, high temperature materials, high strength materials

## 2.1 Introduction

The field of additive manufacturing (AM) is rapidly growing and has fundamentally altered the process of product making. AM offers a number of

**Corresponding author*: jagdishsg@nitc.ac.in

R. Thanigaivelan, Pradeep Kumar Krishnan, Kamalakanta Muduli and Santosh Kumar Tamang (eds.) New Materials, Processing and Manufacturability: Fabrication and Processing of Advanced Materials, (21–34) © 2024 Scrivener Publishing LLC

benefits over traditional manufacturing processes, including the ability to build intricate geometries and specialized parts with high accuracy and precision. One of the key advantages of AM is the ability to build structures layer by layer, which reduces material waste and permits the integration of different materials and functionality [1]. Over the past few decades, AM has seen significant advancements in materials, processes, and applications [2]. A few of the several AM methods that have been developed include vat photopolymerization, material extrusion, binder jetting, and powder bed fusion. The individual requirements of the application determine the best course of action. Each of these methods has advantages and restrictions of its own [3]. The use of cold spray AM (CSAM) has grown in popularity recently. In a process known as CSAM, metallic or composite powders are accelerated before being deposited onto a substrate to produce three-dimensional (3D) structures [4]. Because CSAM does not require high temperatures like other AM processes do, it is perfect for a wide range of materials and applications [5].

The principle behind CSAM is that high-velocity particles impart their kinetic energy to the substrate. This process uses a supersonic gas stream created by a converging-diverging de Laval nozzle [6]. The gas stream is then directed in the direction of the introduction of the feedstock powder through a carrier gas. The high-velocity gas particles contact and distort the powder particles, causing them to flatten and adhere to the substrate [7]. The accumulation of successive layers of powder leads to the creation of a 3D structure [8]. CSAM may be used to repair, modify, and create entirely new electrical, automotive, aerospace, and biological parts. It offers a variety of advantages over traditional manufacturing methods, including quicker lead times, reduced material waste, and the ability to manufacture complex geometries with accuracy and precision. CSAM of powder [9] enables the creation of hybrid structures with a variety of capabilities and the use of cutting-edge materials. Figure 2.1 shows the schematics of cold spray additive manufacturing (CSAM).

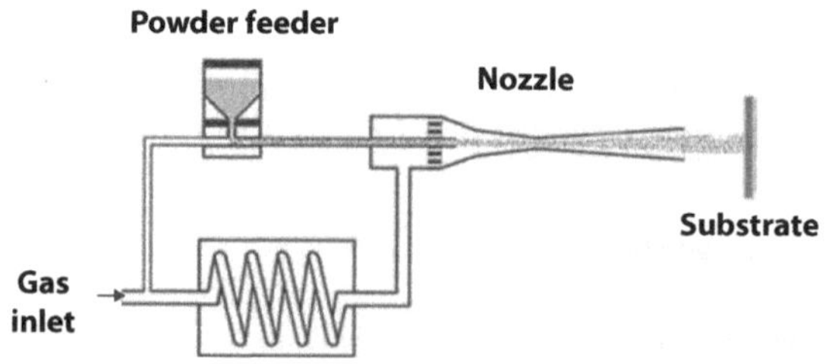


**Figure 2.1** CSAM basic diagram.

Despite this, there are still a number of technological challenges and limitations in the production of cold spray additives. Among these challenges include the effectiveness of powder deposition, interactions between particles and substrates, and surface quality control. The size and speed of the particles, the pressure and temperature of the gas stream, and the nozzle arrangement all have an impact on how well powder deposition works. The properties of the substrate and powder adhesion, impact angle and velocity, and surface topography all affect how the particles interact with the substrate. Surface quality control is affected by the nozzle-to-substrate distance, substrate preparation, and cleaning [10–15].

This chapter's objective is to present a thorough review of CSAM, including its fundamentals, uses, and most current developments.

The impact of gas pressure on the mechanical properties and microstructure of Ti6Al4V coatings deposited using the cold spray method is investigated by Chen *et al.* (2019). The range of gas pressures employed was 0.5 to 1.5 MPa. Microhardness and tensile tests were utilized to evaluate the coatings' mechanical properties, while SEM was employed to examine their microstructure. The results showed that the gas pressure had a substantial impact on the mechanical properties and microstructure of the cold-sprayed Ti6Al4V coatings. At low gas pressures (0.5 MPa), the coatings displayed a substantial amount of porosity and poor mechanical properties. With increasing gas pressure, the mechanical properties increased and the porosity decreased, peaking at 1.0 MPa. Raising the gas pressure resulted in a loss in mechanical properties, most likely as a result of the particles' faster speed producing excessive deformation and fragmentation upon collision. The authors claim that 1.0 MPa is the appropriate gas pressure for applying Ti6Al4V coatings using the cold spray approach. Under this pressure, the coatings exhibited minimal porosity, robust mechanical properties, and a smooth and uniform microstructure. In order to achieve the required coating features and performance, the study also highlights how vital it is to adjust cold spray technology process parameters such as gas pressure.

The principles of particle acceleration and impact, critical process parameters, and the various varieties of cold spray systems have all been covered in detail by Pathak and Saha (2017) in their study of cold spray technology. The current state of CSAM is then examined, including new advancements in the development of novel materials, the use of multi-material deposition, and the integration of cold spray with other production methods. The authors go into great detail on the creation of intricate structures and geometries, the creation of functional coatings, the repair and refurbishment of high-value components, and other applications for CSAM. One of the key advantages they point out is the cold spray technique's capacity to produce high-density, high-strength coatings with minimum distortion or degradation of the substrate material. Additionally, they discuss the unique challenges associated with CSAM, including the necessity of carefully controlling the process variables for optimal particle acceleration and deposition. Rokni *et al.* (2017) have used X-ray diffraction, transmission electron microscopy, and scanning electron microscopy (SEM) to describe the microstructure of cold spray deposits. They also studied the hardness, tensile strength, and ductility of cold spray deposits as mechanical attributes. They reported that changes in the critical impact velocity are brought on by particle size, which affects particle deformation. As a result of variations in heat capacity, thermal conductivity, cooling speed, and shear level impact instability for various particle sizes. The substrate/coating interface's temperature gradients and frictional heating are influenced by changes in spray angle, which has an impact on how particles deform.

Xie *et al.* 2020 go into great detail about in situ synthesis of metal matrix composites (MMCs) utilizing CSAM. One of the primary contributions of this research is the extensive analysis of the microstructure and properties of the resulting MMCs. SEM, energy-dispersive X-ray spectroscopy, and X-ray diffraction are used to analyze the MMCs' microstructure and phase composition. Mechanical testing is also used to evaluate the MMCs' mechanical qualities, including hardness, tensile strength, and ductility. By using this technique, MMCs with specialized microstructures and properties may be created, and the matrix and reinforcing components can be connected strongly.

According to Lu *et al.* (2020), CSAM at low pressure was used to produce thick copper coatings. Using a low-pressure cold spray technique, they coated a copper substrate with pure copper. The aim of the study was to investigate the effects of several process factors on the properties of copper coatings, including gas pressure, standoff distance, and nozzle diameter. The powder had a spherical shape and had an average particle size of 50 m. The scientists claim that the copper coatings were later sprayed using a low-pressure cold spray technique. The system consisted of a supply of nitrogen gas, a feeder for powder, an injection nozzle for powder, a mixing chamber, and a substrate holder. After the powder was inserted, the nitrogen gas was blended with the powder in the mixing chamber. The gas-powder mixture was then accelerated using the nozzle as it was being deposited onto the substrate. The investigation found that the gas pressure had a substantial influence on the coating quality. When the gas pressure was elevated owing to an increase in particle velocity, the coating thickness and density increased. It was found that the standoff distance and nozzle diameter also had an impact on the coating quality. Reducing the standoff distance and increasing the nozzle diameter resulted in thicker coatings with higher densities. The study also found that the properties of the coating were influenced by the powder feed rate. The microstructure consisted of a mixture of equiaxed and elongated grains, and it was found that grain size increased with increasing gas pressure. The coatings had high microhardness values, and these values increased as gas pressure and powder feed rate increased. The exceptional microhardness was attributed by the researchers to the high dislocation density and grain refinement in the coatings.

Functionally graded materials (FGMs) are a class of materials with compositions, microstructures, and properties that vary spatially. They are well known for their ability to provide mechanical, thermal, and electrical properties that are tailored and optimized for specific applications. CSAM, which can deposit a wide variety of materials with precise control over their composition and microstructure, is a practical technique for producing FGMs. Ma *et al.* (2020) provide an in-depth analysis of the most recent advancements in the CSAM of FGMs. The authors go over the many fabrication techniques that have been employed to create FGMs utilizing cold spray, such as powder mixing, powder feeding, and multilayer deposition. Along with discussing recent developments in the field, the authors also list some of the current difficulties and constraints of CSAM, such as the need for further process parameter optimization, the constrained range of materials that can be deposited, and the expensive equipment [9–12].

## 2.2 Phenomena and Factors Behind CSAM

A new method for creating metallic and composite materials called CSAM has gained popularity as a productive, affordable, and adaptable way to create parts with outstanding surface quality, high density, and exceptional mechanical qualities. The fundamental idea behind the CSAM method is to deposit metallic or composite powders onto a substrate by accelerating the powder particles to supersonic speeds (up to 1200 m/s) using a high-pressure gas stream, commonly nitrogen or helium. Without the use of any external heat sources or melting of the material, the high kinetic energy of the powder particles upon collision with the substrate causes the development of a thick and tightly bound layer of material.

A complex interplay of several physical and chemical processes, including gas dynamics, powder-particle interactions, solid-state deformation, and adhesion mechanisms, governs the CSAM process. The deposition efficiency, which is determined by the mass of deposited powder to the mass of powder injected into the gas stream, is one of the important variables that affect the quality of the deposited material. The gas pressure, temperature, and velocity; the particle size, shape, and composition; the substrate material and surface preparation; and the nozzle design and geometry have all been studied as elements that affect the CSAM process' deposition efficiency. The development and evolution of the particle-substrate interface, which is a critical factor in determining the bonding strength and quality of the deposited layer, are other significant features of the CSAM process. The deformation and fragmentation of the particle upon impact, followed by the plastic deformation and flow of the substrate material surrounding the particle, are what drive the mechanical interlocking and cold welding processes that control the interfacial bonding in the CSAM process. The interdiffusion and solid-state interactions between the particle and substrate materials, which can result in the creation of intermetallic compounds and other phases, can have an impact on the bonding strength.

The particle size, shape, and distribution; the degree of deformation and work hardening; the interparticle bonding and porosity; and the residual stress and strain distribution all affect the microstructural and mechanical properties of the deposited material in the CSAM process. The CSAM technique has been shown to have the ability to create components with distinct microstructures and characteristics, such as FGMs, composite materials, and nanostructured materials, in a number of studies. In addition, the CSAM process has a number of benefits over traditional manufacturing methods, including high deposition rates, less temperature input, little distortion, and the capacity to fix and recover harmed components.

## 2.3 Different CSAM Techniques

There are several different CSAM techniques that can be used to deposit or repair materials. Some of the most common techniques include the following [13–15]:

- Low-pressure cold spray: Compared to regular cold spray, this approach uses a lower-pressure gas system, often less than 10 MPa. This makes it more cost-effective for particular applications because it enables the use of smaller equipment and uses less gas. Decreased coating densities and mechanical qualities, however, might also be a consequence of the decreased pressure.

- High-pressure cold spray: This method, which is the most used for applying cold spray, typically uses gas pressures between 10 and 40 MPa to spray particles onto a substrate. This method is appropriate for a variety of materials and produces high-density coatings with good mechanical qualities.
- Cold ultrasonic spray: This method accelerates particles onto a substrate using high-frequency ultrasonic vibrations. Denser coatings with improved mechanical properties are produced as a result of the vibrations' assistance in dispersing agglomerated particles and enhancing adherence to the substrate.
- Kinetic metallization: It is a modified method of cold spray that produces a more controlled particle impact by using a bigger particle size and lower gas pressure. Polymers and metals can be deposited with this process, among many other materials.
- High-velocity oxy-fuel (HVOF) cold spray: This method combines cold spray with a HVOF thermal spray procedure. Prior to being launched onto the substrate, the particles are heated during the HVOF process. This improves their adhesion and produces denser coatings with better mechanical properties.
- Hybrid cold spray: This process combines cold spray with additional AM techniques, including 3D printing or laser cladding. The substrate is covered with a small layer of material using the cold spray method, and the remaining material is added using another AM procedure.

## 2.4 Advantages of CSAM

- Dense, high-strength coatings and repairs can be made using the CSAM technology. Due to the fact that the process does not involve melting or solidifying the material, it can be done with little heat input. Instead, tiny coating material particles are accelerated to supersonic speeds using high-pressure gas and then deposited onto a substrate.
- Low thermal distortion and high particle-to-particle bonding are produced by the low heat input and high velocity

of the particles, which can lead to coatings with superior strength and hardness. When it comes to some situations, CSAM coatings can have characteristics that are similar to those of the original material, which makes them a great option for repairing worn-out or broken components.

- Wide range of materials: Metals, alloys, ceramics, and polymers are just a few of the materials that can be employed with CSAM. Due to its adaptability, the method is a desirable choice for a variety of industries, including biomedicine, electronics, aerospace, and the automobile industry.
- Effective and economical: CSAM is a very effective and economical process. It does away with the necessity for pricey tooling, molds, and fixturing that other AM procedures could call for. Additionally, rather than having to completely replace worn-out or damaged components, the process can be used to repair or restore them, which can be more environmentally friendly and economically advantageous.
- Enhanced corrosion and wear resistance: CSAM coatings can provide improved thermal and electrical properties in addition to increased corrosion and wear resistance. Utilizing specialized coatings or alloy powders can further improve these qualities, enhancing the final product's performance and durability.
- Precision and customization: Coatings and repairs can be made with accuracy and personalization using CSAM. Particle size, velocity, and temperature can all be precisely controlled through the process to meet the demands of a given application. Furthermore, complex geometries and features can be produced using the process, which may be difficult or impossible to do with other manufacturing methods.

## 2.5 Disadvantages of CSAM

While CSAM offers many advantages, there are also some potential disadvantages to consider:

- High-pressure gas systems: One of the main drawbacks of CSAM is that the particles are often propelled by high-pressure gas systems. Due to the requirement for specialized machinery and supplies of compressed gas, this can be costly to operate and maintain. The high-pressure gas may also lead to safety issues, which need to be carefully managed.

- Controlling and improving the CSAM process can be challenging because so many variables can impact how well the coatings or repairs turn out. The final product's quality and attributes can be affected by a number of variables, including particle size, velocity, temperature, gas pressure, and others. Extensive testing and optimization may be needed to produce consistent and predictable results.
- Limited material selection: CSAM can be utilized with a number of materials; however, compared to other AM techniques, cold spray still has a considerably smaller material selection. The types of materials and applications that can be employed for CSAM may be constrained as a result.
- Slow process: When compared to other AM processes, the CSAM process might be slow. Process time can be increased if many passes are required to build up a coating or repair. Because of this, CSAM might not be as effective in applications where speed is crucial.
- Limitations on coating thickness: The maximum coating thickness that CSAM can produce may have some restrictions. Due to the possibility that some applications may require thicker coatings, this may limit the variety of uses for the procedure.
- Requirements for surface preparation: The substrate surface must be suitably prepared in order to obtain high-quality coatings using CSAM. In addition to other methods, this may entail cleaning, preheating, and sandblasting. These extra preparation processes might lengthen the process and increase costs.
- Limited accessibility to intricate geometries: Although CSAM is capable of producing intricate geometries and features, there might be restrictions on the accessibility of particular regions for coating or repair. This may restrict the variety of components that can be coated or repaired using the method.

## 2.6 Numerical Simulation of CSAM of Ti6Al4V Pellets on Ti6Al4V Substrate

A numerical simulation model of multiple particle impact during the cold spray process of Ti-6Al-4V powder onto a Ti-6Al-4V substrate is presented in the publication "Numerical simulation of multiple particle impact process during cold spray of the Ti-6Al-4V powder onto the Ti-6Al-4V substrate" by Xu Song. The goal of the study is to shed light on the mechanisms that control how particles interact with substrates during cold spraying, which is important for comprehending the procedure and improving coating quality.

The discrete element method (DEM), a particle-based numerical technique that can mimic the motion and interaction of individual particles, served as the foundation for the simulation model. The impact of Ti-6Al-4V particles on the substrate surface was simulated using the DEM model under a variety of parameters, including varying particle velocities, angles of incidence, and surface conditions.

A viscoelastic contact model, which accounts for the elastic and plastic deformation of the particles and substrate during impact, was used to simulate the particle-substrate interaction. To correctly estimate the coating morphology, the model also takes into account how particle size and shape affect impact behavior. Figure 2.2 shows the simulation image of cold spray process.

The authors matched the simulation findings with real data from cold spray tests using Ti-6Al-4V particles in order to verify the simulation model. The coating thickness, porosity, and surface roughness—important

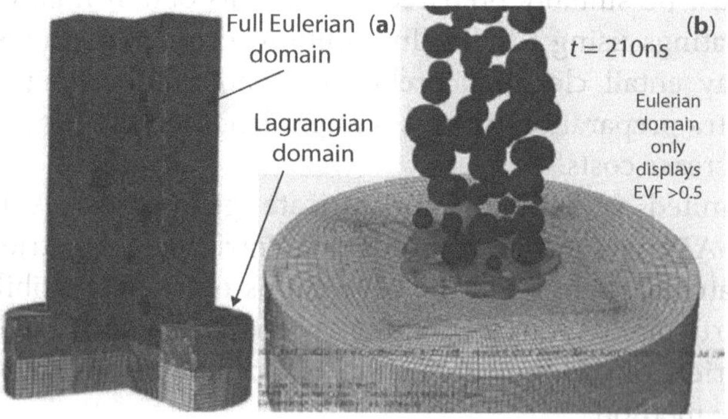


**Figure 2.2** Model (a) assembly and (b) snapshot at time frame t = 210 ns with Eulerian domain display EVF > 0.5 only.

factors that affect the coating quality—were all included in the experimental results.

The simulation model is able to properly anticipate the impact behavior of Ti-6Al-4V particles during cold spray, as evidenced by the simulation findings' strong agreement with the experimental data. The influence of other factors on the coating morphology, such as particle velocity, angle of incidence, and surface roughness, was also examined by the authors using the simulation model. Figure 2.3 shows the simulation image of cold spray deposition.

The simulation findings showed that the particle velocity and angle of incidence had a significant impact on coating thickness and porosity. The particles tend to penetrate the substrate farther at greater particle velocities and smaller angles of incidence, resulting in a thicker and less porous covering. It was discovered that the substrate's surface roughness had a major impact on the coating quality, with smoother substrates producing a coating that was more uniform and included fewer pores.

In conclusion, the numerical simulation model described in the work offers important new perspectives on the processes governing the interaction of particles and substrates during the cold spray process. By anticipating the impact of different factors on the coating morphology, the model may be utilized to improve the coating quality. The simulation findings and experimental data were in good agreement, demonstrating the model's

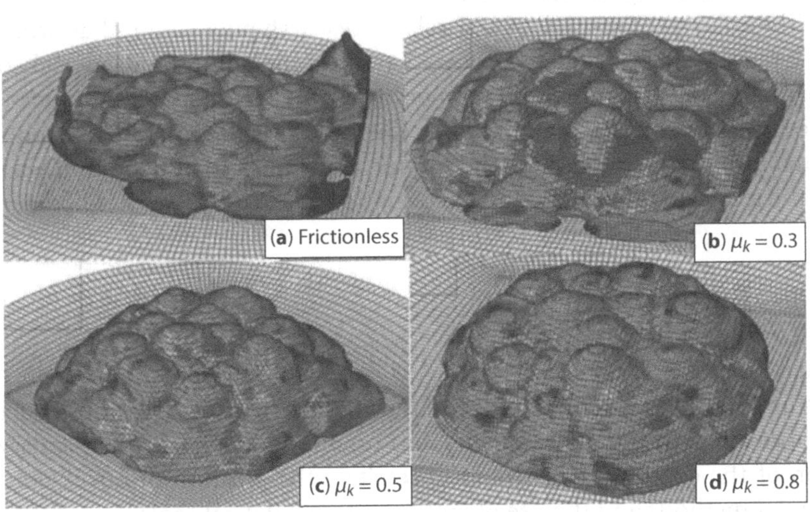


**Figure 2.3** Influence of friction on one hundred particles cold spray deposition process simulation.

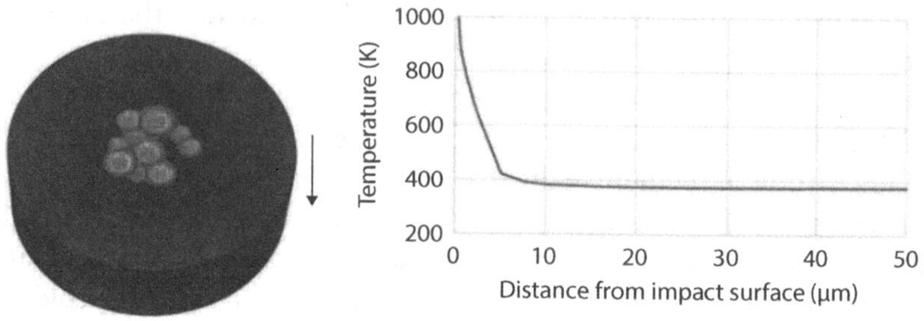


**Figure 2.4** Plot of the nodal temperature vs. depth in the substrate under multiple particle impact.

capability to predict the impact behavior of Ti-6Al-4V particles under cold spray. Overall, the study offers a strong starting point for future investigation into the cold spray method and its uses in diverse sectors. Figure 2.4 shows the subtract with deposited particles and temperature curve.

## 2.7 Conclusion

For the creation of metallic and composite materials, CSAM is a promising approach. This technique has demonstrated considerable promise for the production and repair of many different kinds of components and structures. It can create coatings that are thick and of a high caliber without seriously harming the substrate.

The analysis of several CSAM-related research publications has brought to light the significance of process variables including powder feedstock characteristics, gas pressure, temperature, and velocity, among others, in producing coatings of a high caliber. Furthermore, the creation of sophisticated numerical simulations and models has improved our comprehension of the intricate processes connected to the CSAM process.

## 2.8 Future Scope

There are several areas where CSAM research will be conducted in the future. First, more research is required to determine how various process factors affect the coatings' quality and functionality. For the characterization of the coatings, this would necessitate the creation of novel experimental methods and the use of cutting-edge analytical tools.

Second, there is a lot of interest in the creation of fresh powder feedstock for CSAM. For a variety of applications, the introduction of novel materials, such as nanocrystalline metals and alloys, can offer special features and improved performance.

Third, by combining CSAM with other production techniques like machining and surface treatment, new hybrid processes with improved capabilities and efficiency may be created.

Last but not least, the application of CSAM to the repair and restoration of vital parts in a variety of sectors, including aerospace and defense, may have a substantial positive impact on uptime, dependability, and cost-effectiveness.

In conclusion, CSAM research and development can result in the creation of novel materials, greater functionality, and increased capacity for a range of applications. For a variety of sectors, the use of CSAM for the repair and restoration of vital components as well as its interaction with other production processes may be quite advantageous.

## References

1. Zhang, X. *et al.*, Effect of Gas Pressure on Microstructure and Mechanical Properties of Cold-Sprayed Ti6Al4V Deposits. *J. Therm. Spray Technol.*, 28, 1653–1663, 2019.
2. Guo, Y., Lavernia, E.J., Zhang, J., Cold Spray Additive Manufacturing of Metallic Materials: Principles and Applications. *Mater. Sci. Forum*, 941, 323–348, 2019.
3. Chen, C., Xie, Y., Yan, X., Yin, S., Fukanuma, H., Huang, R., Liao, H., Effect of hot isostatic pressing (HIP) on microstructure and mechanical properties of Ti6Al4V alloy fabricated by cold spray additive manufacturing. *Addit. Manuf.*, 27, 595–605, 2019.
4. Pathak, S. and Saha, G.C., Development of sustainable cold spray coatings and 3D additive manufacturing components for repair/manufacturing applications: A critical review. *Coatings*, 7, 8, 122, 2017.
5. Rokni, M.R., Nutt, S.R., Widener, C.A., Champagne, V.K., Hrabe, R.H., Review of relationship between particle deformation, coating microstructure, and properties in high-pressure cold spray. *J. Therm. Spray Technol.*, 26, 1308–1355, 2017.
6. Xie, X., Ma, Y., Chen, C., Ji, G., Verdy, C., Wu, H., Liao, H., Cold spray additive manufacturing of metal matrix composites (MMCs) using a novel nano-TiB2-reinforced 7075Al powder. *J. Alloys Compd.*, 819, 152962, 2020.
7. Lu, P. *et al.*, Fabrication of dense copper coatings by low-pressure cold spray additive manufacturing. *J. Mater. Process. Technol.*, 278, 116525, 2020.

8. Ma, Y. *et al.*, Cold spray additive manufacturing of functionally graded materials: Process, properties, and applications. *J. Mater. Sci. Technol.*, 48, 39–53, 2020.
9. Trinh, H.T. *et al.*, Cold Spray Additive Manufacturing of Lightweight Metallic Materials: A Review. *Adv. Eng. Mater.*, 21, 1900195, 2019.
10. Schmidt, T. *et al.*, On the Control of Deposition Efficiency in Cold Spray Additive Manufacturing of Ti-6Al-4V. *J. Mater. Process. Technol.*, 265, 29–36, 2019.
11. Vaz, R., Garfias, A., Albaladejo, V., Sanchez, J., Cano, I.G., A Review of Advances in Cold Spray Additive Manufacturing. *Coatings*, 13, 267, 2023.
12. Song, X., Numerical simulation of multiple particle impact process during cold spray of the Ti-6Al-4V powder onto the Ti-6Al-4V substrate. *J. Therm. Spray Technol.*, 22, 7, 1035–1046, 2020.
13. Montavon, G., Coddet, C., Sampath, S., Herman, H., Berndt, C.C., Quality control of the intrinsic deposition efficiency from the controls of the splat morphologies and the deposit microstructure. *J. Therm. Spray Technol.*, 6, 153–166, 1997.
14. Natrayan, L., Balaji, S., Bharathiraja, G., Kaliappan, S., Veeman, D., Mammo, W.D., Experimental Investigation on Mechanical Properties of TiAlN Thin Films Deposited by RF Magnetron Sputtering. *J. Nanomater.*, 2021, Article ID 5943486, 7, 2021, https://doi.org/10.1155/2021/5943486.
15. Natrayan, L., Merneedi, A., Bharathiraja, G., Kaliappan, S., Veeman, D., Murugan, P., Processing and Characterization of Carbon Nanofibre Composites for Automotive Applications. *J. Nanomater.*, 2021, Article ID 7323885, 7, 2021, https://doi.org/10.1155/2021/7323885.

# 3

# Optimization of Gas Metal Arc Welding (GMAW) Cladding Parameters for Enhanced Weld Integrity in Low-Carbon Steel Plates (EN30)

**Vivek Singh[1], Amit Kumar Singh[2], Akash Deep[2], Som Ashutosh[2], M. Chandrasekaran[3] and Bishub Choudhary[3*]**

[1]*Indian Institute of Technology, Indore, Madhya Pradesh, India*
[2]*Department of Mechanical Engineering, AKGEC, Ghaziabad, Uttar Pradesh, India*
[3]*Department of Mechanical Engineering, NERIST, Nirjuli, Arunachal Pradesh, India*

***Abstract***

Weld cladding involves the application of a substantial coating of corrosion-resistant material onto vulnerable substrates with the goal of enhancing their resistance to corrosion. This method provides economic advantages by enabling the utilization of cost-effective and easily workable base materials, which can then be layered with high-quality metals and alloys to attain the desired characteristics in specific parts of the component. Present investigation, gas metal arc welding (GMAW) was utilized to clad low-carbon steel plates (EN30). The cladding parameters considered were voltage (*V*), wire feed rate (WFR), travel speed (*S*), and gas flow rate (*G*). Using a Taguchi L9 orthogonal array, experiments were conducted with three levels for each parameter. ER309L wire with a 1.2-mm diameter and direct current electrode positive polarity were used. A shielding gas of 98% argon and 2% $CO_2$ prevented oxidation during welding. The resulting welds were evaluated by measuring bead profile parameters, including bead width, bead height, depth of penetration, and dilution. Additionally, the weld penetration shape factor (PSF) and weld reinforcement form factor (RFF) were calculated based on these bead profile parameters. Graphical investigation indicated that WFR had the most dominant influence, followed by voltage, on the resultant characteristics. Increasing WFR

**Corresponding author*: bishub.choudhury73@gmail.com

R. Thanigaivelan, Pradeep Kumar Krishnan, Kamalakanta Muduli and Santosh Kumar Tamang (eds.) New Materials, Processing and Manufacturability: Fabrication and Processing of Advanced Materials, (35–56) © 2024 Scrivener Publishing LLC

led to higher welding current, resulting in greater filler material melting per unit length, increased penetration, and higher dilution. To determine the optimal cladding parameters, Taguchi optimization using signal-to-noise *(S/N)* ratio analysis was employed. The minimum dilution percentage was achieved at $WFR_1V_1G_1S_1$ (WFR = 6 m/min, $V$ = 20 V, $G$ = 16 L/min, $S$ = 240 mm/min). The Taguchi method effectively improved the performance of GMAW cladding for EN30, offering valuable insights for future applications.

***Keywords:*** AISI 201LN Gr steel, GMAW, wire feed rate, dilution, Taguchi

## 3.1 Introduction

Cladding is a surfacing method that involves the creation of a substantial layer of corrosion-resistant material, typically stainless steel, on low-carbon steel (also known as mild steel) to enhance its resistance to corrosion. In engineering applications, there is a need for materials that offer both high strength and inherent corrosion resistance to ensure long-term reliability and performance. While low-carbon steel meets the strength requirements for industrial applications, it falls short in terms of corrosion resistance [1]. The primary aim of this process is to economically combine the desirable properties of stainless steel with the base material, where the use of a full alloy construction is not necessary. Stainless steel cladding provides the necessary resistance to corrosion, abrasion, and oxidation, while the base material enhances fabricability, contributes structural strength, and improves the thermal conductivity of the composite [2].

There are various cladding techniques available, including cold roll bonding, hot roll bonding, centrifugal casting, explosive bonding, weld overlaying, and brazing. In addition, adhesive bonding, extrusion, and hot isostatic pressing have been employed for manufacturing clad metals. When using hot/cold explosive bonding, the roll bonding process involves bringing clean metal surfaces into close contact, leading to the sharing of electrons between the metals [3]. Among these methods, gas metal arc welding (GMAW) is often considered one of the most versatile, as it can be operated in semiautomatic, machine, or automatic modes. It is applicable to a wide range of commercially significant metals, including carbon steel, aluminum, stainless steel, titanium, copper, and nickel alloys, and can be employed in various positions.

Verma *et al.* found that cladding methods include thermal spraying, laser-based methods, and arc welding processes. Laser methods offer high-quality cladding but have high setup costs, while arc welding processes are more economical and productive. GMAW allows control over metal transfer variables and techniques like pulsed welding and electrode vibration control droplet characteristics [3]. Chandel *et al.* aimed to identify key factors for selecting stainless steel cladding processes, emphasizing dilution and bead geometry as crucial factors to consider [4]. Pires *et al.* presented experimental data on GMAW and shielding gas mixtures, revealing that thermal conductivity and active component content affect metal transfer modes. Ternary mixtures offer flexibility, and fume formation rates increase with higher $CO_2$ and $O_2$ content [5]. Senthilkumar *et al.* highlighted different welding techniques for surfacing, noting consumable electrode processes' impact on bead dimensions and phase formation. Nonconsumable electrodes and laser welding offer independent filler feed rate control [6]. Mahapatra *et al.* stressed the importance of process parameter selection for high-quality weld cladding, finding that dilution increases with higher welding current and speed, while bead width increases with higher current, distance, and torch angle [7]. Shahi and Pandey concluded that wire feed rate (WFR) has the greatest impact on dilution, and the universal gas metal arc welding (UGMAW) process with auxiliary preheating reduces penetration and achieves lower dilution, providing cost-effective surfacing [8]. Sreeraj *et al.* employed ANN and genetic algorithm (GA) models to optimize bead geometry and minimize dilution in GMAW, outperforming experimental data and regression methods, offering accurate solutions for parameter optimization [9].

Kumar *et al.* emphasized the influence of current, non-destructive testing (NTD), and voltage on weld bead geometry, with current being the most significant parameter affecting dilution. Dilution increases with higher current and voltage, while it decreases with increased NTD [10]. Sreeraj *et al.* demonstrated the effectiveness of a genetic algorithm (GA) for optimizing weld bead dimensions in GMAW cladding, surpassing MINITAB 15 results. The GA approach maximized reinforcement and weld width while minimizing penetration and dilution [11]. Senthilkumar *et al.* used a central composite rotatable design matrix to develop mathematical models for predicting clad bead geometry in GMAW for austenitic steel cladding. Weld bead width and reinforcement height increased with higher WFRs and nozzle-to-plate distances [12]. Terner *et al.* employed response surface methodology to predict bead geometry, hardness, and heat input in low-carbon steel cladding, identifying optimal welding parameters [13]. Chandel *et al.* explored mathematical modeling and optimization techniques for GMAW cladding processes, recommending further investigation into pulse type effects and the inclusion of qualitative factors [14]. Kannan developed a mathematical model using a factorial design matrix to predict weld deposit area in GMAW cladding of austenitic stainless steel, with welding current and nozzle-to-plate distance positively influencing the area [15].

Jha *et al.* identified dilution and bead geometry as crucial factors in cladding. Thermal conductivity and active components of the mixture affect the range of parameters for spray transfer, while oxidant potential influences arc length. The fume formation rate increases with $CO_2$ and $O_2$ content, as well as with arc temperature and instability [16]. Kannan and Jayaraman developed mathematical models using a full-factorial design matrix for predicting clad bead geometry in austenitic steel cladding. Factors such as WFR, welding speed, welding gun angle, and nozzle-to-plate distance were found to influence bead geometry significantly [17]. Sreeraj and Kannan used neural network models to accurately predict the clad bead geometry in stainless steel cladding deposited by GMAW, offering an alternative to traditional calculation methods [18]. Sreeraj *et al.* conducted GMAW experiments on austenitic stainless steel cladding and developed a mathematical model that accurately predicted bead geometry. Welding current, contact-to-plate distances, and welding speed were identified as significant factors affecting bead geometry, dilution, and reinforcement height, respectively [19]. Chaudhari and More studied the influence of process parameters on Vickers hardness and microstructure of weldments. Minitab software was used to analyze the relationship between input parameters and Vickers microhardness, showing variations in weldments attributed to welding process parameters [20].

Marinho *et al.* found that the presence of primary ferrite (FA) in cladding indicates low susceptibility to hot cracking, making it desirable for welded structures subjected to thermal cycles. Pulsed current GMAW resulted in a finer microstructure, while mean current significantly impacted secondary arm spacing [21]. Khara *et al.* highlighted that the cladding technique exhibited good reinforcement and low penetration, with a slightly increased hardness of the clad layer. Microstructural analysis revealed a favorable bond between the base material and cladding, although corrosion resistance was slightly lower compared to duplex stainless steel [22]. Verma *et al.* concluded that a moderate heat input with specific welding parameters resulted in a favorable microstructure and lower corrosion rate for cladding low alloy steel components with duplex stainless steel electrodes using GMAW [23]. Gao and Kao observed that the FeCoCr1.5NiAl coating exhibited a strong metallurgical bond with the matrix and demonstrated significantly increased hardness, leading to reduced wear volume. The cladding layer experienced abrasive wear, while the matrix encountered both abrasive and fatigue wear [24]. Liangyu *et al.* successfully bonded TA1 and 304 stainless steels using explosive welding, creating a

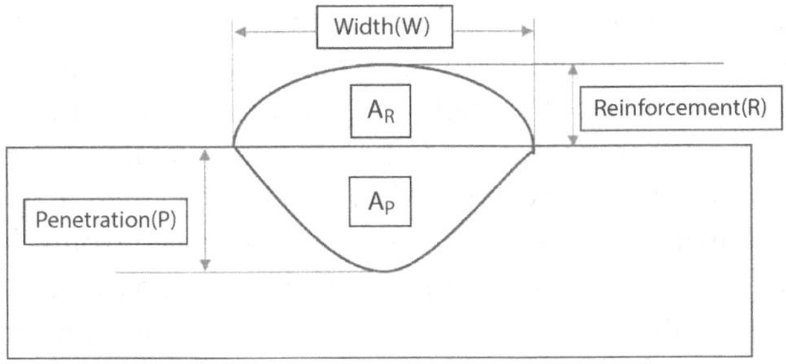


**Figure 3.1** Weld bead geometry.

titanium-stainless steel cladding plate with improved interfacial bonding strength, hardness, and corrosion resistance. Explosive welding proved effective in producing high-performance cladding materials [25].

In this investigation, GMAW was used to clad low-carbon steel plates (EN30) with consideration of voltage (*V*), WFR, travel speed (*S*), and gas flow rate (*G*) as cladding parameters. Experiments were conducted using a Taguchi L9 orthogonal array, with three levels for each parameter. Initially, graphical plots were used to investigate the influence of parameters on the resultant characteristics. Taguchi optimization with signal-to-noise (S/N) ratio analysis was applied further to determine the optimal cladding parameters during GMAW cladding of EN30 with ER309L wire. Figure 3.1 shows the weld bead geometry.

## 3.2 Experimental Work

The equipment and consumables listed were used for conducting an experiment involving GMAW on low-carbon steel plates (EN30). The Fronius TransPuls Synergic 3200 is a specific model of welding machine that provides a constant voltage power source for GMAW. It regulates the welding current and voltage to ensure a stable and controlled welding process. The stainless steel wire with a diameter of 1.2 mm (ER-308L) serves as the filler material in the welding process. It is used to add material to the weld joint and create a strong and durable bond between the low-carbon steel plates. The shielding gas cylinder contains a mixture of 98% argon and 2% carbon dioxide ($CO_2$). This mixture is commonly used as a shielding gas in GMAW. The shielding gas protects the welding area from atmospheric contamination, such as oxygen and nitrogen, which can negatively affect the quality of the weld. These plates are the base metal on which the welding experiment is conducted. Low-carbon steel, such as EN30, is a commonly used material for various applications due to its good weldability, strength, and affordability. By combining the constant voltage GMAW machine, the stainless steel wire filler, the shielding gas mixture, and the low-carbon steel plates, the experiment aimed to study the welding characteristics, joint quality, and other aspects of GMAW on the specified material.

Samples measuring 150 × 150 × 8 mm were obtained from a low-carbon steel plate of EN30 grade. One surface of these samples was thoroughly cleaned to eliminate any oxide and dirt before the cladding process. The cladding was carried out using ER-308 L stainless steel wire with a diameter of 1.2 mm, which was fed through a feeder. The chemical composition of the consumables is detailed in Table 3.1.

The specific experimental conditions for cladding will depend on the WFR, *V*, *G*, and *S* of the cladding process. Proper optimization and experimentation are typically required to achieve the desired properties and performance of the cladded material shown in Table 3.2. The experimental setup, namely, Fronius TransPuls Synergic 3200, is shown in Figure 3.2. Parameters are selected from the Taguchi design method listed in Table 3.3.

**Table 3.1** Chemical elements of LCS(EN30) and ER309L.

| Materials | Si | C | P | Mn | Mo | S | S | Ni | Cr |
|---|---|---|---|---|---|---|---|---|---|
| LCS(EN30) | 0.22 | 0.29 | 0.022 | 0.45 | 0.12 | 0.02 | 0.02 | 4.15 | 1.17 |
| ER309L | 1.0 | 0.03 | 0.045 | 2.0 | - | 0.03 | 0.03 | 10.0 | 19.0 |

**Table 3.2** Experimental condition during cladding.

| S. no. | Input parameters | Symbol | Levels | | |
|---|---|---|---|---|---|
| | | | 1 | 2 | 3 |
| 1 | Wire feed rate (m/min) | WFR | 6 | 7 | 8 |
| 2 | Voltage (V) | *V* | 20 | 22 | 24 |
| 3 | Gas Flow Rate (L/min) | *G* | 16 | 18 | 20 |
| 4 | Travel Speed (mm/min) | *S* | 240 | 260 | 280 |

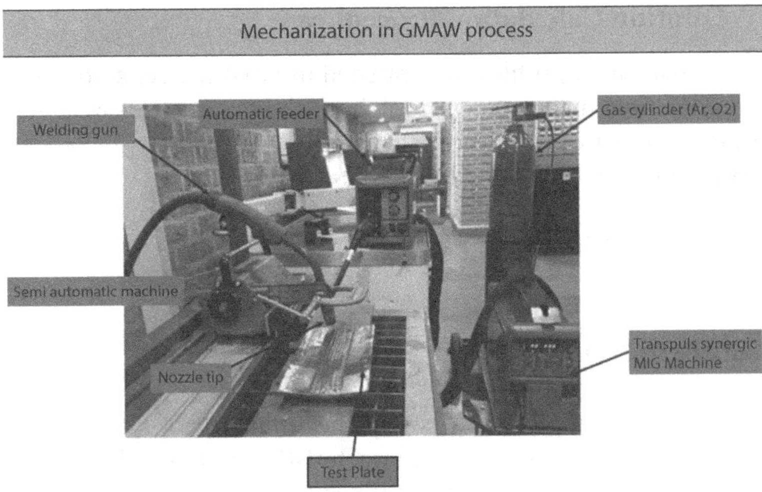


**Figure 3.2** Experimental setup for cladding.

**Table 3.3** Taguchi design with selected parameters.

| | Process parameters | | | | Weld responses | | | | | |
|---|---|---|---|---|---|---|---|---|---|---|
| **S. no.** | **WFR** | ***V*** | ***G*** | ***S*** | ***P* (mm)** | ***W* (mm)** | ***R* (mm)** | ***D* (mm)** | **PSF** | **RFF** |
| 1 | 6 | 20 | 16 | 240 | 0.720 | 8.265 | 3.567 | 10.04 | 11.479 | 2.317 |
| 2 | 6 | 22 | 18 | 260 | 1.247 | 10.326 | 3.297 | 20.19 | 8.281 | 3.132 |
| 3 | 6 | 24 | 20 | 280 | 1.311 | 11.339 | 3.175 | 20.95 | 8.649 | 3.571 |
| 4 | 7 | 20 | 20 | 260 | 1.377 | 9.423 | 3.499 | 18.06 | 6.843 | 2.693 |
| 5 | 7 | 22 | 16 | 280 | 1.543 | 12.686 | 3.332 | 22.70 | 8.222 | 3.807 |
| 6 | 7 | 24 | 18 | 240 | 1.646 | 12.635 | 3.289 | 25.10 | 7.676 | 3.962 |
| 7 | 8 | 20 | 18 | 280 | 2.237 | 10.796 | 3.482 | 28.93 | 4.826 | 3.841 |
| 8 | 8 | 22 | 20 | 240 | 1.925 | 12.583 | 3.374 | 32.20 | 6.537 | 3.964 |
| 9 | 8 | 24 | 16 | 260 | 2.148 | 15.542 | 3.228 | 33.43 | 7.235 | 3.729 |

*P*–penetration (mm), *R*–reinforcement (mm), *W*–bead width (mm), *D*–dilution(%), RFF–reinforcement form factor, PSF–penetration shape factor.

### 3.2.1 Dilution Calculation Procedure

All the bead-scanned profiles were opened in Foxit Reader software. With the help of an area measurement tool, both the penetration area ($A_P$) and the reinforcement area ($A_R$) were calculated and compared (Table 3.4). The formula used [6] to calculate dilution is as follows:

$$\text{Dilution (\%)} = [A_P / (A_P + A_R)] \text{ x } 100 \tag{3.1}$$

## 3.3 Parametric Study

Graphical plots were plotted to study the influence of GMAW process parameters on resultant response characteristics. Initially, the main effect plots were plotted that depict the influence of each welding parameter (*WFR, V, S, G*) on the welding variables on penetration shape factor (PSF), reinforcement form factor (RFF), and dilution (D). Furthermore, contour plots based on the interactive effect of significant parameters are plotted to study the interactive effect of process parameters on the resultant characteristics. Figure 3.3 shows the cross sections of weld bead samples.

**Table 3.4** Dilution calculation in Taguchi design samples.

| **Weld specimen** | **Penetration area ($A_P$)** | **Reinforcement area ($A_R$)** | **Total bead area ($A_P + A_R$)** | **Dilution (%)** |
|---|---|---|---|---|
| 1 | 0.22 | 1.97 | 2.19 | 10.04 |
| 2 | 0.41 | 1.62 | 2.03 | 20.19 |
| 3 | 0.44 | 1.66 | 2.10 | 20.95 |
| 4 | 0.41 | 1.86 | 2.27 | 18.06 |
| 5 | 0.52 | 1.77 | 2.29 | 22.70 |
| 6 | 0.60 | 1.79 | 2.39 | 25.10 |
| 7 | 0.68 | 1.67 | 2.35 | 28.93 |
| 8 | 0.86 | 1.81 | 2.67 | 32.20 |
| 9 | 1.09 | 2.17 | 3.26 | 33.43 |

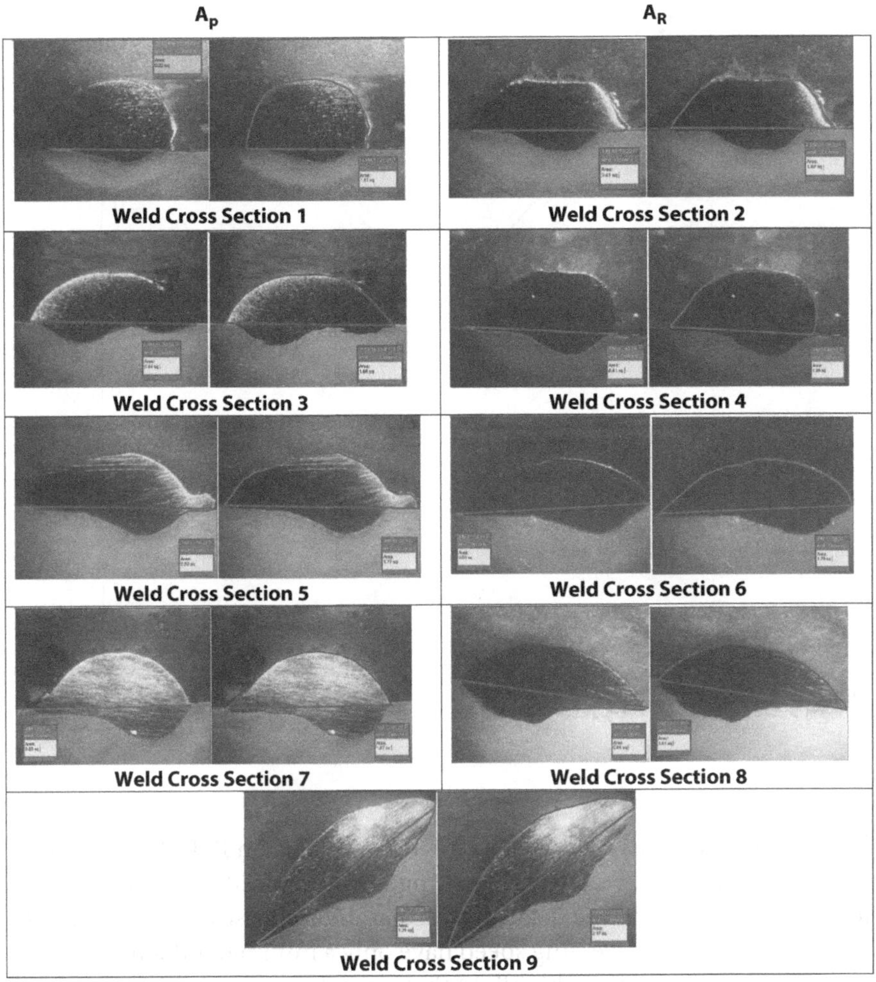


**Figure 3.3** Cross sections of weld bead samples.

### 3.3.1 Influences on PSF

Figure 3.4 illustrates the main effect plot for PSF. As observed, the PSF is decreasing with the increase in the *WFR*, this happens due to an increase in heat input at a higher welding current. In the GMAW process, the increase in *WFR* tends to increase the welding current. The PSF is the ratio of bead width to penetration. As the weld penetration increases, the PSF will decrease. *V* has a very minimal effect on PSF; however, increases in voltage cause wider bead width, which increases the PSF. The higher the welding speed, the lower in PSF S indicates. The gas flow rate indicates higher PSF at a flow rate of 16 lpm.

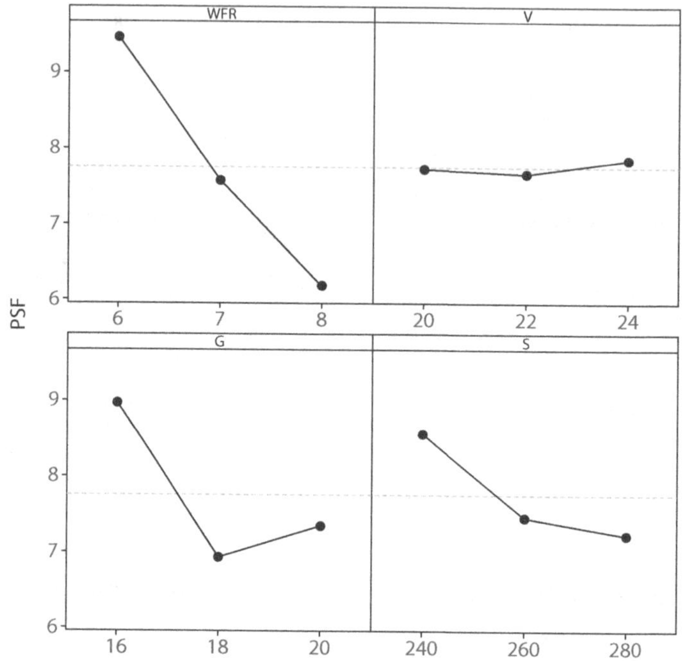


**Figure 3.4** Main effect plot of PSF.

Figure 3.5 shows the contour plots for *WFR, V, S, G* on PSF. An increase in WFR, attributed to greater wire melting and enhanced penetration, leads to increased dilution. The data suggest that dilution rises with higher voltage, while changes in welding speed have only a minimal influence on dilution. Notably, the findings highlight that a gas flow rate of 18 lpm results in a higher degree of dilution.

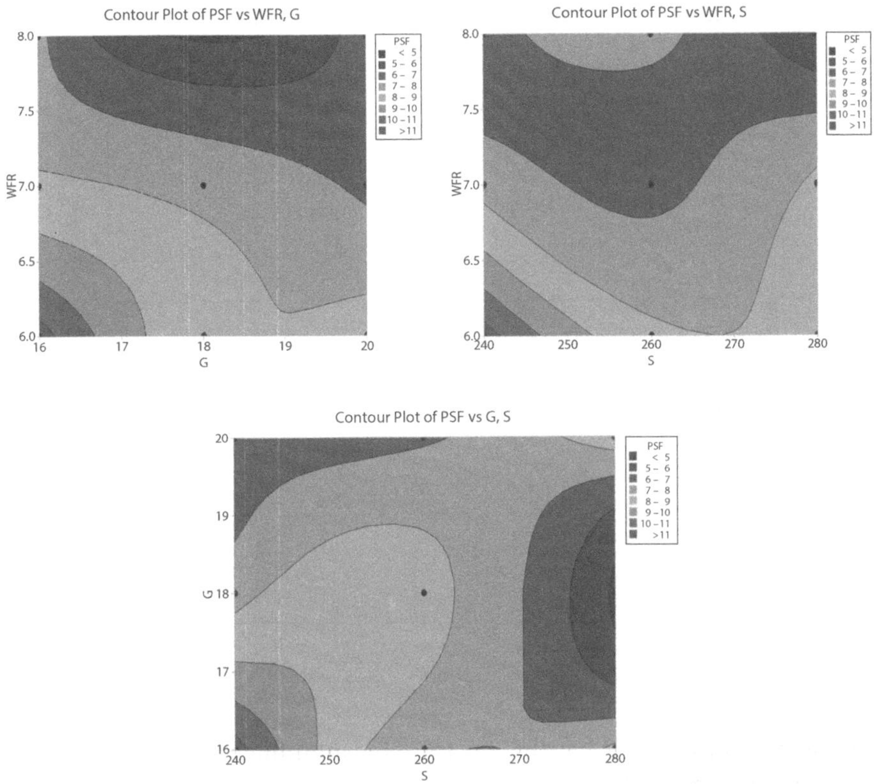


**Figure 3.5** The influence of *WFR*, *V*, *S*, *G* on PSF.

## 3.3.2 Influences on RFF

The influence of *WFR*, *V*, *S*, and *G* on RFF is depicted using a main effect plot (Figure 3.6) and contour plots (Figure 3.7). WFR is identified as the most influencing parameter followed by *V* on resultant RFF. The intermediate effect of *G* and *S* is also observed as evident in Figure 3.6. A lower setting of WFR, *V*, and *G* is found ideal for a lower value of RFF. Furthermore, the RFF is increasing with the increase in the *WFR*, the higher *WFR* leads to high welding current and melting of more filler per unit length. The rise in voltage results in an increase in the RFF due to the widening of the bead. Similarly, an increase in welding speed leads to higher RFF due to the formation of a narrower bead and greater reinforcement.

## 3.3.3 Influence on D

The main effect plot (Figure 3.8) and contour plots (Figure 3.9) illustrate the impact of WFR, *V*, *S*, and *G* on dilution (*D*). Figure 3.8 clearly

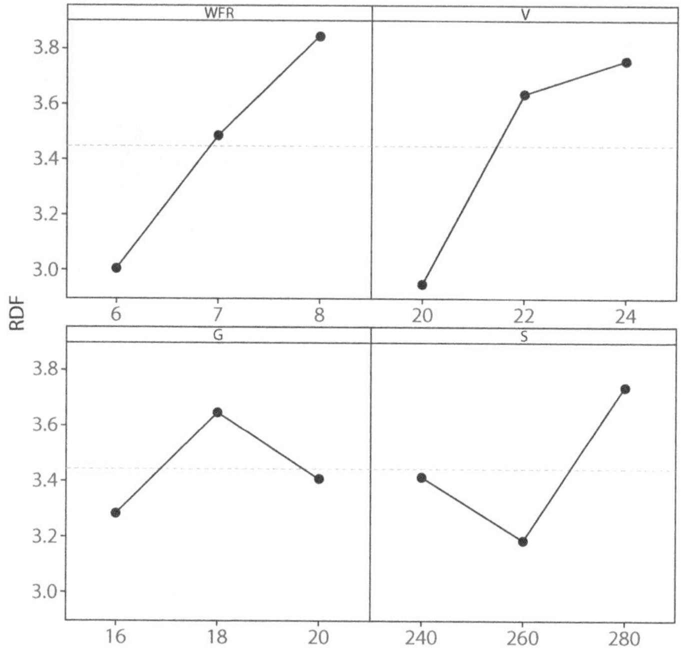


**Figure 3.6** Main effect plot of RFF.

demonstrates that WFR has the most significant influence on the resultant RFF, followed by V. The figure also illustrates the nominal impacts of G and S.

## 3.4 Taguchi Optimization

The Taguchi methodology is a straightforward and resilient approach used for designing experiments and optimizing process parameters to minimize process variation. In the Taguchi optimization technique, the impact of uncontrollable noise factors on process or product variability is mitigated by identifying the optimal control factor levels through the application of robustness measures. The optimal configuration of process variables, aimed at minimizing product variability caused by noise factors, is ascertained by assessing the S/N ratio. Higher S/N ratios are preferred as they signify the best process variable settings for minimizing the impact of noise factors. Taguchi optimization is carried out in two steps [26–30].

1. In this step, the responses are converted into S/N ratios. S/N ratios represent a logarithmic transformation of the desired

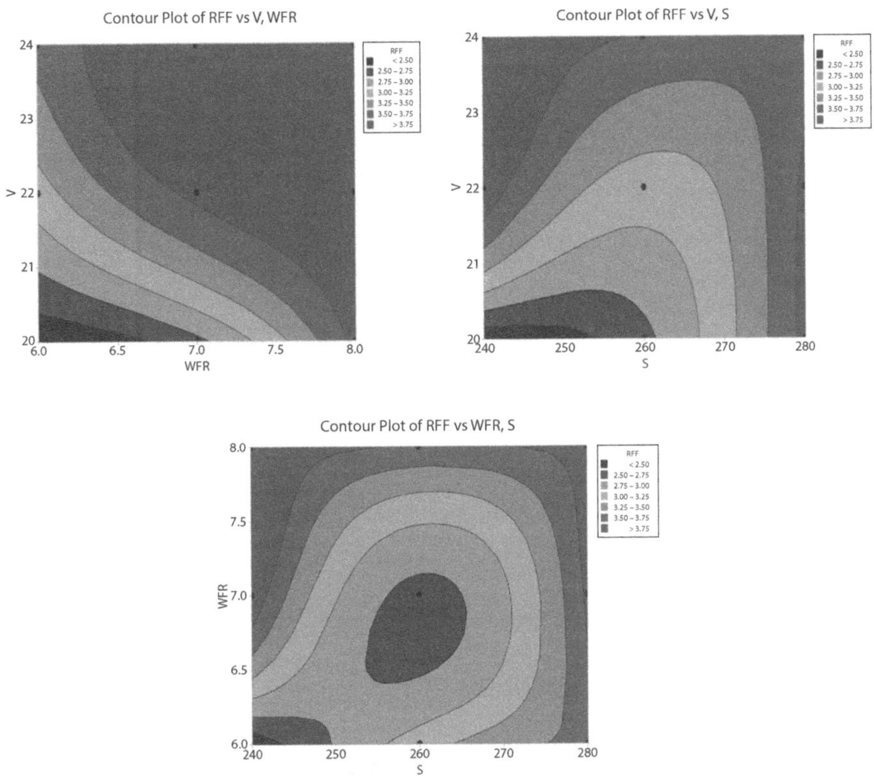


**Figure 3.7** The influence of *WFR*, *V*, *S*, and *G* on RFF.

response and are computed to establish the optimal configuration of process variables. Various types of S/N ratios are chosen, namely, (i) 'the-smaller-the-better,' (ii) 'the-larger-the-better,' and (iii) 'the-nominal-the-best,' based on whether the objective is to minimize, maximize, or achieve a nominal response, respectively.

2. In the second step, an analysis is conducted to identify process variables that shift the mean toward the target while having minimal or no impact on the S/N ratio. In this present investigation, RFF and D are required to be minimized, while the PSF needs to be maximized. Hence, the larger-the-better S/N ratio ($LB_{SN}$) is used for the analysis of PSF, while the smaller-the-better S/N ratio ($SB_{SN}$) is used to analyze RFF and D. Mathematically, $LB_{SN}$ and $SB_{SN}$ are given as in Equations (3.2) and (3.3) respectively

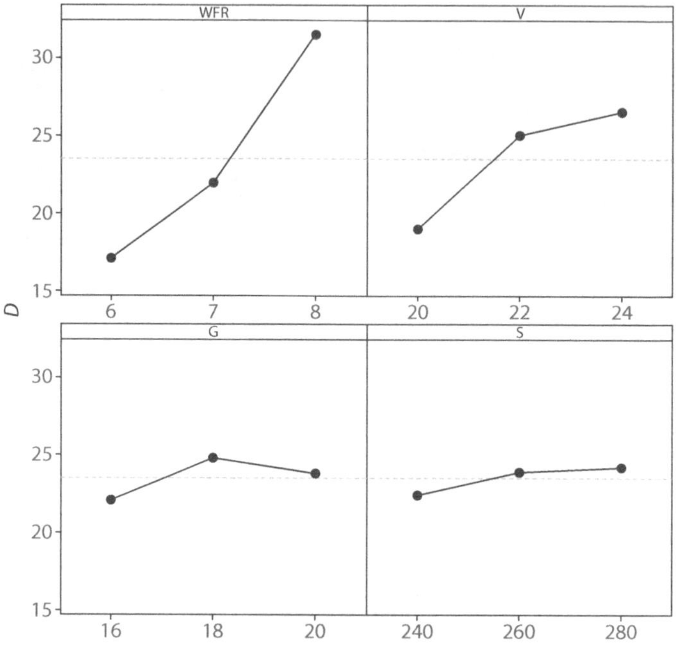


**Figure 3.8** Main effect plot of Dilution (D).

$$LB_{SN} = -10 \times log_{10}\left[\frac{1}{n}\sum_{i=1}^{n}\frac{1}{Y_i^2}\right] \tag{3.2}$$

$$SB_{SN} = -10 \times log_{10}\left[\frac{1}{n}\sum_{i=1}^{n}Y_i^2\right] \tag{3.3}$$

where $Y_i$ and $n$ are $i$th response variable and number of observations, respectively.

The optimal parameter configuration is determined using analysis of means (ANOM), a method for assessing the primary effects of each parameter. The effect of a parameter level is quantified as the deviation it induces from the overall mean response (m), as described in Equation 3.4

$$m = \frac{1}{k}\sum_{k=1}^{k}\eta_k \tag{3.4}$$

where $n$ is the experimental trial, and $\eta$ is the $S/N$ ratio.

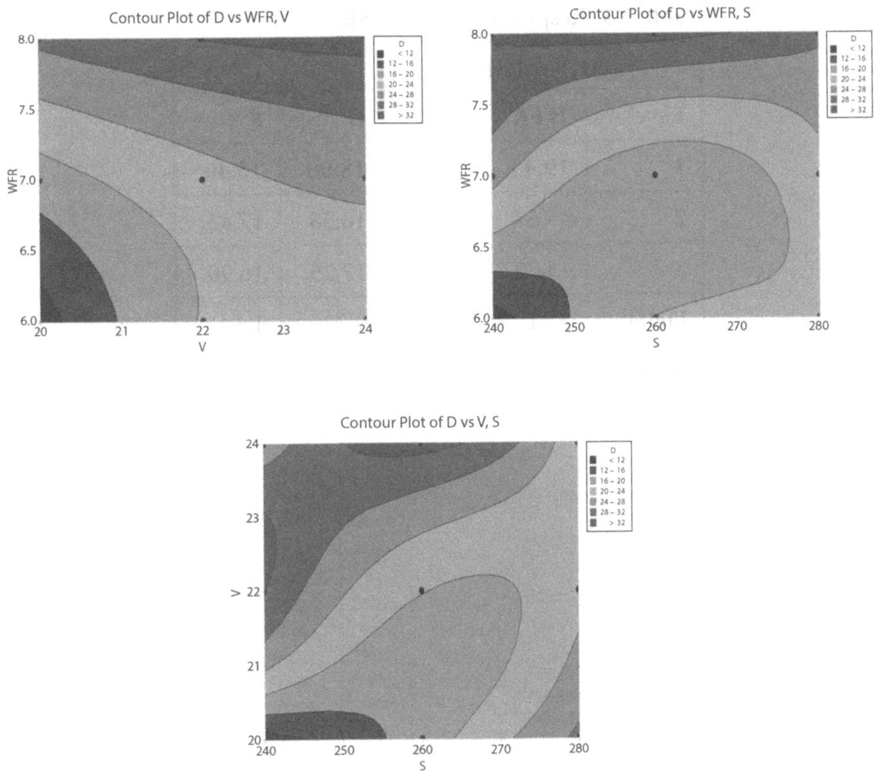


**Figure 3.9** Influence of *WFR, V, S, G* on dilution (D).

### 3.4.1 Determination of Optimal Cladding Conditions

For obtaining the optimal cladding parameters for a minimum RFF and *D* and maximum PSF, the S/N ratio was obtained and the main effect plots of S/N ratios are plotted. The delta statistic and rank-of-factor are also summarized. The delta statistic is calculated by differentiating the maximum and minimum of the mean effect for every parameter. The parameter with the highest delta value is considered as the most significant on the resultant characteristics, while the lowest delta value signifies the least significant process parameter.

### 3.4.2 Maximization of PSF

To maximize the PSF of cladded GMAW weldments, the 'larger-the-better' S/N ratio relation is employed using Equation 3.2. The table presents the obtained S/N ratios, while the corresponding main effect plots of these S/N ratios are depicted in Figure 3.10.

**Table 3.5** Response table for PSF.

| | For PSF | | | |
|---|---|---|---|---|
| **Level** | ***WFR*** | ***V*** | ***G*** | ***S*** |
| **1** | **19.43** | 17.19 | **18.90** | **18.40** |
| **2** | 17.57 | 17.66 | 16.58 | 17.42 |
| **3** | 15.72 | **17.88** | 17.25 | 16.90 |
| **Delta** | 3.71 | 0.69 | 2.32 | 1.50 |
| **Rank** | 1 | 4 | 2 | 3 |

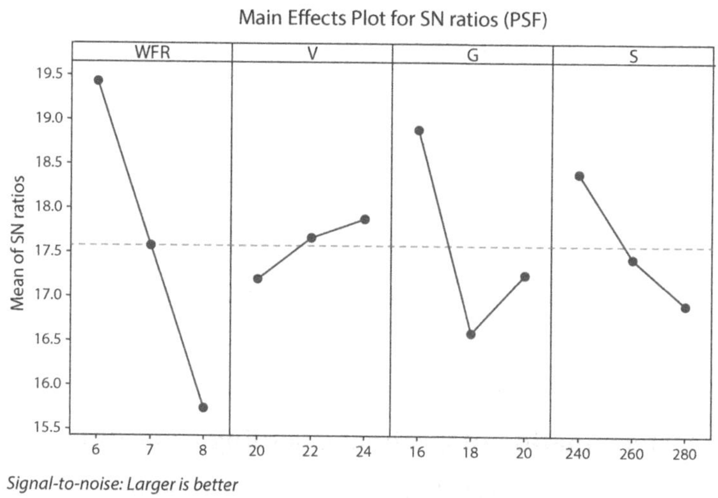


**Figure 3.10** Main effect plot for PSF.

Based on the result stated in Table 3.5, *WFR* and *G* are established as the most significant process variables affecting the S/N ratio for PSF. The next most significant parameter is found as *S* followed by *V*. The optimal parametric setting for a maximum PSF is identified as $WFR_1V_3G_1S_1$ ($WFR$ = 6 m/min, $V$ = 24 V, $G$ = 16 L/min, $S$ = 240 mm/min) based on the factor levels resulting in the highest value of S/N ratio. Figure 3.10 also confirms the optimal parameter setting obtained for PSF.

### 3.4.3 Minimization of RFF

To minimize *RFF* of cladded GMAW weldments, the 'smaller-the-better' S/N ratio relation is employed using Equation 3.3. The S/N ratio obtained

**Table 3.6** Response table for RFF.

| | For RFF | | | |
|---|---|---|---|---|
| **Level** | ***WFR*** | *V* | *G* | *S* |
| **1** | **−9.42** | **−9.20** | **−10.11** | −10.41 |
| **2** | −10.73 | −11.16 | −11.19 | **−9.98** |
| **3** | −11.69 | −11.48 | −10.54 | −11.45 |
| **Delta** | 2.27 | 2.29 | 1.07 | 1.47 |
| **Rank** | 2 | 1 | 4 | 3 |

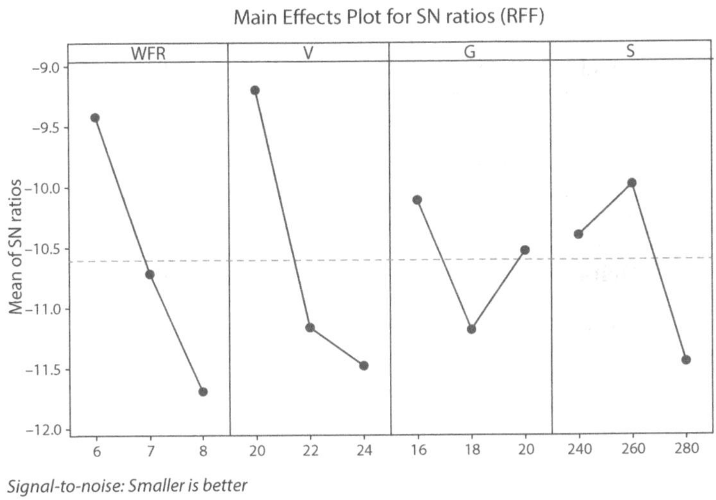


**Figure 3.11** Main effect plot for RFF.

is stated in Table 3.6. The respective main effect plots of S/N ratios are depicted in Figure 3.11.

In terms of the RFF, the most influential process parameter is *V*, with WFR and *S* being the next in significance. On the other hand, *G* is found to have the least impact on the resulting RFF. The optimal parametric setting for a minimum *RFF* is identified as $WFR_1V_1G_1S_2$ (*WFR* = 6 m/min, *V* = 20 V, *G* = 16 L/min, *S* = 260 mm/min) which is again confirmed by the main effect plots depicted in Figure 3.11.

### 3.4.4 Minimization of D

To minimize *D* of cladded GMAW weldments, the 'smaller-the-better' S/N ratio relation is employed using Equation 3.3. Table 3.7 presents the obtained S/N ratios, while the corresponding main effect plots of these ratios are illustrated in Figure 3.12.

Again for *D*, *WFR* is established as the most significant process parameter with a delta value of 5.77 followed by *V*. The next most significant parameter is found as *G* followed by *S*. Based on Taguchi analysis, the optimal cladding parameter setting of $WFR_1V_1G_1S_1$ ($WFR$ = 6 m/min, $V$ = 20

**Table 3.7** Response table for D.

| | **For D** | | | |
|---|---|---|---|---|
| **Level** | ***WFR*** | ***V*** | ***G*** | ***S*** |
| **1** | **−24.19** | **−24.80** | **−25.88** | **−26.06** |
| **2** | −26.75 | −27.79 | −27.77 | −27.24 |
| **3** | −29.96 | −28.30 | −27.24 | −27.59 |
| **Delta** | 5.77 | 3.50 | 1.90 | 1.53 |
| **Rank** | 1 | 2 | 3 | 4 |

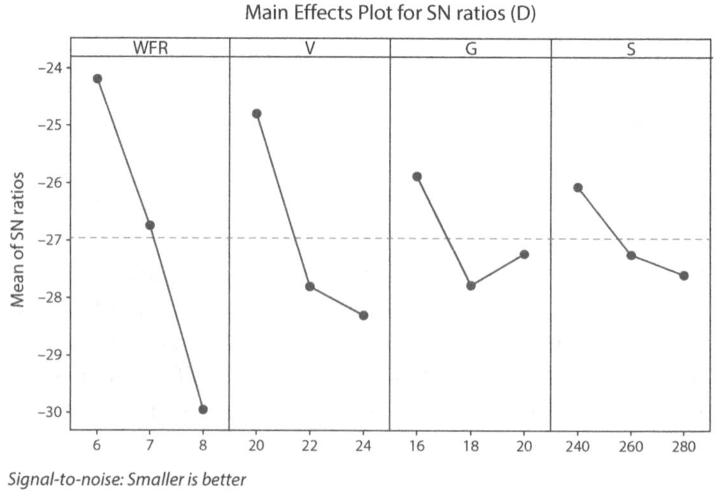


**Figure 3.12** Main effect plot for *D*.

**Table 3.8** Optimal cladding parameters.

| Responses | Factor | Optimal cladding parameters | | | |
|---|---|---|---|---|---|
| | | ***WFR*** | ***V*** | ***G*** | ***S*** |
| PSF | Level | 1 | 3 | 1 | 1 |
| | Value | 6 m/min | 24 V | 16 L/min | 240 mm/min |
| RFF | Level | 1 | 1 | 1 | 2 |
| | Value | 6 m/min | 20 V | 16 L/min | 260 mm/min |
| D | Level | 1 | 1 | 1 | 1 |
| | Value | 6 m/min | 20 V | 16 L/min | 240 mm/min |

V, $G = 16$ L/min, $S = 240$ mm/min) is identified for a minimum percentage dilution. The main effect plot of the S/N ratio given in Figure 3.12 also confirms the obtained optimum cladding conditions as $WFR_1V_1G_1S_1$. Overall, the complete optimization results using Taguchi analysis for maximum PSF and minimum RFF and D are tabulated in Table 3.8 during the cladding of the low-carbon steel plate of grade EN30.

## 3.5 Conclusions

This investigation is based on a cladding investigation using the GMAW process. The conclusions drawn from the investigation are as follows:

1. The WFR is found to be the most dominant factor affecting the performance characteristics. As the WFR (wire feed rate) increases, it results in a higher welding current, causing more filler material to melt per unit length. This, in turn, leads to increased penetration during welding and consequently higher dilution.
2. The Taguchi results reveal that, in the case of PSF, the most influential process variables affecting the S/N ratio are WFR and *G*, with *S* following closely in significance. Likewise, in the case of RFF, *V* is identified as the most crucial process parameter, followed by WFR and *S*. Moreover, concerning *D*, WFR stands out as the most significant process parameter, with a delta value of 5.77, followed by *V*.

3. Utilizing Taguchi optimization, the maximum PSF is achieved at parametric setting $WFR_1V_3G_1S_1$ ($WFR$ = 6 m/min, $V$ = 24 V, $G$ = 16 L/min, $S$ = 240 mm/min). Conversely, the minimum RFF is obtained when the parameters set as $WFR_1V_1G_1S_2$ ($WFR$ = 6 m/min, $V$ = 20 V, $G$ = 16 L/min, $S$ = 260 mm/min) are identified.
4. For achieving the minimum percentage dilution, the optimal parameter setting for cladding is identified as $WFR_1V_1G_1S_1$ ($WFR$ = 6m/min, $V$ = 20 V, $G$ = 16 L/min, $S$ = 240 mm/min).

## Acknowledgement

The author extends heartfelt gratitude to the Director General of AKGEC and the Head of CORE at Ajay Kumar Garg Engineering College, Ghaziabad.

## References

1. Kumar, S. and Shahi, A.S., Effect of heat input on the microstructure and mechanical properties of gas tungsten arc welded AISI 304 stainless steel joints. *Mater. Des.*, 32, 3617–3623, 2011, https://doi.org/10.1016/j.matdes.2011.02.017.
2. Vashishtha, H., Taiwade, R.V., Sharma, S., Patil, A.P., Effect of welding processes on microstructural and mechanical properties of dissimilar weldments between conventional austenitic and high nitrogen austenitic stainless steels. *J. Manuf. Process.*, 25, 49–59, 2017, https://doi.org/10.1016/j.jmapro.2016.10.008.
3. Verma, A.K. and Das, S., A Brief overview on Cladding Techniques with a Reference to Weld Cladding Using GMAW. *Reason-Tech. J.*, X, 45–48, 2011, DOI:10.21843/reas/2011/45-48/108212.
4. Chandel, V.S., Om, H., Goyal, S., Identification of Factors Affecting the Selection of Stainless Steel Cladding Process. *J. Mater. Sci. Mech. Eng.*, 2, 10, 65–69, 2015, http://www.krishisanskriti. org/jmsme.html.
5. Pires, I., Quintino, L., Miranda, R.M., Analysis of the influence of shielding gas mixtures on the GMAW metal transfer modes and fume formation rate. *Mater. Des.*, 28, 1623–1631, 2007, https://doi.org/10.1016/j.matdes.2006.02.012.
6. Senthilkumar, B., Kannan, T., Surendran, P., Influence of Welding Process Parameters on Bead Geometry-A Review. *J. Mech. Mech. Eng.*, 1, 1–13, 2015.

7. Mahapatra, S., Laha, A., Dey, A., Mondal, A., Bhowmik, B., An Overview on Weld Cladding. *Int. J. Sci. Eng. Res.*, 7, 4, 194–203, 2016.
8. Shahi A, S. and Pandey, S., Modelling of the effects of welding conditions on dilution of stainless steel claddings produced by gas metal arc welding procedures. *J. Mater. Process. Technol.*, 196, 339–344, 2008, doi: 10.1016/j.jmatprotec.2007.05.060.
9. Sreeraj, P., Kannan, T., Maji, S., Estimation of Optimum Dilution in the GMAW Process Using Integrated ANN-GA. *J. Eng.*, 2013, 1–17, 2012, http://dx.doi.org/10.1155/2013/285030.
10. Kumar, V., Singh, G., Yusufzai, M.Z.K., Effects of Process Parameters of Gas Metal Arc Welding on Dilution in Cladding of Stainless Steel on Mild Steel. *MIT Int. J. Mech. Eng.*, 2, 2, 127–131, 2012.
11. Sreeraj, P., Kannan, T., Maji, S., Optimization of Weld Bead Geometry in Gas Metal Arc Cladded Austenitic Stainless Steel Plates Using Genetic Algorithm. *Asian Rev. Mech. Eng.*, 2, 1, 7–15, 2013, https://doi.org/10.51983/arme-2013.2.1.2328.
12. Senthilkumar, B., Birundha, P., Kannan, T., Modelling and Simulation of Austenitic Stainless Steel Claddings deposited by GMAW. *Int. J. Sci. Eng. Res.*, 5, 5, 363–370, 2014.
13. Terner, M., Bayarsaikhan, T.-A., Hong, H.-U., Lee, J.-H., Influence of Gas Metal Arc Welding Parameters on the Bead Properties in Automatic Cladding. *J. Weld. Join.*, 35, 1, 16–25, 2017, https://doi. org/10.5781/JWJ.2017.35.1.16.
14. Chandel, V.S., Om, H., Goyal, S., Gas Metal ARC Weld Cladding Process: A Review. *Int. J. Eng. Res. Technol.*, 8, 1, 69–73, 2018.
15. Kannan, T., Effects of Gas Metal Arc Welding Parameters on Weld Deposit Area of Stainless Steel Claddings. *J. Manuf. Sci. Prod.*, 10, 2, 123–135, 2009, DOI:10.1515/IJMSP.2009.10.2.123.
16. Jha, T.K., Bhardwaj, B., Bhagat, K., Sharma, V., Prediction and optimization of weld bead geometry in gas metal arc welding process using d-optimal methodology. *Int. J. Sci. Eng. Technol.*, 2, 7, 1451–1454, 2014.
17. Kannan, T. and Jayaraman, Y., Effect of process parameters on clad bead geometry and its shape relationships of stainless steel claddings deposited by GMAW. *Int. J. Adv. Manuf. Technol.*, 47, 9, 1083–1095, 2009, DOI:10.1007/s00170-009-2226-1.
18. Sreeraj, P. and Kannan, T., Modelling and Prediction of Stainless Steel Clad Bead Geometry Deposited by GMAW Using Regression and Artificial Neural Network Models. *Adv. Mech. Eng.*, 2012, 1–12, 2015, doi: 10.1155/2012/237379.
19. Sreeraj, P., Kannan, T., Subhasismaji, Sensitivity analysis of process parameters in cladding of stainless steel by GMAW. *Mach. Des.*, 5, 1–10, 2013.
20. Chaudhari, P.D. and More, N.N., Effect of Welding Process Parameters on Micro hardness and Microstructure. *Int. J. Eng. Res. Technol.*, 3, 5, 1937–1942, 2014.

21. Marinho, F.L.C., Quintana, F.L.P., Filho, C.S.A., Mota, C.A.M., Barga, E.M., Microstructural analysis of stainless steel cladding deposited with pulsed current and conventional continuous current gmaw processes. *ABCM J.*, 25, 1508, 2005.
22. Khara, B., Mandal, N.D., Sarkar, A., Sarkar, M., Chakarbarti, B., Das, S., Weld Cladding with Austenitic Stainless Steel for Imparting Corrosion Resistance. *Indian Weld. J.*, 49, 1, 74–81, 2016, DOI:10.22486/iwj/2016/v49/ i1/125902.
23. Verma, A.K., Biswas, B.C., Roy, P., De, S., Saren, S., Das, S., On the Effectiveness of Duplex Stainless Steel Cladding Deposited by Gas Metal Arc Welding. *Indian Weld. J.*, 47, 4, 24, 2014, DOI:10.22486/iwj.v47i4.141081.
24. Gao Y, L. and Kao, G.P., Performance analysis of FeCoCr1.5NiAl high entropy alloy coating on H13 steel by laser cladding. *J. Phys.: Conf. Ser.*, 2256, 2022, 012022, 2022, doi: 10.1088/1742-6596/2256/1/012022.
25. Liangyu, L., Yong, S., Jian, C., Yu, F., Xiaoyuan, X., Jin, Y., Study on Microstructure and Properties of TA1-304 Stainless Steel Explosive Welding Cladding Plate. *Mater. Res. Express*, 7, 2, 026557, 2020, https://doi.org/10.1088/2053-1591/ab7357.
26. Maniraj, S. and Thanigaivelan, R., Optimization of electrochemical micromachining process parameters for machining of AMCs with different% compositions of GGBS using Taguchi and TOPSIS methods. *Trans. Indian Inst. Met.*, 72, 3057–3066, 2019.
27. Mythili, T. and Thanigaivelan, R., Optimization of wire EDM process parameters on Al6061/$Al_2O_3$ composite and its surface integrity studies. *Bull. Pol. Acad. Sci.: Tech. Sci.*, 68, 1403–1412, 2020.
28. Saravanan, K.G., Thanigaivelan, R., Soundarrajan, M., Comparison of electrochemical micromachining performance using TOPSIS, VIKOR and GRA for magnetic field and UV rays heated electrolyte. *Bull. Pol. Acad. Sci.: Tech. Sci.*, 69, 5, 1–10, 2021.
29. Gunasekaran, K., Pradeep Kumar, G., Thanigaivelan, R., Arunachalam, R., Shanmugam, V., Optimization of turning parameters of cryogenic soaked AZ91 magnesium alloy using TOPSIS coupled Taguchi technique. *J. New Mater. Electrochem. Syst.*, 24, 1, 49–54, 2021, https://doi.org/10.14447/jnmes.v24i1.a09.
30. Thanigaivelan, R., Arunachalam, R.M., Jerald, J., Niranjan, T., Applications of Taguchi technique with fuzzy logic to optimise an electro- chemical micromachining process. *Int. J. Exp. Des. Process Optim.*, 2, 4, 283–298, 2011.

# 4

# Design and Development of Brake Pedal by Topology Optimization Approach

**Sivashankar N.[1]*, Jagadeesha T.[2], Mothilal T.[3] and Natrayan L.[4]**

*[1]Department of Mechanical Engineering, Kongunadu College of Engineering and Technology (Autonomous), Trichy, Tamil Nadu, India*
*[2]Department of Mechanical Engineering, National Institute of Technology NIT, Calicut, Kerala, India*
*[3]Department of Automobile Engineering, KCG College of Technology, KCG Nagar, Karapakkam, Chennai, Tamil Nadu, India*
*[4]Department of Mechanical Engineering, Saveetha School of Engineering, SIMATS, Chennai, Tamil Nadu, India*

## *Abstract*
This study presents an innovative approach to the design and development of a brake pedal using topology optimization techniques. The objective is to enhance the performance and efficiency of the brake pedal while maintaining structural integrity. Topology optimization is employed as a powerful tool to iteratively refine the geometry of the brake pedal, redistributing material to minimize weight while ensuring that stress and displacement constraints are met. The methodology encompasses the definition of loading conditions, constraints, and material properties, followed by the application of generative design algorithms to derive an optimal structural layout. The proposed approach showcases the practical applicability of advanced optimization techniques in the automotive industry, paving the way for improved component designs with enhanced performance characteristics.

***Keywords*:** Optimization, topology, fusion 360, weight reduction, brake pedal

**Corresponding author*: sivashankar1911@gmail.com

R. Thanigaivelan, Pradeep Kumar Krishnan, Kamalakanta Muduli and Santosh Kumar Tamang (eds.) New Materials, Processing and Manufacturability: Fabrication and Processing of Advanced Materials, (57–70) 

## 4.1 Introduction

Topology optimization is a groundbreaking computational technique that revolutionizes the way engineers and designers approach the problem of creating optimal structures and designs. Rooted in the fields of mathematics, engineering, and computer science, topology optimization is a powerful tool that aims to discover the most efficient and high-performing configuration of materials within a given design space, subject to specific constraints [1]. Traditionally, design processes have relied on intuition, trial and error, and iterative methods to arrive at efficient structures. However, as industries demand ever more lightweight, robust, and innovative solutions, the limitations of conventional design approaches have become increasingly apparent. Topology optimization addresses these limitations by utilizing advanced algorithms and mathematical formulations to automatically generate optimized shapes and layouts, often resulting in designs that exhibit remarkable improvements in terms of structural integrity, weight reduction, and material utilization [2, 3]. At its core, topology optimization explores the distribution of material within a defined space to achieve specific goals, such as minimizing stress concentrations, maximizing stiffness, or enhancing heat dissipation. By formulating the design problem as an optimization task, engineers can guide the algorithm to create designs that outperform traditional handcrafted ones, often leading to unconventional and innovative solutions that might not have been considered through traditional methods. The implementation of topology optimization involves intricate algorithms and computational methods, leveraging principles from structural mechanics, finite element analysis, and optimization theory [4]. These algorithms iteratively adjust the distribution of material while considering factors like load conditions, boundary constraints, and design objectives. The result is a design that can be fine-tuned further by engineers, incorporating their domain knowledge and adjusting for manufacturing feasibility. In a world where industries are continuously pushing the boundaries of what is possible, topology optimization emerges as a transformative tool. Its ability to uncover novel and optimal design solutions while enhancing resource utilization holds promise across sectors ranging from aerospace and automotive engineering to architecture and consumer goods. As technology advances and computational capabilities grow, topology optimization is poised to play an increasingly integral role in shaping the future of design and innovation [5].

Weight reduction is a critical factor in the design and performance of modern bicycle components. As cycling continues to evolve into a pursuit of greater speed, efficiency, and maneuverability, the significance of minimizing the weight of bike parts becomes increasingly evident. Each gram shaved-off component can lead to improved acceleration, enhanced handling, and increased overall efficiency, ultimately translating into a more enjoyable and competitive cycling experience. Bicycles are essentially a delicate balance between the rider's power output and the resistance they encounter. Every extra ounce of weight in the components adds to the overall load the rider has to propel, especially during uphill climbs or rapid accelerations [6]. As such, manufacturers and engineers have recognized the advantages of investing in materials and design methodologies that prioritize weight reduction without compromising structural integrity and safety. In competitive cycling, where marginal gains can determine victory or defeat, the importance of lightweight components is particularly pronounced. Professional cyclists and enthusiasts alike seek the cutting-edge technology that allows them to maximize their potential and achieve peak performance. Whether it is reducing the mass of the frame, wheels, drivetrain, or other critical parts, the cumulative effect of these weight reductions can significantly impact the overall performance of the bicycle [7].

Moreover, weight reduction in bike parts does not just cater to elite athletes. It also benefits recreational riders and commuters who seek more comfortable, efficient, and enjoyable rides. A lighter bike is easier to handle, requires less effort to pedal, and can lead to reduced fatigue during longer rides. Additionally, it can be particularly advantageous for those who rely on their bikes for commuting, as lighter components make it easier to navigate urban environments and tackle inclines. Advancements in material science, such as the use of carbon fiber, titanium, and high-strength alloys, have allowed engineers to push the boundaries of weight reduction while maintaining durability and safety. Computer-aided design and advanced manufacturing techniques further contribute to the creation of intricate and optimized structures that achieve a delicate equilibrium between weight, strength, and performance. In this exploration of the importance of weight reduction in bike parts, we will delve into the various aspects that underscore its significance. From the technological innovations driving lighter component production to the real-world implications for cyclists of all levels, a comprehensive understanding of this aspect of bicycle design promises to enhance the riding experience and push the boundaries of what is achievable on two wheels [8]. Many optimization techniques are currently available in recent days which include TOPSIS techniques, ANOVA analysis, and RSM. All these techniques are used to find the optimum parameter value in machining, casting, and welding [9–15].

Sudin *et al.* [16] have performed topology optimization to reduce the weight of the brake pedal. CATIA V5 was used to design the model of the brake pedal. Altair OptiStruct software was used to do topology optimization. The end result indicated that a 22% weight reduction was achieved. Oğuz *et al.* [17] have performed surface optimization for weight reduction of automobile pedals and achieved nearly 40% accurate design. MATLAB was used to do optimization. Arun *et al.* [18] have studied and performed the optimization to reduce the weight reduction in the brake pedal. They reduced the weight by 21.2% when compared to the original design. Many works were carried out to reduce the weight of the brake pedals. Most of the research works were forced to reduce weight by using MATLAB and Altair OptiStruct software. They mostly achieved a 22% weight reduction. Hence, this work is focused on reducing the weight of the brake pedal with the help of Autodesk Fusion 360. Commonly used brake pedal was used to design the pedal.

## 4.2 Structure Optimization

Structure optimization refers to the process of finding the optimal configuration or arrangement of components within a system or structure. This process is commonly employed in various fields such as engineering, chemistry, materials science, and computer science. The goal of structure optimization is to find a configuration that maximizes or minimizes a specific objective while satisfying certain constraints. Structure optimization can take various forms depending on the nature of the problem and the specific goals. Here are some common types of structure optimization:

Topology Optimization: In topology optimization, the goal is to determine the optimal distribution of material within a given design space to achieve specific performance objectives. This is often used in engineering and manufacturing to create lightweight and efficient structures. The algorithm decides which regions of the design space should be filled with material and which should be left void.

Shape Optimization: Shape optimization involves finding the optimal shape of a given structure while keeping its size and location fixed. This is useful in fields such as aerodynamics, where the goal is to design shapes that minimize drag or improve airflow.

Parameter Optimization: In parameter optimization, specific design parameters are varied to optimize a structure's performance. This could involve adjusting dimensions, angles, or other geometric properties of the components. Genetic algorithms and gradient-based methods are often used for parameter optimization.

Material Optimization: Material optimization focuses on finding the best combination of materials for a structure to achieve desired properties like strength, stiffness, and durability. This is common in materials science and engineering.

Multi-Objective Optimization: Sometimes, a design problem involves multiple conflicting objectives. Multi-objective optimization aims to find solutions that balance these objectives. For example, a bridge design might aim to minimize both cost and material usage.

Constraint-Based Optimization: When designing structures, there are often constraints that must be satisfied. These could be related to factors like stress, displacement, thermal properties, or manufacturability. Constraint-based optimization ensures that the final design meets these requirements.

Discrete Optimization: In discrete optimization, the design parameters can only take on specific discrete values. This is common in problems where components come in standardized sizes or quantities, and the challenge is to find the best combination.

Uncertainty-Based Optimization: Many real-world problems involve uncertainties in factors like loads, material properties, or environmental conditions. Uncertainty-based optimization aims to find designs that are robust and perform well under different possible scenarios.

Coupled Optimization: In complex systems, different aspects might be interdependent. Coupled optimization considers multiple interconnected variables simultaneously. For example, optimizing an aircraft wing design might involve considering both its aerodynamic performance and structural strength.

Multi-Scale Optimization: In problems where structures exhibit behavior at multiple scales (microscopic to macroscopic), multi-scale optimization aims to optimize parameters and designs at each scale to achieve the best overall performance.

Dynamic Optimization: Dynamic optimization involves optimizing structures over time, considering transient, or changing conditions. This is common in control systems and areas where designs need to adapt to varying inputs.

Sequential Optimization: This involves optimizing a structure in multiple stages, often addressing different aspects at each stage. For example, optimizing a car's chassis might involve separate stages for structural integrity, aerodynamics, and ergonomics.

## 4.3 Topology Optimization

This work is aimed at creating a brake pedal with less weight. Hence, a regular automobile brake pedal is selected to create the 3D model. Fusion 360 educational version is used to design the brake pedal as per dimensions. The selected automotive brake pedal is presented in Figure 4.1. For stage optimization, new work was created in Autodesk Fusion 360 and it is presented in Figure 4.2. In order to do topology optimization, stainless steel material and its properties were selected. The brake pedal is pivoted and constrained to do analysis. An applied load of 20 kgf is given on the brake pedal and a reaction force of 10 kg is given at the pivot point. The application of load and reaction force is given in Figure 4.3.

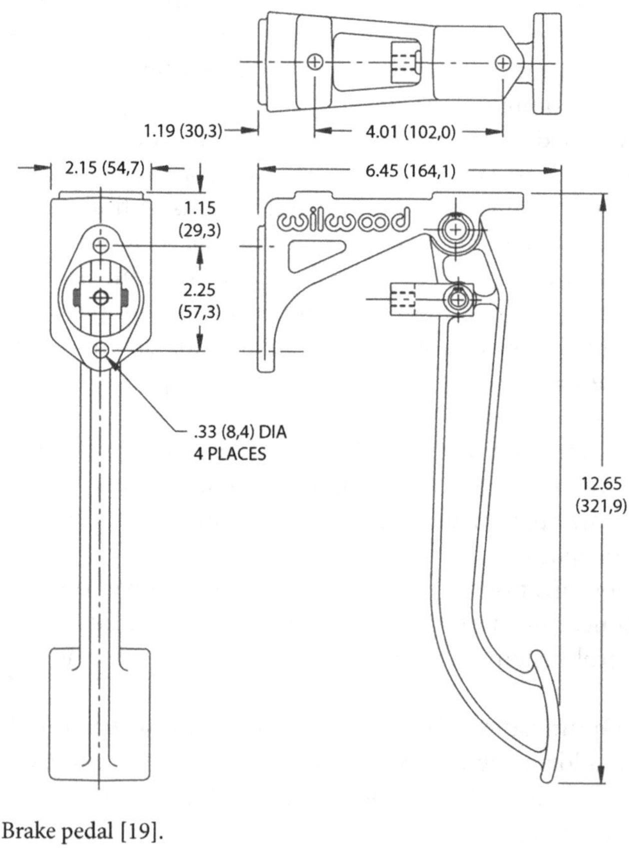


**Figure 4.1** Brake pedal [19].

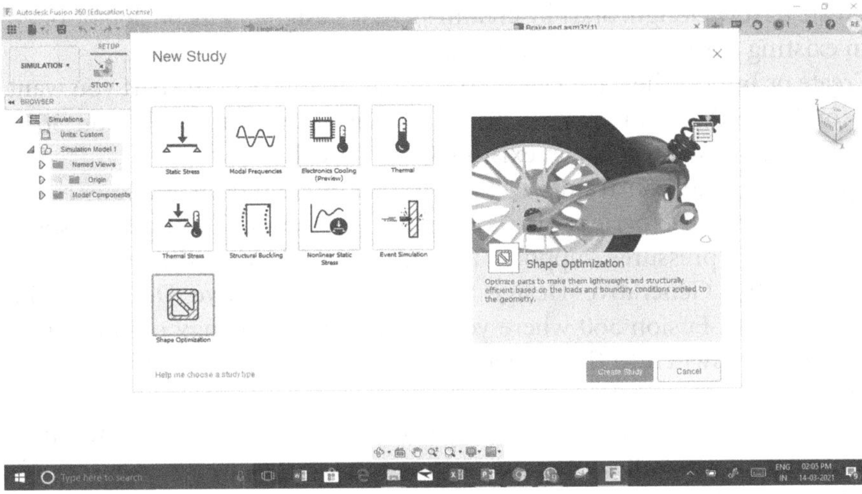


Figure 4.2 Stage optimization.

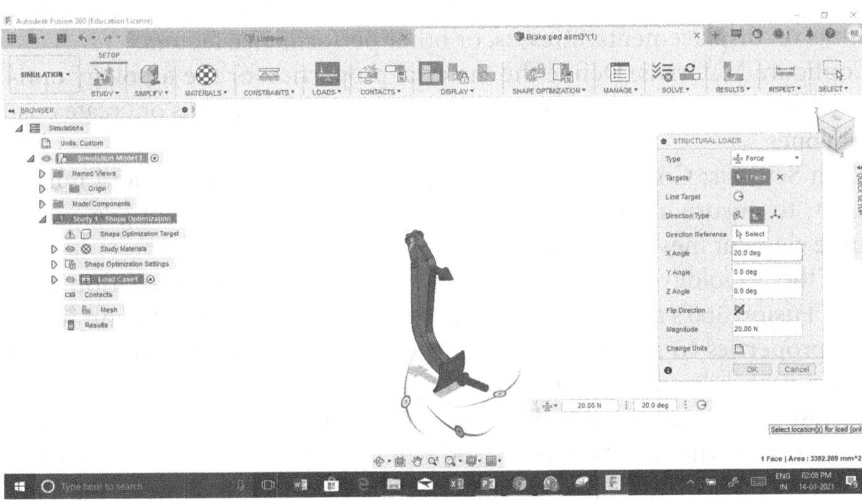


Figure 4.3 Application of working load.

## 4.4 Optimization Results

Topology optimization in Fusion 360 is a process used to design and optimize the structural layout of a part while maximizing its performance and minimizing its weight. Here are the general steps to perform topology optimization in Fusion 360:

Create a New Design: Start by creating a new design in Fusion 360 or open an existing design that you want to optimize.
Create or Import the Part: Design the initial geometry of the part you want to optimize. You can create a new part from scratch using Fusion 360's modeling tools or import an existing CAD model.
Apply Loads and Constraints: Define the loading conditions and constraints that the part will experience in its intended application. This includes specifying forces, pressures, supports, and other boundary conditions.
Activate the Generative Design Workspace: Generative Design is the workspace in Fusion 360 where you will perform topology optimization. To access it, switch to the Generative Design workspace from the workspace switcher.
Set up a New Study: Inside the Generative Design workspace, create a new study. This study will contain the settings and parameters for your topology optimization.
Define Objectives and Constraints: Specify your design objectives and constraints. Objectives might include minimizing mass, maximizing stiffness, or reducing stress concentrations. Constraints might involve setting limits on displacements, stresses, or other performance factors.
Configure Materials: Define the material properties for the topology optimization. You can choose from various predefined materials or create custom ones.
Mesh Settings: Configure mesh settings, such as element size and refinement, to ensure accurate simulation results. Fusion 360 will generate a finite element mesh for analysis.
Run the Topology Optimization: Initiate the topology optimization process. Fusion 360's algorithms will analyze the loads, constraints, and material properties to generate an optimized shape that meets your specified objectives while adhering to constraints.
Review and Interpret Results: After the optimization process is complete, review the results. Fusion 360 will display an optimized geometry that represents the suggested material distribution to achieve the best performance.
Refine and Iterate: Analyze the optimized geometry and make adjustments as needed. You might need to adjust objectives, constraints, or other parameters to achieve a design that meets your requirements.
Export the Result: Once you are satisfied with the optimized design, export the resulting geometry back to your main design workspace in Fusion 360. This geometry can be used as the basis for further detailing and refinement.

Based on the above steps, the created brake pedal model is analyzed in Fusion 360 and the test results are presented below. The created model is imported into the simulation section of Fusion 360 in order to do an analysis. The load is applied and meshed to find the possibility of a surface that is not affected by the applied load. The load application and meshing are presented in Figures 4.4 and 4.5, respectively.

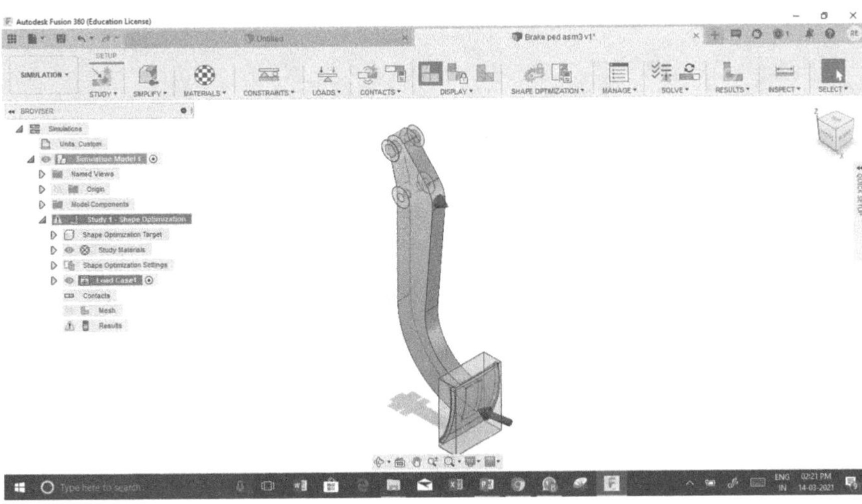


**Figure 4.4** Lower load case in the brake pedal.

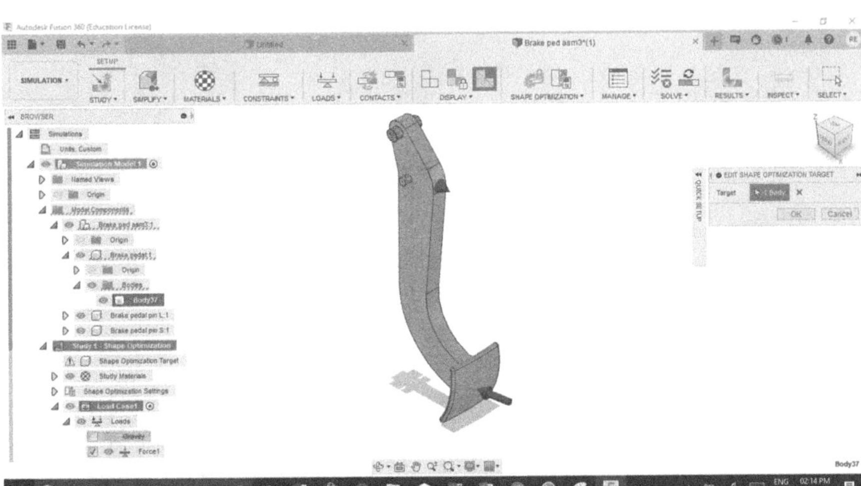


**Figure 4.5** Meshing of the brake pedal.

Figure 4.6 indicates the predicted analysis of the brake pedal. From the figure, it is found that the predicted weight of the brake pedal based on the material input is 1.252 kg. Due to the applied load and pivoted point, the pedal area and pivoted region were stressed. The center portion of the pedal did not undergo any stress due to the applied load and reaction force. Hence, the possibility of material removal is done in the center part of the

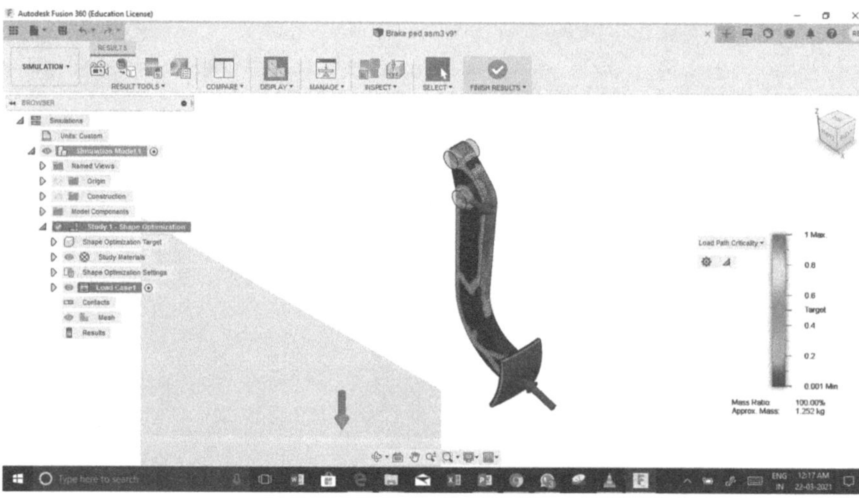


**Figure 4.6** Analysis of brake pedal.

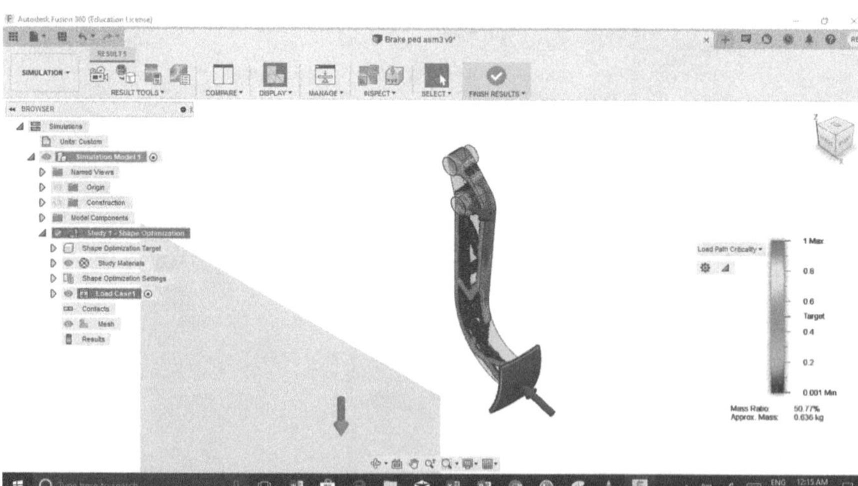


**Figure 4.7** Optimized brake pedal.

brake pedal. The optimized brake pedal with less weight is given in Figure 4.7. From Figure 4.7, it is very clear that the Fusion 360 topology optimization was done and the weight of the pedal was reduced to 0.636 kg. The comparison of brake pedals is given in Figure 4.8. The induced stress on before and after optimization is given in Table 4.1.

To validate the optimized results, the designed pedal was fabricated by using 3D printing. Instead of steel, initially, it was fabricated by using polymers. The fabricated 3D-printed brake pedal is given in Figure 4.9. From the figure, it is clear that weight reduction was achieved accurately and it is ready to fabricate in original material.

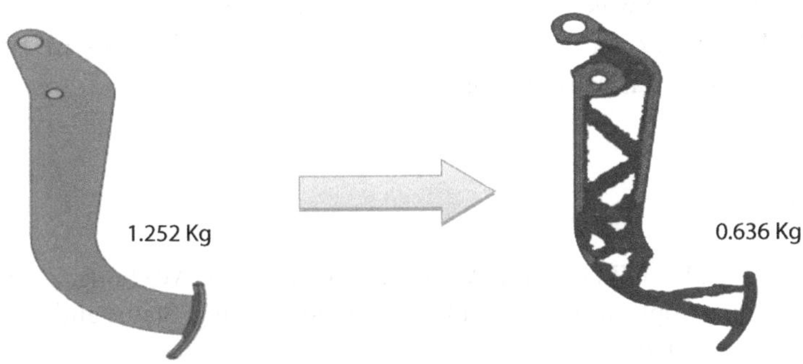


**Figure 4.8** Comparison of brake pedal.

**Table 4.1** Test results.

| Parameter | Original design | Optimized design |
|---|---|---|
| Mass | 1.252 kg | 0.636 kg |
| Von stress [max] | 13.849 MPa | 13.437 MPa |

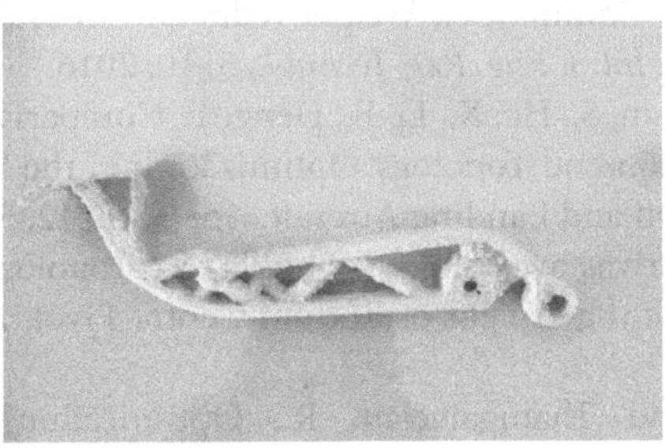

**Figure 4.9** 3D-printed brake pedal.

## 4.5 Conclusion

In this work, topology optimization is done to reduce weight in automobile brake pedals. Based on the topology optimization, the following concussions were found and pointed out below.

1. Fusion 360 was used to create the CAD model of the brake pedal and force was applied on pivoted end and front end.
2. The computed pedal weight was found as 1.252 kg in the original design and 0.636 kg in the optimized design.
3. Weight reduction of the brake pedal was done by 50.7% and von stress was identified as 13 MPa for both designs.
4. This work successfully achieved weight reduction in the brake pedal using topology optimization.

## References

1. Jihong, Z., Han, Z., Chuang, W., Lu, Z., Shangqin, Y., Weihong, Z., A review of topology optimization for additive manufacturing: Status and challenges. *Chin. J. Aeronaut.*, 34, 1, 2021.
2. Jewett, J.L. and Carstensen, J.V., Topology-optimized design, construction and experimental evaluation of concrete beams. *Autom. Constr.*, 102, 59–67, 2019.
3. Deng, X., Chen, H., Xu, Q. *et al.*, Topology optimization design of three-dimensional multi-material and multi-body structure based on irregular cellular hybrid cellular automata method. *Sci. Rep.*, 12, 5602, 2022.
4. Wu, J., Sigmund, O., Groen, J.P., Topology optimization of multi-scale structures: a review. *Struct. Multidisc. Optim.*, 63, 1455–1480, 2021.
5. Andrea, M., Andrea, G., Elena, B., Sorin Ion, M., Bogdan, G. *et al.*, Weight reduction by topology optimization of an engine subframe mount, designed for additive manufacturing production. *Mater. Today: Proc.*, 19, 3, 2019.
6. Swapnil, T. and Chinmay, S., Weight reduction in brake disc using topologyoptimization. *Int. J. Eng. Res. Technol.*, 5, 10, 2016.
7. Zeng, Y., Li, J., Lin, S., He, X., Li, B., Deng, T., Comparison of Manual Setting Weight Reduction and Topology Optimization of the Wing Tips of Electric Vertical Take-Off and Landing Aircraft. *Appl. Sci.*, 12, 5548, 2022.
8. Prashant, K., Srivastava, Simant, Shukla, S., Topology Optimization: An Effort to Reduce the Weight of Bottom Centre Pivot. *Int. J. Recent Technol. Eng.*, 8, 2, 2019.
9. Maniraj, S. and Thanigaivelan, R., Optimization of Electrochemical Micromachining Process Parameters for Machining of AMCs with Different

% Compositions of GGBS Using Taguchi and TOPSIS Methods. *Trans. Indian Inst. Met.*, 72, 3057–3066, 2019.
10. Thanigaivelan, R., Arunachalam, R.M., Jerald, J., Niranjan, T., Applications of Taguchi technique with fuzzy logic to optimise an electrochemical micromachining process. *Int. J. Exp. Des. Process Optim.*, 2, 4, 283–298, 2011.
11. Thanigaivelan, R., Arunachalam, R.M., Kumar, M., Dheeraj, B.P., Performance of electrochemical micromachining of copper through infrared heated electrolyte, Mater. *Manuf. Process.*, 33, 4, 383–389, 2018.
12. Sivashankar, N., Thanigaivelan, R., Saravanan, K.G., Electrochemical micromachining and parameter optimization on AZ31 alloy—ANN and TOPSIS techniques. *Bull. Chem. Soc. Ethiop.*, 37, 5, 1263–1273, 2023.
13. Sivashankar, N. and Thanigaivelan, R., Electrochemical micromachining of magnesium AZ31 alloy using minimum quantity electrolyte, Mater. *Manuf. Process.*, 38, 11, 1406–1415, 2023.
14. Sivashankar, N. and Thanigaivelan, R., Performance of Electrochemical Micromachining of Magnesium Alloy through Sodium Nitrate Electrolyte, in: *Advances in Modern Machining Processes. Lecture Notes in Mechanical Engineering*, M.S. Shunmugam, B. Doloi, R. Ramesh, A.S. Prasanth, (Eds.), Springer, Singapore, 2023.
15. Maniraj, S., Thanigaivelan, R., Gunasekaran, K., Saravanan, K.G., Optimization of Process Parameters in Electrochemical Micromachining of AMCs by Using Different Techniques of Weight Evaluation. *Adv. Mater. Sci. Eng.*, 1–9, 2023.
16. Sudin, M.N., Musthafah, M.T., Faiz, R.R., Shamsul, A.S., Topology Optimization in Automotive Brake Pedal Redesign. *Int. J. Eng. Technol.*, 398–402, 2014.
17. Oğuz, D., Onur, C.K., Emirhan, K., Fatih, K., Optimum Design of Brake Pedal for Trucks Using Structural Optimization and Design of Experiment Techniques. *Int. J. Automot. Sci. Technol.*, 4, 272–280, 2020.
18. Arun, J.K., Rao, C.S.P., Cherian, J., Topology optimization of lattice structure on a brake pedal. *Mater. Today: Proc.*, 47, 15, 2021.
19. https://www.wilwood.com/pedals/pedalprod?itemno=340-1290.

5

# Enhancing Resistance Spot Welding Weld Quality: A Comprehensive Analysis of Influencing Factors and the Role of Modeling and Optimization for Improved Quality

**Kaushal Jha[1], Santosh Kumar Tamang[2], Rajeev Kumar[3] and Bishub Choudhury[1]***

*[1]Department of Mechanical Engineering, Dronacharya Group of Institutions, Greater Noida, Uttar Pradesh, India*
*[2]Department of Mechanical Engineering, NERIST, Nirjuli, Arunachal Pradesh, India*
*[3]Department of Mechanical Engineering, GNIOT, Greater Noida, Uttar Pradesh, India*

***Abstract***

Resistance spot welding (RSW) is a widely used welding technique in the automobile and aerospace industries that joins sheet metal by applying heat and pressure. The quality, metallurgical properties, and mechanical aspects of the welded joint rely on numerous input factors involved in the RSW process. This literature review provides valuable insights into the factors that affect weld quality, emphasizing the need to understand the welding parameters and their impact on the welding procedure to achieve reliable and superior welds. It examines how various factors, including welding current, welding duration, surface conductivity, surface cleanliness, electrode shape, and applied pressure, impact the quality of welds in different materials. Furthermore, the review underscores the significance of utilizing modeling and optimization in RSW to achieve precise control and the creation of high-quality welds. The findings reveal that higher welding currents result in larger and more consistent weld nuggets, leading to improved joint strength and

**Corresponding author*: bishub.choudhury73@gmail.com

R. Thanigaivelan, Pradeep Kumar Krishnan, Kamalakanta Muduli and Santosh Kumar Tamang (eds.) New Materials, Processing and Manufacturability: Fabrication and Processing of Advanced Materials, (71–94) © 2024 Scrivener Publishing LLC

quality. Insufficient welding time results in weak joints, while excessive time causes overheating and deformation. Additionally, surface resistivity, preparation, and cleanliness play crucial roles in achieving strong welds, as understanding surface conditions enhances adhesion properties and mechanical performance. Adequate electrode force is vital in RSW, as it minimizes resistance, concentrates heat, and ensures uniform pressure, thereby contributing to high-quality joints. The choice of the electrode tip's shape and design plays a crucial role in determining the quality and strength of the weld, underscoring the importance of selecting the right tip design. The review also emphasizes the importance of utilizing modeling and optimization tools to achieve precise control and produce high-quality welds.

***Keywords*:** Resistance spot welding (RSW), weldability, resistivity, optimization, weld preparation

## 5.1 Introduction

Resistance spot welding (RSW) is commonly utilized for joining sheet metal components in various industries, including automobile manufacturing, aerospace structures, space shuttle fabrication, furniture and home appliance fabrication, railway industries, and many other industries [1]. More than 90% of the assembly in the automation industries is done using RSW in manufacturing plants and factories, with typical modern vehicles having 2000 to 5000 spot welds. The high production rate of RSW is attributed to its automation and robotization capabilities [2]. In recent years, significant advancements have been made in RSW technology, including improvements in design, weldability, resistance, welding current, and the use of new technologies for analysis. However, despite these advancements, the complex interactions between various RSW parameters make it challenging to control and ensure proper weld quality for industrial applications [1–3].

The RSW process involves numerous elements, including the RSW machine's design, and welding parameters such as time, current, resistivity, frequency, material properties, electrode configuration, pressure, material type, and surface preparation before welding. These factors contribute to a multifaceted interplay of thermal, electromagnetic, fluid dynamics, mechanical, and metallurgical processes at the point where the materials make contact. Precise control of these variables is of utmost importance to maintain the structural integrity of the welded joint and improve welding productivity. Thus, this paper focuses on five prime factors, i.e., design of the RSW machine; analysis of weldability-related issues and precautionary measures; the influence of welding current; effect of surface resistivity, surface preparation, etc.; and advanced control techniques, process optimization, monitoring, and control systems [1, 3, 4]. By doing so, this review paper will provide a comprehensive understanding of the recent advancements in RSW and help in making future RSW machines more portable, efficient, and cost-effective.

In the conventional RSW process, a mixture of external pressure and resistance thermal energy is used to execute welding. Typically, two or more sheets of base metal are compressed together by electrode force. The heat at the metal interface is produced by the electrical energy carried by the welding current as it flows through the workpiece, coupled with the electrical resistance encountered between the two workpieces [5]. The heat generated by the current flow is expressed as:

$$Q = I^2Rt$$

Thermal energy ($Q$) in joules can be calculated as a function of the current ($I$) in amperes, time ($t$) in seconds, and resistance ($R$) in ohms. Here, the resistance is composed due to several factors that include contact resistance between workpieces, electrode, and workpieces, the resistance of electrode as well as workpiece. This resistance is further governed by the current flows through the electrode and workpieces. Thus, optimized resistance with an optimized current is essential to guarantee high-quality welded structures. Implementation of optimization techniques, implementation of the design of experiments (DOE) by replacing conventional trial and error methods, implementation of quality control measures, and *in situ* monitoring are mainly suggested by various researchers to overcome weldability issues and to maintain high-quality weld joint [6, 7]. Furthermore, factors such as welding time, electrode force, and shape are found to have a significant impact on RSW performance.

Overall, this review paper provides a valuable viewpoint regarding the factors influencing the quality of RSW and emphasizes the importance of comprehending these welding factors and their effects on the welding process in order to achieve consistently strong and reliable welds. The significance of this review paper lies in its ability to provide a reference for engineers and researchers working in the field of RSW to better understand the complex interplay between different parameters affecting RSW [1, 3, 7]. The optimal welding parameters and techniques may vary depending on the material being welded, highlighting the need to consider this factor during the welding technique. Additionally, research is required to develop a more portable, cost-efficient, energy-efficient, and durable RSW machine that can meet the needs of various industries.

This thorough analysis explores the impact of different factors on the quality of welds in RSW. These factors comprise welding current, time, surface resistivity, surface cleanliness, electrode pressure, and geometry. Furthermore, the review emphasizes the significance of modeling and optimization in RSW to achieve precise control and ensure high-quality welds. The paper concludes by summarizing the key findings of each section and presenting an overall conclusion that takes into account all the parameters affecting RSW. Overall, this review paper provides a comprehensive overview of recent advancements in RSW technology and serves as a valuable resource for researchers and engineers working in this field.

## 5.2 Influence of Welding Current and Time

As per the available studies, the welding current has a pivotal influence on defining the effectiveness of RSW. Higher current levels are generally associated with the production of larger and more consistent weld nuggets, resulting in improved weld quality and greater strength of the welded joints. Figure 5.1 presents a diagram of an RSW machine, showing the route of electrical current as it travels from one electrode to the other through the workpiece being joined. In the welding operation, the maximum welding current is limited by the electrical system's capacity and the resistance of the load supply [3, 4, 8].

Another important RSW factor to consider in terms of overall weld quality comprising physical, metallurgical, and mechanical characteristics

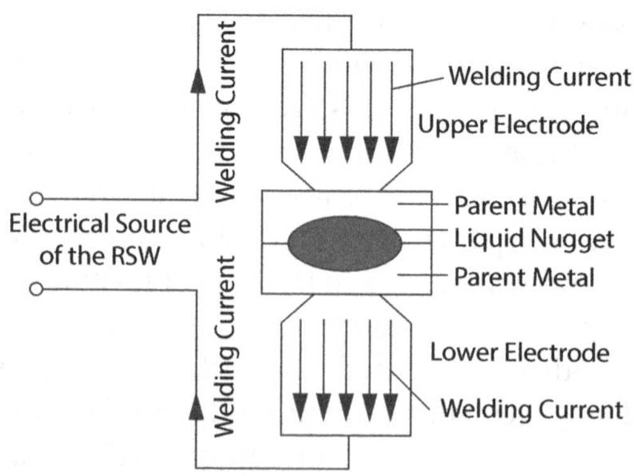


**Figure 5.1** RSW schematic [8].

is welding time. The duration of welding time affects the formation of the weld nugget. Insufficient welding time may result in an underdeveloped nugget, leading to weak joints [5, 8]. Meanwhile, excessive welding time can cause overheating, excessive growth of the nugget, and potential material deformation. A representation of the growth of the welding nugget with respect to time is depicted in Figure 5.2.

Several research investigations have explored both the separate and combined impacts of welding current and time on the structural soundness of RSW joints in various materials. A significant portion of these studies has primarily concentrated on scrutinizing the impact of welding current on the weld integrity. Wan *et al.* (2014) explored the significance of welding current on the weldability of advanced high-strength DP600 steels through RSW. The experiments conducted within 6 to 12 kA revealed minimal influence of welding current on microstructure, but a significant impact on nugget size. At the maximum current of 12 kA, expulsion occurred and unsatisfactory partial interfacial failure was observed [10]. Zhang *et al.* extensively examined the RSW of CR590T/340YDP galvanized dual-phase steel, a prevalent automotive material. They observed a decline in weld quality and appearance as the welding current increased, resulting in a zinc island on the weld surface. The current of 10.5 kA was identified as

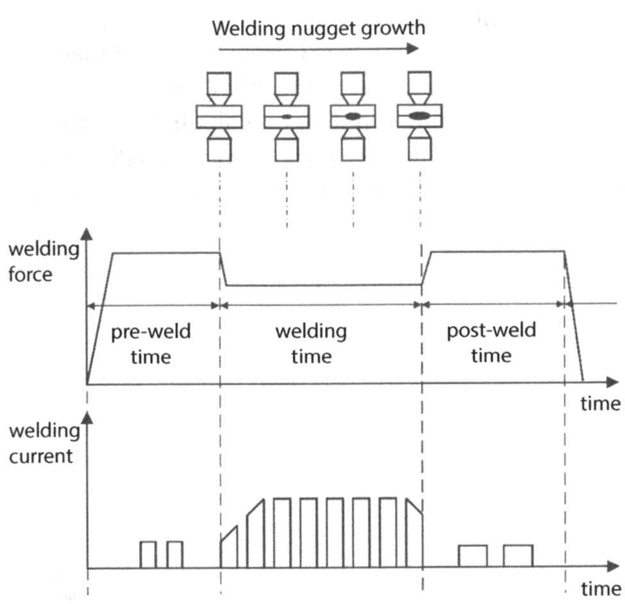


**Figure 5.2** Welding nugget growth with respect to time [9].

the ideal setting, leading to the highest tensile shear strength. Figure 5.3 illustrates how varying the current influences both the nugget width and the tensile shear strength [11]. The fluctuations in welding current on both the residual stress and the dimensions of the weld nugget were studied by Ariffin *et al.* (2014). With an increase in time and current, residual stresses within the outer and inner portions of the weld nugget expand, though they exhibit a slight reduction at the edges of the weld nugget [12].

Aslanlar *et al.* (2007) explored how welding current can affect the quality of RSW of galvanized chromate steel sheets. The maximum tensile peel and tensile shear strengths were recorded at 11 kA for 10 cycles and 10 kA for 15 cycles, respectively while preserving a high level of surface quality [13]. In 2021, Taufiqurrahman *et al.* conducted a study to explore on impact of electrode force and welding current on heat input, as well as the mechanical and physical characteristics of a joint between SS316L and Ti6Al4V, with the inclusion of an aluminum interlayer. The greatest tensile shear load, measuring 8.71 kN, was attained when employing 11 kA of welding current with a 3-kN electrode force [14]. Khan *et al.* conducted research on the impact of welding current on the strength of RSW welded 6061-T6 aluminum joints. Across the range of currents investigated, the tensile shear load indicates an upward trend with elevated current at specific welding time, resulting in the growth of nugget size. The maximum tensile shear load achieved was approximately 2.75 kN, observed at 20 kA and eight-cycle welding time [15]. To reduce the impact of heat input in terms of welding current, Schulz *et al.* (2021) proposed medium frequency direct current (MFDC) systems during RSW of AA6016-T4 aluminum alloy. Both experimental and simulated results indicate that in this alloy, MFDC-RSW can decrease the necessary energy, specifically heat input, for producing strong welds, all without the need to raise the welding current.

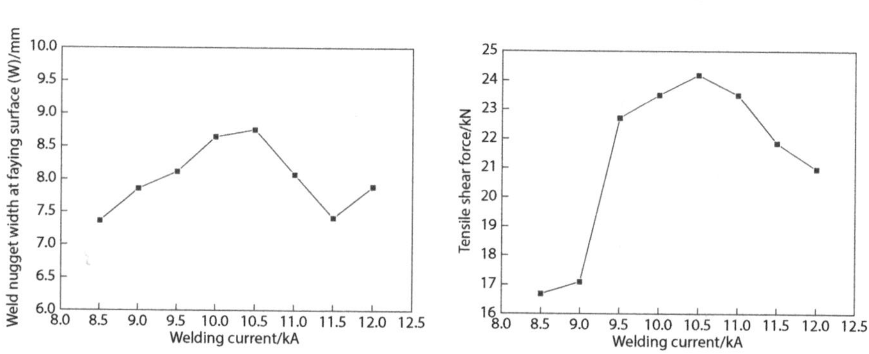


**Figure 5.3** Influence of welding current [11].

The welding process also exhibited distinct phases of weld nugget nucleation and growth [16]. In the investigation of Muhammed, the focus was on the RSW of DP1200 steel. The investigation examined a range of factors, such as the fracture patterns, microstructure, microhardness, and the extent of softening in the HAZ. The findings highlighted the substantial influence of current and electrode pressure, on the load-displacement characteristics and fracture modes [17]. Zhang *et al.* conducted research to assess how changes in welding current affect the evolution, composition, and mechanical properties of CR590T/340YDP galvanized dual-phase steel for automotive production. The findings suggest an ideal welding current value for achieving peak tensile strength while also indicating a decline in weld quality with higher current settings [18].

The combined effect of welding current and time has been investigated by some researchers. Moshayedi and Sattari-Far conducted a study to examine how time along with the current affects residual stresses in spot-welded specimens. Their study revealed that as both welding time and current were elevated, the rate of weld nugget growth decreased. Additionally, the extent and strength of residual stresses increased and intensified with longer welding time and higher welding current, causing the maximum residual stress to shift away from the center of the weld [19]. Pouranvari investigated the impact of various factors such as welding current and time that affect the process regarding the efficiency of RSW welds on low-carbon steel. Exerting too much pressure on the electrode can significantly reduce the peak load and maximum energy [20]. Badkoobeh *et al.* explored the properties of spot welds between two types of stainless steel (AISI 430 and AISI 321). The microstructure of the welds was analyzed and found to comprise ferrite, austenite, and martensite phases. Enhancing the welding current and duration led to improved maximum load capacity and failure energy. However, the specific optimal times for achieving these improvements varied. Regardless of weld parameters, the welds exhibited a pullout failure mode [21]. In a separate study, Sivaraj *et al.* investigated the impact of welding parameters on the quality of RSW joints in low-carbon steel sheets. Increasing the current and time resulted in a larger weld nugget diameter and higher tensile strength, but applying a greater electrode force reduces the tensile strength [22]. Hasanbaşoğlu *et al.* (2007) explored the effect of peak current and time on the weld quality of dissimilar materials joined by RSW. The maximum joint strength was achieved by finding the optimal welding parameters. Excessive grain growth due to improper current in the interstitial-free (IF) steel was found to be the prime reason for the weakening of the weldment [23].

Overall, understanding and controlling welding current and time parameters are crucial for achieving high-quality RSW joints with sufficient strength and integrity in different materials. From the above study, the following important remarks were drawn:

1. The quality and integrity of RSW joints are notably affected by the welding current. Studies have shown that higher welding currents can result in larger and more consistent weld nuggets, improving weld quality and joint strength. However, excessively high currents can lead to expulsion, unsatisfactory interfacial failure, and reduced weld appearance.
2. The time of welding is found equally essential in shaping the weld nugget. Inadequate time leads to weak joints with underdeveloped nuggets, while excessive time causes overheating, excessive nugget growth, and potential material deformation. To achieve the desired weld quality, it is essential to find the optimal welding duration.
3. The interaction between welding current and time has been examined as well. Increasing time and employing a higher welding current can delay weld nugget expansion and raise residual stresses. Achieving the best balance between current and time is vital to maximize joint strength without jeopardizing weld integrity.

## 5.3 Influence of Resistivity, Surface Preparation, and Cleanliness

Various studies in the literature have placed great emphasis on the significance of surface resistivity, surface preparation, and surface cleanliness to ensure high strength and better metallurgical properties. Surface resistivity is directly affected by the material used for the electrode and workpiece, and it plays an essential role in anticipating the heat produced by the contact of the surfaces [5, 6].

The resistance welding machine's secondary circuit and the weld specimen form a sequence of resistances as shown in Figure 5.4 [8]. The cumulative resistance encountered by the current influences its intensity. In a welding operation, there are typically about five resistances linked together in series, collectively influencing the distribution of temperature.

The combined resistance of these elements is denoted as R, and is expressed as follows:

$$R = R_2 + R_3 + R_4 + R_5 + R_6$$

Here, $R_4$ determines the resistance required in the formation of spot weld and a higher value of it implicates higher weldability. The degree of resistance is contingent on the cleanliness of the electrode and work piece surface, electrode force, and the size and shape of the electrode [8, 9].

Han carried out an investigation into how internal oxidation and zinc diffusion impact the RSW of advanced high-strength automotive steel sheets (AHSS). Their study underscores the significance of surface conditions in relation to electrical contact resistance and introduces an electrical dynamic resistance approach based on heat input to identify the appropriate welding current for AHSS. It is concluded that heat treatment time from 4 to 10 min can establish an effective RSW schedule for producing strong welds, and dynamic electrical resistance can be used as a monitoring tool for weld quality [24]. Zhang *et al.* (2020) explored the use of surface treatment methods like silane coupling agent (SCA) and Ti/Zr conversion treatment for the improvement of adhesion characteristics of AA5182-O alloys in RSW. The study investigates how the weldability of AA5182-O can be affected by protective layers and storage conditions. Ti/Zr conversion coating improves weld quality and mechanical performance than the SCA surface treatment [25].

Hamedi and Atashparva (2017) conducted a study on interface resistance in RSW. Despite numerous models developed for interface resistance, discrepancies in measurements and the complexity of the phenomenon

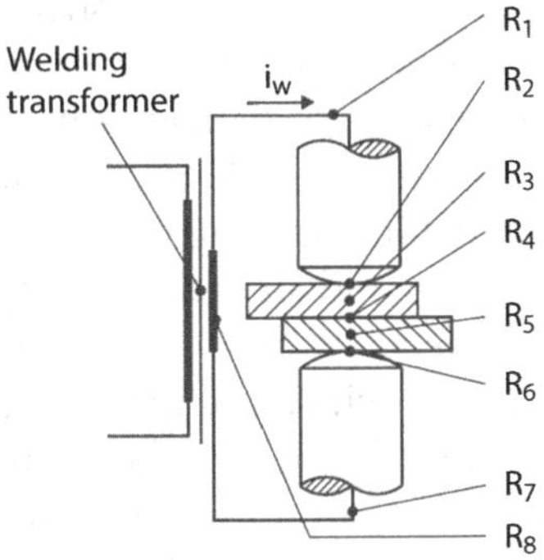


**Figure 5.4** The resistance occurred in RSW [9].

have hindered the creation of a comprehensive model encompassing all important parameters. The study reviews existing works, emphasizing the requirement for a systematic and quantitative model of dynamic resistance [26]. Naimi *et al.* (2015) examined the challenges faced in the RSW of aluminum alloys due to the oxide layer, which leads to inconsistent weld quality. The study investigated the impact of pretreatment on the weld quality of AA1050 sheets with different thicknesses. The presence of a thick oxide film resulted in higher electrical contact resistance in the as-received sheets. Glass-blasting, which increased surface roughness, reduced electrical contact resistance by aiding in the breakdown of the oxide film. Pickling in NaOH improved strength and reduced strength variation [27].

Kim *et al.* studied how surface roughness and electrode force affect the RSW of aluminum 6061 alloys. They discovered that rough electrode surfaces facilitated oxide layer removal, leading to decreased resistance heat, smaller weld nugget size, and lower tensile shear strength. While high electrode force improved sticking for as-received electrodes, it had no effect on abraded electrodes. Increasing electrode surface roughness reduced sticking by slowing nugget growth in the thickness direction during continuous welding [28]. Ghatei-Kalashami *et al.* (2022) examined how surface characteristics and the initial microstructural composition resulting from the continuous galvanizing process (CGP) affect the failure behavior and mechanical properties of AHSS when subjected to RSW. The study found that during the annealing and coating process, the development of internal oxides caused a decrease in heat input during RSW, which led to reduced tensile shear peak load values and a smaller nugget size [29]. The RSW of AZ31 Mg alloy sheets with varying surface conditions (as received-black (AR-B) having a black surface, as received-white (AR-W) having a shiny surface and acid clean) was examined by Liu *et al.* (2009). The variations of surface resistance at different surface conditions are shown in Figure 5.5. The contact resistance was higher for the sheets in their received state, leading to serious expulsion and internal defects, while acid-cleaned sheets exhibited better strength with less damage on electrode tip faces. This phenomenon occurred because the elevated contact resistance resulted in lower current thresholds for initiating the weld and forming a smaller nugget size causing expulsion and surface porosity [30]. Finally, Wei and Wu (2012) explored the effects of local electrical interface resistance on transport variables, the rate at which the material cools, the way in which the solute is distributed, and the shape of the nugget, during RSW. The study highlights the importance of understanding electrical resistance for monitoring and achieving good weld quality [31].

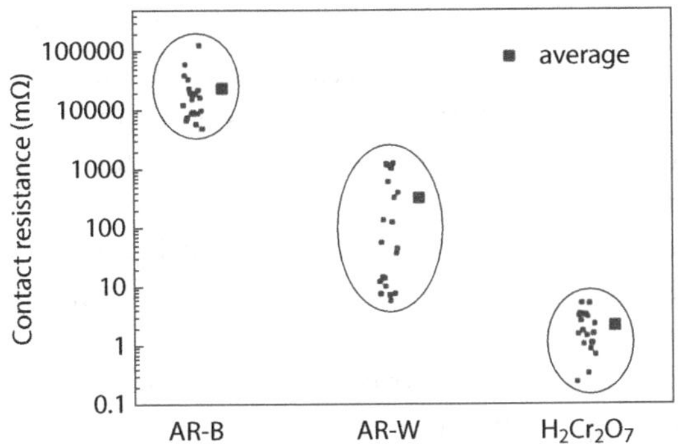


**Figure 5.5** Effect of surface condition on contact resistance [30].

Janardhan *et al.* investigated the weldability of dissimilar IF and high-strength low-alloy steels (HSLA) using RSW. The study examines how the nugget diameter is affected by dynamic contact resistance, along with the mechanical characteristics of spot-welded samples in configurations of tensile shear and coach peel. The load-carrying ability of the IF–HSLA joint is lower than that of the HSLA–HSLA joint but closer to that of the IF–IF joint [32]. Dobosy *et al.* investigated the weldability of high-strength 7075 aluminum alloy, commonly used in the automotive industry, through RSW tests. Various surface pretreatment methods were tested on sheets that were 1 mm thick and in T6 condition. Results showed that chemical etching improved joint strength and quality, decreased electrode wear, and can be integrated into the production line for improved reliability [33].

Overall, these studies explored how surface resistivity, preparation, and finish can affect the quality of RSW in various materials. Based on the studies, the following remarks were made:

1. Surface resistivity, surface preparation, and surface cleanliness are critical factors influencing the strength and metallurgical properties of RSW joints. Surface resistivity is affected by the electrode and workpiece materials and plays a vital role in identifying the heat generated during surface contact. Cleanliness of the electrode and workpiece surfaces, as well as surface preparation techniques, is crucial in achieving higher weldability and optimal resistance values.
2. Surface conditions like internal oxidation, zinc diffusion, and oxide layers impact electrical contact resistance and weld quality. Heat treatment and surface treatment techniques

improve adhesion properties in RSW. Effective treatment times and suitable surface treatments enhance weld quality, mechanical performance, and adhesion properties.

3. Interface resistance and surface characteristics such as roughness and initial microstructural composition also impact RSW. Research on dynamic resistance, oxide layer effects, and surface conditions aids in optimizing welding parameters and ensuring good weld quality. Understanding electrical and contact resistance is crucial for monitoring and ensuring reliable weld quality.

## 5.4 Effect of Electrode Force and Electrode Tip Geometry

The characteristics of the joint are predominantly determined by the electrode force, which affects both contact resistance and contact area. Inadequate electrode force results in excessive heat production, elevated contact resistance, expulsion, and results in the formation of low-quality joints [4, 5]. Moreover, the shape and design of the electrode tip have a significant impact on weld quality. Proper electrode tip geometry ensures adequate heat concentration, uniform pressure distribution, and efficient current flow. Improper electrode tip design can lead to uneven heating, inadequate nugget formation, and reduced joint strength [8, 9].

Wei and Wu (2013) examined how electrode geometry influences weld quality in RSW. The impact of electrode shape, size, and material on weld quality was investigated, including factors such as nugget size, tensile strength, and electrode wear. They found that a concave electrode shape resulted in a larger nugget size and higher tensile strength compared to a flat electrode shape [34]. Mahmud *et al.* investigated how electrode deterioration influences the occurrence of Zn-assisted liquid metal embrittlement (LME) cracking during the RSW of galvannealed TRIP steel. The primary factor impacting LME cracking was the physical deterioration of the electrode, particularly the enlargement of the curvature radius. Metallurgical degradation had a minor impact. LME crack severity remained stable in the first 80 welds but significantly decreased in 80 to 200 welds, underscoring the significance of electrode condition for weld quality [35]. Jo *et al.* investigated the weldability of AA6061-T6 considering the effect of electrode surface roughness and force. Using an abraded electrode with a rough surface reduced contact resistance, effectively removed oxide film, minimized heat generation, and slowed down weld nugget growth. The abraded electrode had less sticking compared to the as-received electrode [36]. Similarly, Wang *et al.* focused on analyzing how electrode tip morphology, specifically electrode pitting (EP) and electrode tip diameter enlargement (ETDE), affects the quality of RSW in DP590 dual-phase steel. The study identifies three different nugget shapes and two failure modes linked to various electrode tip structures. EP shapes cause inadequate fusion at the center of the workpiece contact area, while ETDE shapes reduce nugget size, resulting in decreased lap shear fracture load (LSFL). A representation of different electrode shapes used by the researcher is shown in Figure 5.6 [37].

Safari and Mostaan (2016) investigated how welding parameters affect the RSW of stainless steel. Elevating both the welding current and duration led to larger weld nuggets and enhanced strength. Conversely, elevating the electrode force results in a reduction in weld strength [38]. Similarly, Afshari *et al.* investigated the influence of RSW factors on failure load and nugget diameter of aluminum 6061-T6 weldments. The impact of electrode force is found vital affecting failure load and nugget diameter as evident from Figure 5.7. With an incremental rise in electrode force, going from 3666 to 4583 N, there is a gradual expansion in the nugget diameter, subsequently increasing the failure load from 3 to 4.4 kN.

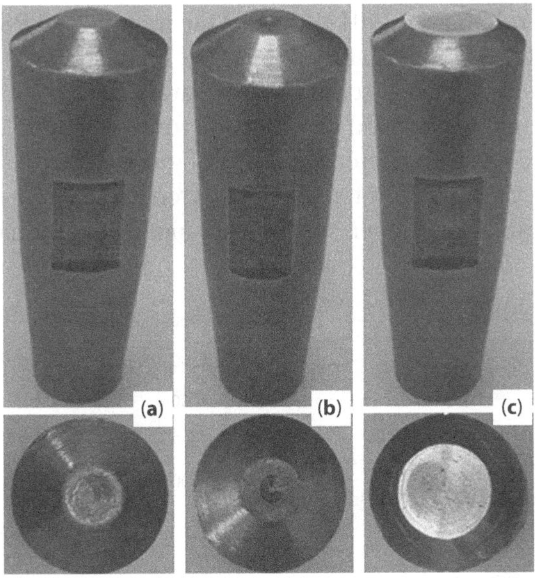


**Figure 5.6** Electrode tip geometries: (a) Conventional, (b) EP, and (c) ETDE [37].

However, increasing the electrode force further reduces both the nugget diameter and the failure load, implying the need for proper control and optimization [39]. Manladan *et al.* (2017) presented a summary of the RSW process for aluminum alloys. The authors addressed the difficulties encountered when welding aluminum alloys, including the creation of intermetallic compounds, and how welding parameters like current, time, and force can impact the quality of the weld. They also highlighted the importance of pre-weld surface preparation and post-weld cleaning to achieve good weld quality [40].

Several research investigations have been carried out to examine the cumulative impact of welding parameters on RSW across a range of materials. Özyürek (2008) focused on austenitic stainless steel and analyzed the effects of RSW parameters on weld strength, nugget size, and electrode wear. They observed that higher current and elevated time led to larger nugget size and increased strength, while higher electrode force reduced electrode wear [41]. Similarly, Wei *et al.* (2014) studied high-strength steels and found that elevating time and current resulted in larger nugget size and higher tensile strength, while an increase in electrode force led to a decrease in both parameters [42]. In a study by Ozsarac (2012), an examination of galvanized steel sheets revealed that greater current and duration resulted in greater hardness. However, increased electrode force led to smaller nuggets and more significant electrode wear [43]. A similar observation was noted by Kolařík *et al.* (2014) in the RSW of low-carbon steel [44]. Sreehari and Bhaskar (2019) studied the influence of RSW parameters while welding for AISI 409 stainless steel. The highest tensile shear strength was achieved with the ideal set of process parameters: a current of 9000A, a duration of eight cycles, and an electrode force of 4500 N. Among these parameters, the welding current, followed by the electrode force, had the most pronounced impact on tensile shear strength [45]. Chen *et al.* (2016) examined the RSW of copper sheets. They noted that enhancing current

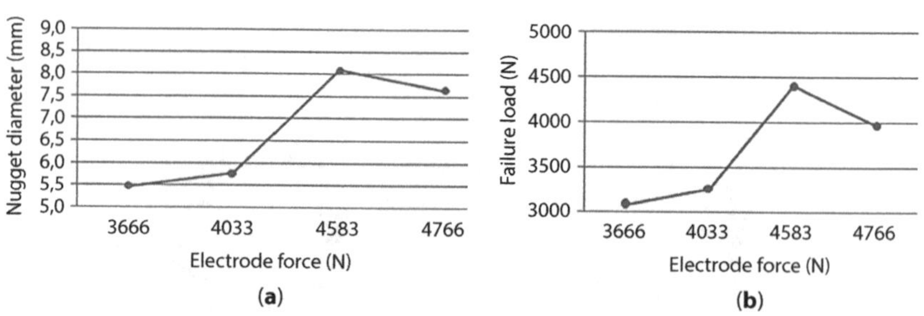


**Figure 5.7** Electrode force influence on (a) nugget size and (b) electrode force.

and duration led to improved copper joint strength, while electrode force had an adverse effect on weld strength. These studies collectively provide insights into the influence of welding parameters on weld quality, nugget size, strength, and electrode wear for different materials in RSW [46].

Overall, the literature reviews discussed here all examine the impact of electrode force and electrode geometry on the weld quality of an RSW joint of various materials. Based on the study, the following key conclusions were made:

1. The electrode force in RSW plays a crucial role in determining joint properties. Sufficient force ensures lower contact resistance, proper heat concentration, and uniform pressure distribution, resulting in high-quality joints.
2. Additionally, the shape and design of the electrode tip significantly impact weld quality, with appropriate geometry leading to adequate nugget formation and optimal weld strength.
3. Numerous studies have investigated how welding parameters RSW weld quality for diverse materials. Typically, raising welding current and duration yields larger nuggets and greater tensile strength, but increased electrode force can diminish weld strength. Understanding the combined effects of these parameters is essential for optimizing weld quality, nugget size, and electrode wear in RSW.

## 5.5 Modeling and Optimization of RSW Parameters

Modeling and optimization tools have a crucial role in optimizing RSW parameters for achieving high-quality welds and precise control. These tools allow researchers and engineers to simulate and analyze the welding process, accurately predicting how different parameters impact weld quality and control [47]. Through techniques like finite element analysis (FEA) and computational modeling, the welding behavior can be simulated, enabling the detection of potential defects such as insufficient fusion, nugget size variations, or excessive heat generation. Optimization algorithms are then employed to identify the best combination of parameters that maximize weld quality and control, considering factors like welding current, time, electrode force, and tip geometry [47, 48]. This approach reduces the need for trial-and-error experiments, resulting in cost and time savings during the optimization process. By utilizing modeling and optimization tools, manufacturers can achieve enhanced weld quality, defect reduction, improved process efficiency, and adherence to desired specifications and standards in RSW applications.

In the welding industry, several modeling and optimization tools are utilized to enhance welding processes. Studies often employ statistical modeling and optimization methods like multi-linear regression, response surface methodology (RSM), Taguchi method, and gray relational analysis (GRA). Soft computing–based approaches, such as fuzzy logic, neural networks, and evolutionary algorithms, are increasingly employed for welding process optimization [48]. These techniques enable intelligent decision-making, adaptive control, and automated optimization, resulting in improved weld quality and efficiency. Furthermore, optimization algorithms like genetic algorithms (GA), ant colony optimization (ACO), particle swarm optimization (PSO), and simulated annealing (SA) are frequently applied to identify optimal parameter combinations that maximize weld quality and control [47, 48].

Karthikeyan and Balasubramaian (2017) implemented RSM to develop a correlation for predicting the tensile shear fracture load (TSFL) of RSW-welded AA2024-T3 alloy. Through optimization using the RSM approach, a maximum TSFL of 6.59 kN was achieved [49]. Similarly, Rostamiyan *et al.* (2015) developed a novel hybrid welding technique, which combines ultrasonic vibration and friction stir spot welding, with the aim of enhancing the lap shear force and hardness of AA6061. The effects of process factors and optimal parameters were analyzed using Taguchi experimental design and GRA. The findings revealed that enhancing weld quality is highly dependent on ultrasonic vibration. The optimal parameters for high lap shear force and hardness were identified [50]. Rajarajan *et al.* (2022) and their study presented the optimization of process parameters for RSW-welded AHSS joints using RSM to fabricate thin-walled automotive structural frames. The use of RSM helps to develop mathematical models and optimize process parameters to enhance the strength of spot joints [51]. Aizuddin *et al.* (2016) employed the Taguchi robust method to fine-tune spot weld parameters for different thicknesses of dissimilar materials in automotive applications, resulting in successful improvements in weld quality and efficiency. Previous literature has also reported success in using the Taguchi method for parameter optimization in spot welding applications [52].

Artificial neural networks (ANNs) are computational models that take inspiration from the organization and functioning of biological neural networks. These networks are composed of interconnected nodes, referred to as "neurons," which perform information processing and communication

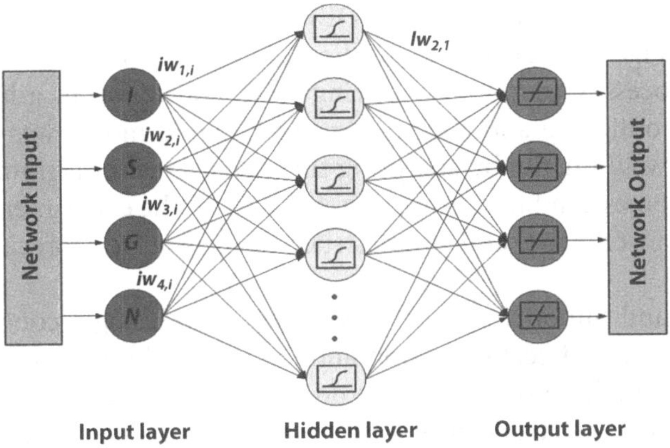


**Figure 5.8** ANN architecture.

tasks. ANNs have been widely used in various fields, including pattern recognition, machine learning, and data analysis, due to their ability to learn from data and make predictions or classifications. An ANN typically comprises three layers: the input layer, the hidden layer, and the output layer [47, 48, 53]. In Figure 5.8, the architecture illustrates the presence of neurons in the hidden layer and the output layer responsible for processing information and producing output, respectively. Another soft computing–based modeling approach, fuzzy logic (FL), finds extensive application in the manufacturing industry for process control, quality assurance, and optimization. It enables the modeling of complex and uncertain systems, facilitating intelligent decision-making and improving efficiency in production processes [47].

Arunchai *et al.* (2014) explored the use of ANN to predict and optimize the parameters of RSW for aluminum alloys in the automotive industry. The results indicate that ANNs are effective in determining welding parameters and achieving high weld quality while reducing the time required for setting and sample testing. The attained outcomes for tensile shear strength yielded a mean square error (MSE) of 0.054 and an accuracy of 95% [53]. A summary of the effects of welding input parameters on the quality of RSW is provided by Jadav and Gandhi (2014). The use of evolutionary algorithms, DoE, and computational networks to optimize these parameters is also discussed. GRA was used to determine optimal parameters for RSW based on tensile strength. The review concludes that while there have been many studies on optimizing RSW for certain materials, the use of GRA is not yet widely employed for this purpose [54]. Haghshenas and Moshayedi (2019) presented a novel algorithm for monitoring RSW using a displacement of electrode (DE) diagram derived from high-speed camera footage. The displacement diagram can provide valuable insights into the processes taking place during RSW and serves as a helpful tool for RSW monitoring and control. An ANN-based model was developed to forecast weld quality by analyzing the displacement diagram. The predicted outcomes exhibit a strong correlation with the experimental results, indicating an effective and nondestructive method for forecasting weld quality [55].

Podržaj and Simončič (2010) developed a fuzzy logic control system for RSW that can detect expulsion and stop the welding process to avoid unnecessary high stresses on electrodes, which increases their life. The problem of short electrode life is particularly prominent when coated materials are welded [56]. Martín and De Tiedra (2022) provided a summary of recent developments in quality control and enhancement techniques for RSW due to the competitive nature of the industrial landscape. Studies can be assisted by machine learning, and new strategies may be based on the process, base metal, or equipment used [57]. Cao *et al.* implemented an RSM-GA–based hybrid optimization approach to optimize RSW parameters for optimal shear strength of Al-Si–coated boron steel. The optimization was found effective and improved the shear strength of the weldment [58]. In their study, Zhao *et al.* introduced a combined approach of the entropy weight method and regression analysis to optimize process parameters in RSW. This approach successfully optimized the welding parameters while simultaneously minimizing the heat input during the welding process [59]. Pashazadeh *et al.* utilized a combined approach of ANN and GA to optimize RSW parameters, specifically targeting optimal nugget height and diameter. By employing this algorithm, a range of 591 to 615 weld spots was found that should be completed before conducting the electrode tip dressing operation [60].

In conclusion, these papers cover a range of topics related to RSW, including the use of modeling and simulation techniques for process optimization and the development of real-time monitoring and control systems. The papers emphasize the importance of understanding welding parameters and their role in optimizing RSW parameters for precise control and high-quality welds. From various studies conducted on this subject, the following observations have been made:

1. Numerous studies have effectively utilized modeling and optimization techniques in RSW. These encompass the application of ANNs to predict and optimize welding parameters, the utilization of GRA to identify optimal parameters based

on tensile strength, the development of monitoring systems using displacement diagrams and high-speed camera footage, and the implementation of fuzzy logic control systems to detect expulsion and enhance electrode lifespan.

2. The combination of different approaches, such as the entropy weight method with regression analysis or the integration of ANN with GA, has also demonstrated promising outcomes in the optimization of RSW parameters and the achievement of optimal weld quality.

## 5.6 Conclusion

This comprehensive review discusses the influence of RSW parameters, namely, welding current, welding time, surface resistivity and surface cleanliness, electrode pressure, and geometry on the weld quality of various materials. Additionally, the practice of modeling and optimization in RSW is highlighted for enabling precise control and high-quality welds. Based on carried review, the following important conclusions were drawn:

1. Higher welding currents lead to larger and more consistent weld nuggets, improving weld quality and joint strength. Insufficient welding time can result in weak joints, while excessive welding time can cause overheating and material deformation.
2. Surface resistivity, preparation, and cleanliness are vital for strong RSW welds. Various surface treatment techniques and understanding the impact of surface conditions contribute to improving weld quality, adhesion properties, and mechanical performance in RSW.
3. Adequate electrode force is essential as it contributes to minimizing contact resistance, facilitating proper heat concentration, and ensuring uniform pressure distribution, all of which are vital for achieving high-quality joints.
4. The geometry of the tip is important as it affects the formation of the nugget and ultimately influences the strength of the weld. Therefore, selecting an appropriate electrode tip design is crucial for achieving optimal weld quality.
5. Increasing current and time yield larger nuggets and higher strength, but higher electrode force may reduce weld

strength. Understanding these parameters optimizes weld quality, nugget size, and electrode wear in RSW.

6. Modeling and optimization tools are of great importance in the optimization of RSW parameters, facilitating precise control and the attainment of high-quality welds. Modeling tools like RSM, ANN, and FZ were found highly efficient and have accurate predictions of the welding process, allowing intelligent decision-making, adaptive control, and automated optimization in RSW.

## References


1. Robert, J., Messler, W., Jou, M., Review of control systems for resistance spot welding: past and current practices and emerging trends. *Sci. Technol. Weld. Join.*, 1, 1, 1–9, 1996.
2. Pouranvari, M. and Marashi, S.P.H., Critical review of automotive steels spot welding: process, structure and properties. *Sci. Technol. Weld. Join.*, 18, 5, 361–403, 2013.
3. Williams, N.T. and Parker, J.D., Review of resistance spot welding of steel sheets part 1: Modelling and control of weld nugget formation. *Int. Mater. Rev.*, 49, 2, 45–75, 2004.
4. Stavrov, D. and Bersee, H.E.N., Resistance welding of thermoplastic composites –An overview. *Compos. A: Appl. Sci. Manuf.*, 36, 1, 39–54, 2005.
5. Singh, R., *Applied Welding Engineering Processes, Codes, and Standards*, Second Edition, pp. 131–162, Butterworth-Heinemann, Oxford, 2016.
6. Patricio, F., *Mendez New trends in welding in the aeronautic industry*, MIT, MA, 02139, USA, 2008.
7. Dickinson, D.W., Franklin, J.E., Stanya, A., Characterization of spot welding behavior by dynamic electrical parameter monitoring. *Weld. J.*, 59, 6, 170s–176s, 1980.
8. Zhou, K. and Yao, P., Overview of recent advances of process analysis and quality control in resistance spot welding. *Mech. Syst. Signal Process.*, 124, 170–198, 2019, doi: 10.1016/j.ymssp.2019.01.041.
9. Podržaj, P., Polajnar, I., Diaci, J., Kariž, Z., Overview of resistance spot welding control. *Sci. Technol. Weld. Joining*, 13, 3, 215–224, 2008, DOI: 10.1179/174329308X283893.
10. Wan, X., Wang, Y., Zhang, P., Modelling the effect of welding current on resistance spot welding of DP600 steel. *J. Mater. Process. Technol.*, 214, 11, 2723–2729, 2014, doi: 10.1016/j.jmatprotec.2014.06.009.
11. Zhang, X., Yao, F., Ren, Z., Yu, H., Effect of Welding Current on Weld Formation, Microstructure, and Mechanical Properties in Resistance Spot

Welding of CR590T/340Y Galvanized Dual Phase Steel. *Materials*, 11, 11, 2310, 2018, https://doi.org/10.3390/ma11112310.

12. Ariffin, A.K., Mohamed, M.F., Ariffin, M.K.A., The effect of welding current on the resistance spot welding of aluminium alloy. *J. Manuf. Processes*, 23, 135–142, 2016, DOI: 10.1016/j.jmapro.2016.06.026.
13. Aslanlar, S., Ogur, A., Ozsarac, U., Ilhan, E., Demir, Z., Effect of welding current on mechanical properties of galvanized chromided steel sheets in electrical resistance spot welding. *Mater. Des.*, 28, 1, 2–7, 2007, doi: 10.1016/j.matdes.2005.06.022.
14. Taufiqurrahman, I., Ahmad, A., Mustapha, M., Lenggo Ginta, T., Ady Farizan Haryoko, L., Ahmed Shozib, I., The Effect of Welding Current and Electrode Force on the Heat Input, Weld Diameter, and Physical and Mechanical Properties of SS316L/Ti6Al4V Dissimilar Resistance Spot Welding with Aluminum Interlayer. *Materials*, 14, 5, 1129, 2021, doi: 10.3390/ ma14051129.
15. Khan, M.F., Sharma, G., Dwivedi, D., Weld-bonding of 6061 aluminium alloy. *Int. J. Adv. Manuf. Technol.*, 78, 5-8, 863–873, 2015.
16. Schulz, E., Wagner, M., Schubert, H., Zhang, W., Balasubramanian, B., Brewer, L.N., Short-pulse resistance spot welding of aluminum alloy 6016–T4—Part. *Weld. J.*, 100, 41–51, 2021.
17. Elitas, M., Effects of welding parameters on tensile properties and fracture modes of resistance spot welded DP1200 steel. *Mater. Test.*, 63, 2, 124–130, 2021, https://doi.org/10.1515/mt-2020-0019.
18. Zhang, X., Yao, F., Ren, Z., Yu, H., Effect of Welding Current on Weld Formation, Microstructure, and Mechanical Properties in Resistance Spot Welding of CR590T/340Y Galvanized Dual Phase Steel. *Materials*, 11, 11, 2310, 2018, doi: 10.3390/ma11112310.
19. Moshayedi, H. and Sattari-Far, I., Resistance spot welding and the effects of welding time and current on residual stresses. *J. Mater. Process. Technol.*, 214, 11, 2545–2552, 2014.
20. Pouranvari, M., Effect of Welding Parameters on the Peak Load and Energy Absorption of Low-Carbon Steel Resistance Spot Welds. *ISRN Mech. Eng.*, 2011, 1–7, 2011, doi: 10.5402/2011/824149.
21. Badkoobeh, F., Mostaan, H., Rafiei, M. *et al.*, A Study on Phase Evolutions and Tensile-Shear Performance of Dissimilar Resistance Spot Welds Formed Between AISI 430 Ferritic Stainless Steel and AISI 321 Austenitic Stainless Steel. *J. Mater. Eng. Perform.*, 32, 5028–5042, 2022, https://doi.org/10.1007/s11665-022-07451-.
22. Sivaraj, P., Seeman, M., Kanagarajan, D., Seetharaman, R., Influence of welding parameter on mechanical properties and microstructural features of resistance spot welded dual phase steel sheets joint. *Mater. Today: Proc.*, 22, 3, 558–562, 2020, https://doi.org/10.1016/j.matpr.2019.08.201.
23. Hasanbaşoğlu, A. and Kaçar, R., Resistance spot weldability of dissimilar materials (AISI 316L–DIN EN 10130-99 steels). *Mater. Des.*, 28, 6, 1794–1800, 2007, doi: 10.1016/j.matdes.2006.05.013.

24. Han, X., Effect of surface condition on resistance spot welding of advanced high strength steel. UWSpace, 37–51, 2019, http://hdl.handle.net/10012/15149.
25. Zhang, Q., Huang, M., Lv, T., Lou, M., Li, Y., Effect of surface treatments and storage conditions on resistance spot weldability of aluminum alloy 5182. *J. Manuf. Processes*, 58, 30–40, 2020, doi: 10.1016/j. jmapro.2020.08.002.
26. Hamedi, M. and Atashparva, M., A review of electrical contact resistance modeling in resistance spot welding. *Weld. World*, 61, 269–290, 2017, https://doi. org/10.1007/s40194-016-0419-4.
27. Naimi, I.K.A., Al Saadi, M.H., Daws, K.M., Bay, N., Influence of surface pretreatment in resistance spot welding of aluminum AA1050. *Prod. Manuf. Res.*, 3, 1, 185–200, 2015, doi: 10.1080/216932 77.2015.1030795.
28. Jo, K., Kang, P., Kim, Effects of Surface Roughness and Force of Electrode on Resistance Spot Weldability of Aluminum 6061 Alloy. *Appl. Sci.*, 9, 19, 3958, 2019, doi: 10.3390/app9193958.
29. Ghatei-Kalashami, A., Zhang, S., Shojaee, M., Midawi, A.R.H., Goodwin, F., Zhou, N.Y., Failure behavior of resistance spot welded advanced high strength steel: The role of surface condition and initial microstructure. *J. Mater. Process. Technol.*, 299, 117370, 2022, doi: 10.1016/j. jmatprotec.2021.117370.
30. Liu, L., Zhou, S.Q., Tian, Y.H., Feng, J.C., Jung, J.P., Zhou, Y.N., Effects of surface conditions on resistance spot welding of Mg alloy AZ31. *Sci. Technol. Weld. Joining*, 14, 4, 356–361, 2009, doi: 10.1179/136217108x394726.
31. Wei, P.S. and Wu, T.H., Electrical contact resistance effect on resistance spot welding. *Int. J. Heat Mass Transfer*, 55, 11-12, 3316–3324, 2012, doi: 10.1016/j.ijheatmasstransfer.2012.01.040.
32. Janardhan, G., Mukhopadhyay, G., Kishore, K. *et al.*, Resistance Spot Welding of Dissimilar Interstitial-Free and High-Strength Low-Alloy Steels. *J. Mater. Eng. Perform.*, 29, 3383–3394, 2020, https://doi.org/10.1007/s11665-020-04857-z.
33. Dobosy, Á., Gáspár, M., Török, I., Resistance Spot Welding of 7075 Aluminium Alloy. *Veh. Automot. Eng.*, 2, 679–693, 2018, doi: 10.1007/978-3-319-75677-6_58.
34. Wei, P.S. and Wu, T.H., Numerical study of electrode geometry effects on resistance spot welding. *Sci. Technol. Weld. Joining*, 18, 8, 661–670, 2013, doi: 10.1179/174329313x13759662488396.
35. Mahmud, K., Murugan, S.P., Cho, Y., Ji, C., Nam, D., Park, Y.-D., Geometrical degradation of electrode and liquid metal embrittlement cracking in resistance spot welding. *J. Manuf. Processes*, 61, 334–348, 2021, https://doi.org/10.1016/j.jmapro.2020.11.025.
36. Jo, H., Kim, D., Kang, M., Park, J., Kim, Y.-M., Effects of Surface Roughness and Force of Electrode on Resistance Spot Weldability of Aluminum 6061 Alloy. *Appl. Sci.*, 9, 19, 3958, 2019, https://doi.org/10.3390/app9193958.

37. Wang, B., Hua, L., Wang, X. *et al.*, Effects of electrode tip morphology on resistance spot welding quality of DP590 dual-phase steel. *Int. J. Adv. Manuf. Technol.*, 83, 1917–1926, 2016, https://doi.org/10.1007/s00170-015-7703-0.
38. Safari, M. and Mostaan, H., Experimental investigation of the effects of process parameters on the strength of eutectoid steel (AISI 1075) sheet resistance spot welds. *Metall. Res. Technol.*, 113, 3, 305, 2016, doi: 10.1051/metal/2016005.
39. Afshari, D., Sedighi, M., Barsoum, Z., Peng, R.L., An approachin prediction of failure in resistance spot welded aluminum 6061-T6 under quasi-static tensile test. *Proc. Inst. Mech. Eng., Part B: J. Eng. Manuf.*, 226, 6, 1026–1032, 2012.
40. Manladan, S.M., Yusof, F., Ramesh, S. *et al.*, A review on resistance spot welding of aluminum alloys. *Int. J. Adv. Manuf. Technol.*, 90, 605–634, 2017, https:// doi.org/10.1007/s00170-016-9225-9.
41. Özyürek, D., An effect of weld current and weld atmosphere on the resistance spot weldability of 304L austenitic stainless steel. *Mater. Des.*, 29, 3, 597–603, 2008, doi: 10.1016/j.matdes.2007.03.008.
42. Wei, S.T., Liu, R.D., Lv, D., Lin, L., Xu, R.J., Guo, J.Y., Lu, X.F., Weldability and mechanical properties of similar and dissimilar resistance spot welds of three-layer advanced high strength steels. *Sci. Technol. Weld. Joining*, 20, 1, 20–26, 2014, doi: 10.1179/1362171814y.0000000250.
43. Ozsarac, U., Investigation of Mechanical Properties of Galvanized Automotive Sheets Joined by Resistance Spot Welding. *J. Mater. Eng. Perform.*, 21, 5, 748–755, 2012, doi: 10.1007/ s11665-012-0189-0.
44. Kolařík, L., Sahul, M., Kolaříková, M., Sahul, M., Turňa, M., Resistance Spot Welding of Low Carbon Steel to Austenitic CrNi Stainless Steel. *Adv. Mater. Res.*, 875-877, 1499–1502, 2014, doi: 10.4028/www. scientific.net/amr.875-877.1499.
45. Sreehari, M. and Bhaskar, G.B., Experimental investigations on resistance spot welding for producing indentation free joints on AISI 409M grade stainless steels. *Mater. Res. Express*, 6, 4, 046527, 2019, doi: 10.1088/2053-1591/aafa97.
46. Chen, F., Tong, G.Q., Ma, Z., Yue, X., The effects of welding parameters on the small scale resistance spot weldability of Ti-1Al-1Mn thin foils. *Mater. Des.*, 102, 174–185, 2016, doi: 10.1016/j.matdes.2016.04.001.
47. Rao, R.V., *Advanced Modeling and Optimization of Manufacturing Processes*, Springer, London Dordrecht Heidelberg New York, 2011.
48. Davim, J.P., *Statistical and Computational Techniques in Manufacturing*, Springer-Verlag, Berlin Heidelberg, 2012, DOI: 10.1007/978-3-642-25859-6.
49. Karthikeyan, R. and Balasubramaian, V., Optimization of Electrical Resistance Spot Welding and Comparison with Friction Stir Spot Welding of AA2024-T3 Aluminum Alloy Joints. *Mater. Today: Proc.*, 4, 2, 1762–1771, 2017, doi: 10.1016/j.matpr.2017.02.018.

50. Rostamiyan, Y., Seidanloo, A., Sohrabpoor, H., Teimouri, R., Experimental studies on ultrasonically assisted friction stir spot welding of AA6061. *Arch. Civ. Mech. Eng.*, 15, 2, 335–346, 2015, doi: 10.1016/j.acme.2014.06.005.
51. Rajarajan, C., Sivaraj, P., Sonar, T., Raja, S., Mathiazhagan, N., Resistance spot welding of advanced high strength steel for fabrication of thin-walled automotive structural frames. *Forces Mech.*, 7, 100084, 2022, https://doi.org/10.1016/j.finmec.2022.100084.
52. Aizuddin, Z.A.Z., Aminudin, B.A., Sanda, P.S., Zetty, R.M.S., Resistance Spot Welding Process Optimization Using Taguchi Robust Method for Joining Dissimilar Material. *Appl. Mech. Mater.*, 835, 248–253, 2016, DOI: 10.4028/www.scientific.net/AMM.835.248.
53. Arunchai, T., Sonthipermpoon, K., Apichayakul, P., Tamee, K., Resistance Spot Welding Optimization Based on Artificial Neural Network. *Int. J. Manuf. Eng.*, 2014, 1–6, 2014, doi: 10.1155/2014/154784.
54. Jadav, S.K. and Gandhi, J., Experimental Investigation of Resistance Spot welding: A Literature Review. *Int. J. Eng. Res. Technol. (IJERT)*, 3, 3, 1503–1506, March 2014, DOI: 10.17577/IJERTV3IS031251.
55. Haghshenas, N. and Moshayedi, H., Monitoring of Resistance Spot Welding Process. *Exp. Tech.*, 44, 99–112, 2019, doi: 10.1007/s40799-019- 00341-z.
56. Podržaj, P. and Simončič, S., Resistance spot welding control based on fuzzy logic. *Int. J. Adv. Manuf. Technol.*, 52, 9-12, 959–967, 2010, doi: 10.1007/s00170-010-2794-0.
57. Martín, Ó. and De Tiedra, P., Advances in the Control and Improvement of Quality in the Resistance Spot Welding Process. *Metals*, 12, 1810, 2022, https:// doi.org/10.3390/met12111810.
58. Cao, X., Li, Z., Zhou, X., Luo, Z., Duan, J., Modeling and optimization of resistance spot welded aluminum to Al-Si coated boron steel using response surface methodology and genetic algorithm. *Measurement*, 171, 108766, 2021.
59. Zhao, D., Ivanov, M., Wang, Y. *et al.*, Multi-objective optimization of the resistance spot welding process using a hybrid approach. *J. Intell. Manuf.*, 32, 2219–2234, 2021, https://doi.org/10.1007/s10845-020-01638-2.
60. Pashazadeh, H., Gheisari, Y., Hamedi, M., Statistical modeling and optimization of resistance spot welding process parameters using neural networks and multi-objective genetic algorithm. *J. Intell. Manuf.*, 27, 549–559, 2016, https://doi.org/10.1007/s10845-014-0891-x.

# 6

# Shot-Peening Influence on Corrosion Behavior of SiC Particle Reinforced Aluminum Composite

**J. Venumurali[1]* and S. Rambabu[2]**

[1]*Department of Mechanical Engineering, Annamacharya Institute of Technology and Sciences (Autonomous), Tirupati (AP), Tirupati, Andhra Pradesh, India*
[2]*Department of Mechanical Engineering, Vignan's Institute of Information Technology, Visakhapatnam, Andhra Pradesh, India*

***Abstract***

This paper focused on the effect of surface modification through shot peening (SP) on corrosion resistance of AA6061-$SiC_p$ metal matrix composite (AMMC). The AMMC was produced using stir casting technique, then the cut samples treated by SP with different durations such as 70 and 140 seconds (s) and SP is repeated to obtain second set of specimens for corrosion test. Then X-ray diffraction (XRD) is utilized to evaluate the surface characteristics of AMMC specimens before and after SP. The residual stress (RS) measurements performed by XRD, which reveals the worthy insights in terms of Debye-Sherrer (DS) rings and are also known as concentric diffraction circles. The increase of RS in AMMC at near the surface taken place after shot peeing and DS rings provided with 2D and 3D visual illustration are more evident for it when compared with AMMC before SP. The maximum value of compressive RS of 23 MPa in AMMC obtained after 140 s duration. The microhardness values of measured depth wise from the surface of the AMMC before and after the SP. The micro-hardness of the AMMC increased after SP due to the induction of work hardening and obtained the maximum value of 143 HV by 140 s duration from 89 HV of AMMC before SP.

The surface roughness ($R_a$) of AMMC before and after SP is measured, which is increased after SP due to induction of plastic deformation from strain

**Corresponding author*: murali.iskapalem502@gmail.com

R. Thanigaivelan, Pradeep Kumar Krishnan, Kamalakanta Muduli and Santosh Kumar Tamang (eds.) New Materials, Processing and Manufacturability: Fabrication and Processing of Advanced Materials, (95–122) © 2024 Scrivener Publishing LLC

hardening and it is observed to be 9.13 μm and 10.9 μm over the surface of 70 s and 140 s treated specimens respectively. To observe the effect of $R_a$ on corrosion behavior of AMMC treated by SP, the $R_a$ value of second set of shot peened specimens reduced to 2.4μm by via grinding. The corrosion test was conducted on the AMMC surface after SP, SP + ground and before SP in 3.5 wt% NaCl solution. The SP + ground specimens exhibited lowest corrosion rate among all the specimens tested. The presence of rich oxygen detected by EDAX acts as passive layer in the surface of SP + ground specimens and enhanced corrosion resistance. Also, there is combined effect of density dislocations, $R_a$ and RS in formation of passive film and to produce corrosion resistance in the surface region of SP + ground specimen.

***Keywords*:** AA6061-$SiC_p$ MMC, shot peening, surface roughness, dislocation density, nanocrystalline structure, residual stress, corrosion resistance

## 6.1 Introduction

Metal matrix composites (MMCs) with different reinforcements such as SiC, $B_4C$, $Al_2O_3$, TiC, and graphene (particles/whiskers) have the potential for the automotive and aerospace industries [1]. Stir casting is one of the most efficient manufacturing techniques for MMC among liquid-state methods [2] to give excellent mechanical properties; also, it is low-cost, highly productive, and simple to operate. The performance of MMCs majorly depends on the size and concentration of reinforcement which could alter the microstructure of the matrix. The modulus and ductility of the MMC are affected by the size and quantity of the reinforcement. During solidification, the agglomeration occurs by pushing the particles into crystallite boundaries in the case of nanoscaled particles which is considered to be a negative effect [1, 3], whereas microscaled particles can improve strength, randomize the texture during load transmission, and stimulate the recrystallization [4, 5].

On the other hand, the majority of failures including fatigue, corrosion, and wear are experienced in almost every engineering system and originate from material surface [6]. Such failures can be bypassed by enhancing the surface properties of materials through severe plastic deformation (SPD). The surface properties have been improved by the changes made in the microstructure of the materials. SPD is related to a group of processing techniques conveyed to stuff a material to high strains, causing significant changes in its crystal structure and other properties. The theme of SPD is to refine the grain size by fragmentation of grains into tiny nuclei and induce a high density of defects in the material, which can lead to enhanced mechanical properties such as strength and hardness. There are many SPD methods that can show great potential for engineering applications [7]. The SPD process can yield materials with ultrafine-grained structures, improved mechanical properties, and better performance in applications such as structural components, aerospace materials, and biomedical implants. Such SPD methods are like (i) equal channel angular pressing which involves pressing a billet material through a channel that has two intersecting channels. The billet material is subjected to SPD as it passes through the channels, resulting in a significant reduction in grain size. (ii) High-pressure torsion: In this, a sample is located between two anvils that are opposite to each other and subjected to high-pressure torsion deformation. The sample experiences repeated shearing as it is rotated and deformed, finally leading to grain refinement. (iii) Accumulative roll bonding: It involves bonding multiple layers of metal sheets or these foils and then rolling them together carefully. The process is continuously repeated multiple times, and then inducing SPD and grain refinement and (iv) shot peening, which involves bombarding the surface of a target material with tiny spherical particles, known as shots, with high velocities. The impact of the shot on the surface of the material induces plastic deformation and the development of compressive residual stresses, which can further enhance the mechanical properties and resistance to fatigue failure of the material.

Among these SPD methods, the shot-peening process is one of the most common and preferred peening techniques since the equipment is cost-effective and can be used easily. During this process, the material surface is bombarded repeatedly and causes a work-hardening layer, induces surface residual stress, increases surface roughness, and reduces crystallite size/forms ultrafine crystallites (UFG)/creates a nanocrystalline (NC) structure. In addition, SP has many advantages in terms of structural, tribological, mechanical, fatigue, and biocompatibility properties of materials, apart from this the corrosion resistance is considered to be an essential property [7].

The study of the corrosion behavior of the material by SP was investigated in different research studies. The corrosion study of materials with NC structure in the surface region has gained the attention of various researchers [8, 9]. The corrosion resistance could be improved significantly with an NC-structured surface developed on the surface of 1Cr18Ni9Ti stainless steel by shot peening [10]. Moreover, very recently, the corrosion resistance of Ti-13Nb-13Zr alloy increased through ultrasonic shot peening as one of the SPD methods [11] and it is attributed to the crystallite refinement and the associated compressive residual stresses in the surface region.

It is observed that the SPD of the surface induced by optimum peening leads to shifting of the southeast because of the formation of a smooth surface, dense structure, and compressive residual stresses [12]. The shot peening with induction of compressive residual stresses surely enhanced the mechanical properties such as yield strength and hardness of the material due to refinement hardening and dislocation hardening [13].

Residual stress measurement after shot peening is a very important method. Also, the strain measurement by XRD is a very common method, the compressive residual stress before and after shot peening is analyzed often using XRD [14]. The experimental results observed from the XRD give a very clear understanding of the induced compressive residual stresses, which are the primary property for the study of other mechanical properties of the metals.

This process describes according to Bragg's law analysis that XRD depends on the principle of Bragg's law, which shapes that when X-rays strike a crystal lattice as seen in Figure 6.1 [15], they can be diffracted at definite angles determined from the spacing of the crystal planes. By calculating the positions and intensities of the diffracted X-ray peaks, it is then possible to determine the strain inside the crystal lattice. Changes that occurred in lattice spacing are mainly due to strain.

When the X-ray beam strikes a crystal lattice at a specific angle of incidence ($\theta$), the X-rays are to be diffracted at an angle ($\theta$) as seen in the figure which satisfies the relationship. This could result in constructive interference between the X-ray waves scattered by the crystal lattice planes, producing a diffraction peak.

The DS ring, which is also known as the DS ring, is in a circular pattern observed in the X-ray diffraction pattern. It will be formed when the X-rays are diffracted by numerously oriented crystallites, which are random (small crystals) in a polycrystalline material that overlap on a flat photographic film or detector. Every point observed on the DS ring corresponds to a different set of crystal planes which could satisfy Bragg's Law at a definite angle of diffraction.

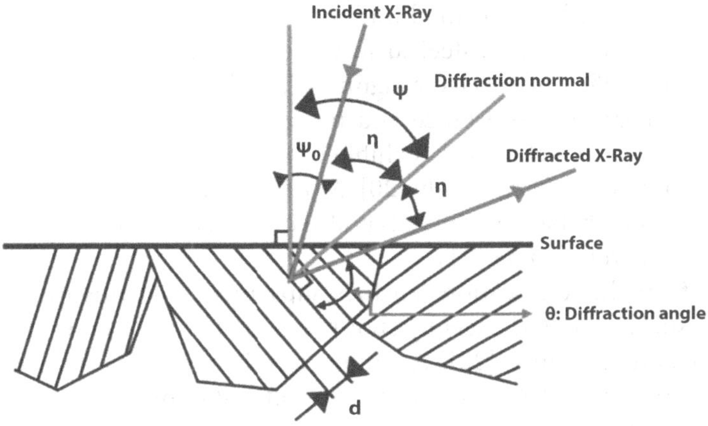


**Figure 6.1** Schematic of residual stress measuring using XRD [15].

The DS ring obtained after the XRD results provides information about the orientation distribution of crystallites inside the material and it can be also used for determining the average crystallite size.

The surface modification by SP treatment favors in creation of a dense passive film over the surface of the material, which protects the surface from corrosion. In general, the passive film refers to a thin, protective oxide layer that forms naturally on the surface of certain metals, most notably stainless steel and aluminum, and when open to oxygen in the atmosphere. This film offers corrosion resistance and plays a crucial role in the material's performance and durability. The formation of passive film is very much influenced by the NC structure developed on the material surface. NC structure has better corrosion resistance by the formation of stable passive film when compared to coarse crystallized structure in aluminum alloys [16].

Apart from the NC structure, surface roughness developed by shot peening shows an influence on corrosion characteristics. The pitting corrosion of carbon steel was inhibited by the dense passive film formed after shot peening [17]. The corrosion tests were conducted in LiBr aqueous solution after shot peening, then the results showed that passive film formation occurred due to an increase in surface roughness by the treatment. Similarly, Toloei *et al.* [18] have reported that the corrosion resistance of mild steel increased with increasing surface roughness. The corrosion rate has been decreased due to the hindrance of corrosive ions diffusion out of the grooves on the mild steel surface. However, it is still a controversial issue that the effect of surface roughness is induced by the shot-peening process on corrosion resistance and passive film formation Bagherifard *et al.* [19]. Similar behavior is exhibited by AZ31 Mg alloy subjected to shot peening in a different study [20]. Moreover, the effect of shot peening on the corrosion behavior of metals is also influenced by surface roughness and structural refinement [21].

Apart from this, dislocation densities, slip bands, and twins are improved over the surface of 304L steel by shot peening and cause deterioration of the corrosion resistance [22]. A similar trend in reducing the corrosion resistance of the 2024 Al alloy is observed after shot peening [23]. In spite of the NC structure induced on the surface of the 7075 Al alloy by ultrasonic shot peening, the corrosion resistance has been decreased due to dislocation density and surface roughness [24]. Hence, it is believed that the NC structure is not alone in governing the corrosion characteristics of the materials treated by SPD techniques. The other governing factors such as surface roughness, dislocation density, and the passive film also play vital roles.

The investigation is carried out to know the corrosion behavior of AA6061-$SiC_p$ MMC by the influence of NC structure obtained due to induction of residual stress, roughness, and dislocations density developed by shot peening with different durations. For this, the MMC was fabricated via a stir casting route and then shot-peened for about 70- and 140-s duration. One set of peened specimens was subjected to grinding to reduce surface roughness and then all the prismatic specimens were tested in 3.5 wt % NaCl solution at room temperature to obtain the corrosion results. The discussion on corrosion behavior is made on the surface of MMC specimens.

## 6.2 Experimental Procedure

### 6.2.1 Fabrication of Composite

The AA6061-$SiC_p$ MMC was cast using the stir casting process and its setup is shown in Figure 6.2. 6061 Al alloy strips were placed in a graphite crucible and heated to above its melting temperature around 720°C using a muffle furnace and its elemental composition is shown in Table 6.1. After attaining molten metal, the stirrer was placed in it to rotate at a speed of

**Table 6.1** The elemental composition of AA6061 aluminum alloy.

| Element | Cu | Mg | Zn | Mn | Si | Fe | Cr | Ti | Al |
|---|---|---|---|---|---|---|---|---|---|
| Wt % | 0.33 | 1.24 | 0.41 | 0.39 | 0.78 | 0.31 | 0.3 | 0.24 | Bal. |

**Figure 6.2** Stir casting setup.

700 rpm to develop a vortex. Silicon carbide particles ($SiC_p$) having an average particle size of 48 μm reinforced by 10 wt % and 1 wt % of Mg were also added as wetting agents. The molten metal is then poured into a steel die as shown in Figure 6.2. The density of the composite is determined after casting which is about 2.78 g/cm$^3$.

### 6.2.2 Microstructure Analysis

The specimens were cut from the casting to study the microstructure using optical microscopy. Before this, the specimens were polished by following a metallographic procedure and etched with H.F. Solution and Keller's reagent.

### 6.2.3 Shot-Peening Process

Shot peening was performed on the surface of composite specimens (8 mm × 8 mm × 3 mm) as shown in Figure 6.3. The steel shots (S-230) were selected according to the SAE-AMS-2431/1 standard with a diameter ranging from 0.45 to 0.6 mm, and a hardness of about 45 to 52 HRC [25]. The peening process is continued for two steps such as 70-s (first step) and 140-s (second step) times by following parameters such as 50-mm

**Figure 6.3** (a) Shot-peening setup and (b) working principle.

distance between nozzle and specimen surface, 4.5-bar air pressure, and 90° impingement angle. The untreated specimen is designated as SP0, whereas those subjected to shot peening for a duration of 70 and 140 s are designated as SP1 and SP2, respectively.

### 6.2.4 Surface Characteristics

The surface morphology of SP0, SP1, and SP2 specimens was examined by using the scanning electron microscope (SEM, TESCAN Vega-3). The phase constitution, size of crystallite, micro-strain, and dislocation density were determined by using the X-ray diffractometer with Cu-Kα radiation (Brucker, Advance D-8). Average surface roughness (Ra) was measured using the Surftest portable surface roughness tester (Mitutoyo, SJ-210), which is shown in Figure 6.4.

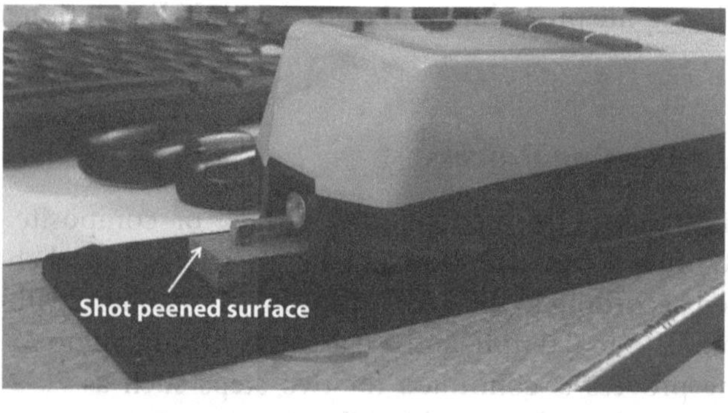


**Figure 6.4** Position of the specimen during surface roughness measurement.

### 6.2.5 Residual Stress Measurement

XRD was employed in this study to determine the exposure of the image to omnidirectional X-rays for the purpose of generating the DS rings. This method [26] serves as a replacement for existing methods such as incremental hole drilling and the $\sin^2 \Psi$ method [27]. The results are determined by taking into account the fixed incident angle of the X-ray at 35° and the (2,2,2) (h,k,l) plane. Finally, the residual stress is determined by the distortion that occurs in the Debye rings.

### 6.2.6 Microhardness Measurement

The microhardness on the surface before and after shot peening and the cross section of SP0, SP1, and SP2 specimens were measured by using the Vickers microhardness tester under the load of 100 gf and dwell time of 10 s.

### 6.2.7 Immersion Corrosion Test

Before the corrosion test as mentioned earlier, one set of shot-peened specimens such as SP1 and SP2 were slightly ground mechanically to eliminate higher superficial peaks developed during shot-peening and designated by SP3 and SP4, respectively. The grinding operation was performed on the surface of another set of shot-peened specimens to reduce Ra values and attained 2.4 μm. If it continues, there may be a loss of shot-peening effect on the specimen surface, therefore stopping the grinding after a slight reduction in the Ra value. The interest in reducing Ra is to check the roughness effect on corrosion behavior. Before the static immersion tests in 3.5 wt % NaCl (pH 7) solution, the specimens SP0, SP1, SP2, SP3, and SP4 were masked except for the surface to be tested, the exposed surface was then cleaned with acetone and weighed using a digital balance with an accuracy of 0.1 mg. After immersion for an interval of 24 hours, the specimens were taken out and washed with distilled water, dried for 30 min at 70°C in the oven, and then finally weighed.

The corrosion rate is then evaluated by using the weight loss method [28]. Specimen photos were taken by a camera after the corrosion tests and also the surface morphology on each specimen was studied by using SEM.

## 6.3 Results and Discussion

### 6.3.1 Microstructure of AA6061- $SiC_p$ MMC

The optical micrographs of stir-cased AA6061-$SiC_p$ MMC are taken at 100x and 200x magnification as shown in Figure 6.5. The distribution of SiC particles in the melt is discernible in etched condition from Figure 6.5(a) with a size range of 40 to 45 μm. An isolated SiC particle in the melt with close proximity at 200x is shown in Figure 6.5(b). The rough surface with a rough interface exhibited by SiC-isolated particles can increase bonding strength to a considerable degree. Such features also contributed to the gripping of reinforcement with the matrix phase and cover the pores when the melt shrinks radially during solidification.

### 6.3.2 Surface Morphology

The surface morphology of the shot-peened specimens exhibits the SPD in the surface of the material due to the random nature of the shot stream coming from the nozzle and nonuniformity in shot size [29]. The surface morphology of the specimens before and after shot peening is shown in Figure 6.6. The SEM photograph of the un-SP specimen in Figure 6.5(a) at 1000x magnification has a smooth surface compared with that of shot-peened specimens. The MMC surface after 70- and 140-s peening generated with distinct dimples and extruded ridges around the indentations created by each shot under the peening action is seen in Figures 6.6(b and c) at 100x magnification. These changes are attributed to the highest energy exerted by the shot media during the process. In general, the formation of dimples over the specimen surface during shot peening is affected by various essential factors such as shot size, pressure, impact angle, and material properties. The size and shape of the dimples can vary depending on these

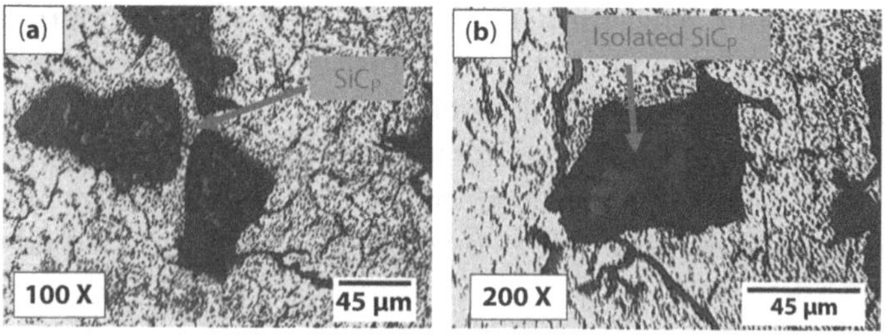


**Figure 6.5** Optical micrographs: (a) Distribution of SiC particles and (b) isolated SiC particles.

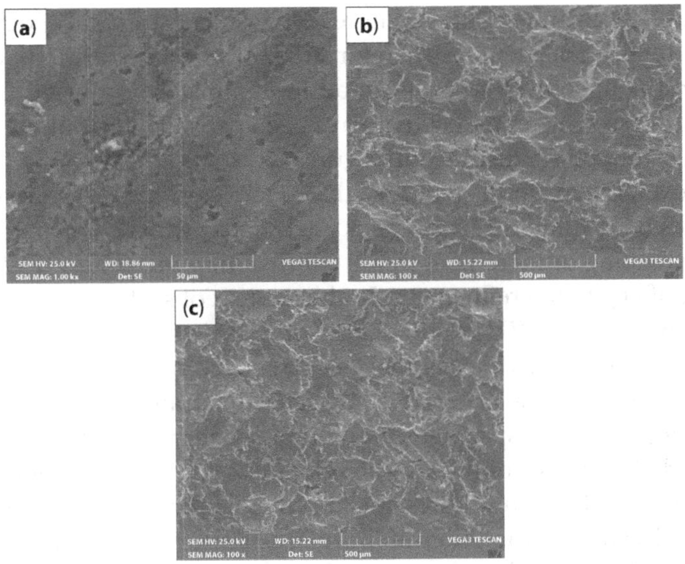


**Figure 6.6** SEM photograph of AA6061-$SiC_p$ MMC: (a) SP0, (b) SP1, and (c) SP2.

factors and the specific characteristics of the material being treated. It is worth noticing that shot-peening parameters must be cautiously controlled to ensure the desired surface properties without causing excessive material removal or negative effects like surface cracking or roughness. Moreover, the number of dimples moved and approached very close, resulting in an increasing number of dimples per unit area and extending the duration of peening from 70 to 140 s as observed in Figure 6.6(c). A similar type of behavior on 316L stainless steel subjected to shot peening is observed and proposed the same in a different study by Azar *et al.* [30].

### 6.3.3 X-Ray Diffraction

The XRD pattern of SP0, SP1, and SP2 specimens was presented in Figure 6.7. The SP0, SP1, and SP2 specimens show the same kind of peaks and also match with JCPDS documents (JCPDS 04-0787). The XRD peaks of un-SP, SP70, and SP140 are presented at an angle (2θ) of Al phases are 38.47°(111), 45.12°(200), 66.23°(220), and 79.42°(311), which displays polycrystalline structure. The peak shift from its original position to the lower angle side occurs in SP1 and SP2, which is caused by peening action and is observed clearly from a magnified view of the Al (311) peak. The maximum of this shift can be seen from the SP2 specimen as a result of more plastic deformation and compressive residual stresses [7] developed in 140-s duration of peening. Therefore, the thickness of the modified layer shown in the upcoming section can be approximately estimated from the microhardness measurements that have been carried out from the surface to the interior depth illustrated in Figure 6.7.

Moreover, the magnified view of Al (311) reveals that the Bragg diffraction peaks of SP1 and SP2 specimens are broader when compared to SP0. Peak broadening or increment in full width at half maximum (FWHM) is an index of crystallite refinement. In the context of surface characteristics measurements, the FWHM broadening of XRD peaks can be also related to the strain-induced changes that occurred due to shot peening in the crystal lattice. When a material is subjected to strain, a slight distortion will develop in lattice spacing, leading to a broadening of the diffraction of FWHM. The degree of FWHM broadening can offer information about the magnitude of the strain presented in the material. The FWHM broadening of XRD peaks is an index of the microstructural effects in a material, dislocation density, and its impact on the crystal lattice.

Also, the plastic deformation accompanied by the lattice defect by shot peening caused for increase in the FWHM [31]. Wang *et al.* [10] show that the NC structure developed over the shot-peened surface of the material is attributed to the effect of FWHM broadening. Table 6.2 shows the FWHM values of SP1 and SP2 specimens which are increased with increasing

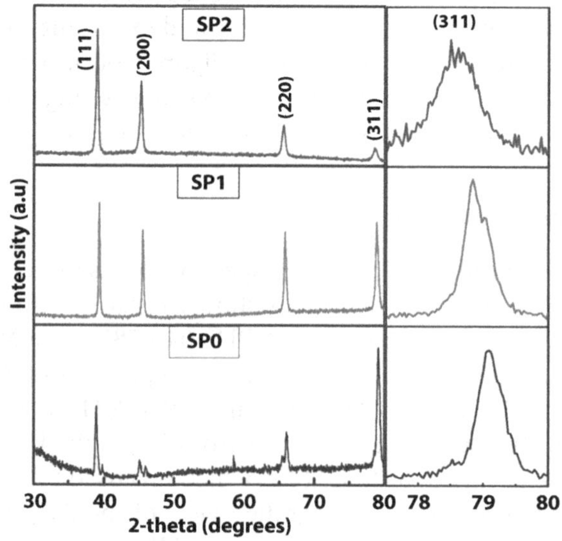


**Figure 6.7** XRD patterns of the AA6061-$SiC_p$ MMC before and after SP.

shot-peening duration. Lai *et al.* [31] reported that the change in FWHM value was also evident to assess the surface coverage during shot peening. The higher the coverage maximum, the higher FWHM broadening, exposure time, and Almen intensity [32].

Based on FWHM values obtained in this study, the average crystallite size of AA6061-$SiC_p$ MMC before and after the treatment was estimated from four Bragg reflection peaks such as (111), (200), (220), and (311) by using Scherrer's equation (6.1) [24]. Since then, Scherrer's equation has been extensively used in X-ray diffraction studies to estimate the average size of crystallites in a material. It relates the FWHM of a diffraction peak to the crystallite size. Moreover, the equation adopts that the peak broadening is merely due to the size of the crystallites and is not influenced by other factors such as micro-strain or lattice strain.

$$D = \frac{0.94\lambda}{\beta.\cos\theta_\beta} \tag{6.1}$$

where D is the average crystallite size, $\lambda$ is a wavelength of Cu-K$\alpha$, $\beta$ is the FWHM, and $\theta$ is the Bragg angle. The micro-strain was estimated by using Williamson Hall equation (6.2) [24].

$$B\cos\theta = \left[\frac{0.94\lambda}{D}\right] + [4\varepsilon\sin\theta] \tag{6.2}$$

where $\varepsilon$ is the root mean square of the micro-strain. Based on the average crystallite size and micro-strain, the dislocation density of the composite before and after the treatment is also estimated by using Equation 6.3 [24].

$$\delta = \frac{2\sqrt{3}}{b}\frac{<\varepsilon^2>^{1/2}}{D} \tag{6.3}$$

where $\delta$ is the dislocation density and b is the Burgers vector. Dislocations are generally known as line defects in the crystal lattice structure of a material. They are considered to be extra or missing planes of atoms within the lattice. Plastic deformation created by the shot peening can lead to the movement of dislocations observed from the XRD results. The values of average crystallite size, micro-strain, and dislocation density of the

**Table 6.2** Surface characteristics by XRD.

| Specimens | FWHM (°) | Crystallite size (nm) | Dislocation density ($\times 10^{14}/m^2$) | Micro-strain ($\times 10^{-3}$) |
|---|---|---|---|---|
| SP0 | 0.18 | 55.4 | 8.1 | 1.6 |
| SP1 | 0.24 | 38.8 | 13.4 | 1.9 |
| SP2 | 0.29 | 33.7 | 19.2 | 3.2 |

composite are presented in Table 6.2. From Table 6.2, it is observed that the crystallite size decreased from 55.4 to 33.7 nm after shot peening for 140-s duration and there was a raise in micro-strain after treatment. Also, dislocation density was increased after the treatment. All these characteristics were increased with increasing peening time.

## 6.4 Residual Stress

During the shot-peening process, the pressurized shots coming through the nozzle collide with the specimen surface resulting in a rapid change in momentum and a transfer of energy. The influence of these factors creates compressive stresses in the surface layer of SP0, SP1, and SP2 and further leads to local plastic deformation. The analyzed residual stresses from XRD clearly show that DS rings are observed in X-ray diffraction patterns of polycrystalline materials and reflect the random orientation distribution of crystallites. From Figure 6.8, it is observed that the concentricity of DS rings is higher in SP2 (Figure 6.8(c)) when compared to SP0 and SP1 specimens, which is a result of the random distribution of crystallites in the specimen. The concentricity of the DS ring increased with increasing shot-peening time. The DS rings appear as concentric circles in the diffraction pattern because the scattering angles and intensities depend on the orientation of the crystallites. Residual stress can distort the circular shape of the DS rings as seen in Figure 6.8(a) for SP0, leading to changes in their ovality or loss of its circularity. Figure 6.8(b) is so informative to show the difference between SP0 and SP2 as it possessed medium ovality. These features clearly show the changes in stresses from the tensile nature to the compressive nature as increases the shot-peening time. The shape changes can be quantified by measuring the eccentricity or aspect ratio

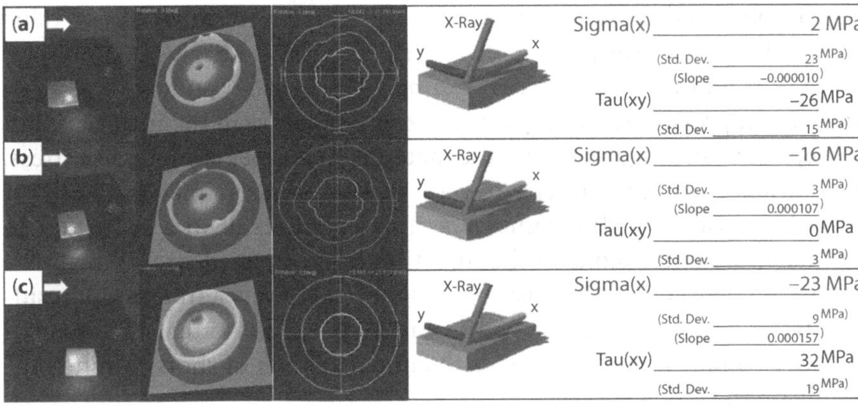


**Figure 6.8** Photo of the measured specimen, 3D DS ring, 2D DS ring, and the resulting diagram of (a) SP0, (b) SP1, and (c) SP2.

of the rings. These shape changes can provide insights into the anisotropy of the residual stress. The quantified result of residual stress is presented in Figure 6.9. The specimen before shot peening (SP0) possessed residual stress +2 MPa, and after 70-s shot-peening duration, the nature of stress changed from positive to negative and had −16 MPa. On further increase of peening duration from 70 to 140 s, the SP2 exhibits −23 MPa. Both the shot-peened specimens show compressive residual stresses.

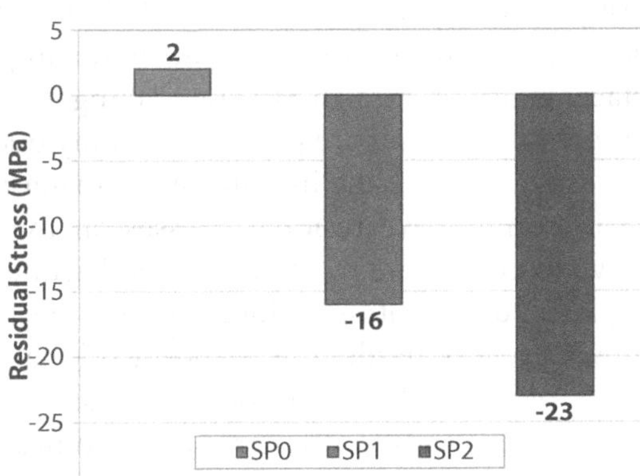


**Figure 6.9** Residual stresses vs. specimens.

It is observed from Figure 6.9 that a negative residual stress value obtained in shot-peened specimens SP1 and SP2 indicates compressive stress. It is known that the stress can be either tensile (pulling or stretching) or compressive (pushing or squeezing) in nature in the materials. But during shot-peening treatment, the material undergoes compressive stress, it experiences forces that tend to push or compress the material together over the substrate of the composite. This could result in a reduction in the material's dimensions in the direction of the applied stress which is a common phenomenon in plastic deformation. However, in terms of residual stress, negative values shown in Figure 6.9 indicate that the material is under compressive stress under the impact action of shots coming randomly on the target surface.

Negative residual stress values −16-MPa SP1 and −23-MPa SP2 are known to be compressive residual stresses and are beneficial in some cases, and they can enhance the material's resistance to cracking, improve fatigue life, or increase its load-bearing capacity.

## 6.5 Hardness Variation

The variation of microhardness along the depth from the surface of SP0, SP1, and SP2 specimens is shown in Figure 6.10. The microhardness was increased greatly in the surface region by shot peening, which is due to the co-action of the crystallite refinement + work-hardening [33] and dislocation densities [34]. Recently, the microhardness of the A356-SiC composite improved greatly by shot peening and it resulted up to a certain depth due to higher dislocation densities [35]. It is due to the impact of shots on the material surface, which creates an SPD by making localized compressive stresses. Compressive stresses are produced mainly due to the surface being deformed plastically while the interior of the material resists deformation resulting in compressive residual stresses on the surface. These stresses really strive to contribute to the enhanced hardness of the material surface. The plastic deformation induced by shot peening develops a phenomenon known as work hardening. Hence, work hardening leads to an increase in the hardness and strength of a material by virtue of plastic deformation. The surface layer after these results becomes more strengthened and resistant to further deformation finally resulting in increased hardness. It is also understood that the repeated action by impacts of the shots can break up or fragment and cause refined grains, leading to a finer microstructure. This finer microstructure commonly results in improved hardness and strength properties of the material. Therefore, it is observed

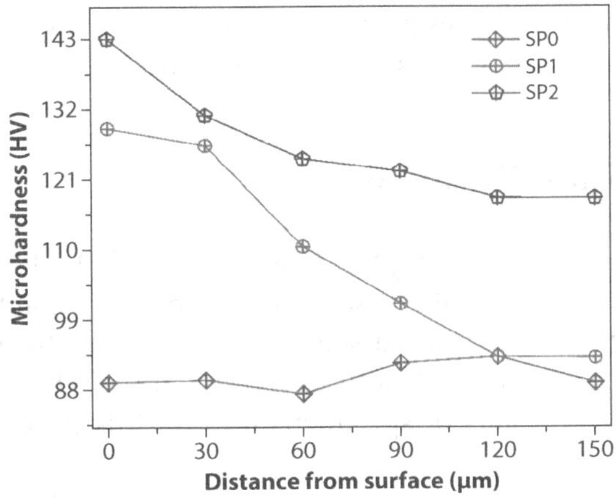


**Figure 6.10** Microhardness variation in AA6061-SiC$_p$ MMC before and after SP with distance from surface to substrate.

from the results of the present study that the microhardness value at the top surface is about 89 HV for SP0, 129 HV for SP1, and 143 HV for SP2 specimens. With the increase in depth, the HV value gradually decreased, which can be seen in Figure 6.10.

## 6.6 Surface Roughness

The average surface roughness (Ra) of the entire specimen series selected for the corrosion test in the present investigation is presented in Table 6.3. Shot peening induces the SPD on the surface of the composite which alters the morphology and average surface roughness. The results from the Softest roughness tester show that the Ra value is higher in the case of SP2 because of the abrupt change in height of the extruded ridges and dimples as shown in Figure 6.6, which is due to excessive work hardening by high impacts. Due to the development of dimples or indentations over the surface of the material by the effect of shot peening, the surface roughness has been increased. These dimples can cause a temporary increase in the roughness of shot-peened composites, the values of Ra of specimens used in this work are shown in Table 6.3. Before shot peening, the Ra value is 0.8 μm only, which is almost smooth and, after shot peening with 70-s duration, the Ra value is about 9.13 μm (SP1) and, on further increase of duration to 140 s, the value of Ra is about 10.9 μm (SP2). The shot-peened specimens SP1 and SP2 after the grinding process brought for Ra value of about 2.4 μm, which are designated by SP3 and SP4, respectively for easy identification from the results of the corrosion test.

## 6.7 Corrosion Behavior

The specimens of AA6061-$SiC_p$ MMC such as SP0, SP1, SP2, SP3, and SP4 with different surface conditions subjected to the corrosion tests and their corrosion resistance can be discussed. Figure 6.11 shows the representative surfaces of specimens after 24-hour immersion in 3.5 wt % NaCl solution. The corrosion test results of specimens used in this study are shown in Table 6.3.

According to Table 6.3, the corrosion resistance of specimens after shot peening has been increased by the formation of the NC structure. It is understood that during the shot-peening process, the shots impinge the surface, which further creates localized plastic deformation. This plastic deformation could have been inducing compressive stresses which are measured in the surface layer of the specimens (SP0, SP1, and SP2), these changes help to counteract tensile stresses and enhance its resistance to fatigue and stress corrosion cracking in general. In this work, the specimens

**Figure 6.11** Photograph of the representative surfaces after 24-hour immersion in 3.5 wt % NaCl solution.

**Table 6.3** Surface roughness values and corrosion rate of the specimens.

| Specimens | Surface roughness (μm) | Corrosion rate (mm/year) |
|---|---|---|
| SP0 | 0.8 | 0.164 |
| SP1 | 9.13 | 0.103 |
| SP2 | 10.9 | 0.123 |
| SP3 | 2.4 | 0.062 |
| SP4 | 2.4 | 0.082 |

with peened + ground condition improved the corrosion resistance and exhibited minimum corrosion rate. The SP3 specimen possessed the highest corrosion resistance among all the specimens and the corrosion rate decreased by 62.52%, 40%, 50.04%, and 25.09% when compared with SP0, SP1, SP2, and SP4, respectively.

Figure 6.12 shows the SEM photographs of corroded surfaces. As seen in Figure 6.12(a), the surface of SP0 is severely attacked by corrosion and combining the surface with numerous pits and wide cracks. The surface region is almost covered by micro-cracks. A smaller number of cracks with small size is found on the surface of peened series SP1 and SP2 from Figures 6.12(b and c). Whereas in peened + ground specimens, such cracks are further reduced, defending the corrosion attack over the surface of the specimens and increasing the uniformity in the formation of corrosion as seen from Figures 6.12(d and e). Moreover, the intergranular type corrosion is observed from the magnified SEM photographs (Figures 6.13(a and b)) of SP3 and SP4. Proneness to the corrosion increases with SP duration and is supported by similar behavior identified from the results obtained in different investigations [24].

This discussion is very important to identify the corrosion behavior by the performance of shot peening. It is discussed with respect to the surface roughness and dislocation densities obtained on the surface of the specimens. Figure 6.14 shows the influence of surface roughness on the corrosion rate. It is very important to observe the variation of the corrosion

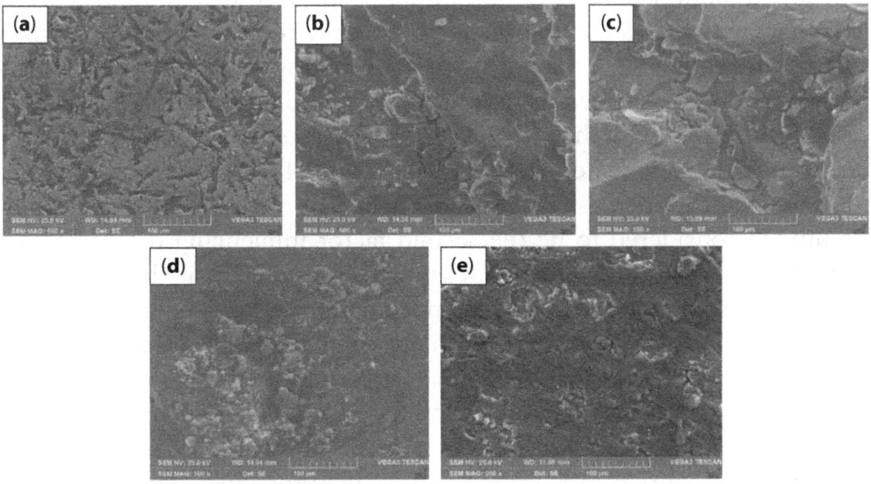


**Figure 6.12** SEM photographs of AA6061-$SiC_p$ composite after 24-hour immersion in 3.5 wt % NaCl solution: (a) SP0, (b) SP1, (c) SP2, (d) SP3, and (e) SP4.

**Figure 6.13** Magnified SEM photographs: (a) SP3 and (b) SP4 specimens.

rate corresponding to the surface roughness of the specimen. The rougher surface causes poor corrosion resistance in spite of the surface NC structure. The increase in roughness leads to variations in the current density during anodic polarization and a higher re-passive current density, which destroys the surface integrity of the stainless steel subjected to the shot peening [21]. The authors reported that the structural refinement obtained by shot peening and by the effect of compressive residual stress induced in the stainless steel is beneficial in enhancing the corrosion resistance, but an increase in roughness has negative effects. The MMC treated for 140-s duration showed the highest corrosion rate among peened specimens in this study. However, after reducing its Ra values by grinding, the corrosion rate is lowered again. Because of that, a rough surface could release the electrons more easily during the specimen being immersed in corrosive media. The smooth surface can minimize the surface area of the specimen in contact with the NaCl solution and reduce the corrosion rate. Therefore, surface morphology plays a key role in controlling the corrosion rate and the same concept has been supported by Ye *et al.* [36]. A similar type of result is discussed with the influence of surface roughness on the corrosion behavior of metallic materials by shot peening [24, 37].

Moreover, the increase of dislocation density with a duration of shot peening is considered to be the second factor influencing the corrosion behavior in the present study. Increasing dislocation density by SPD techniques shows an increase in corrosion formation and caused deterioration of corrosion resistance of metallic materials reported in different studies [24, 38–40]. Because of that, high dislocation density in the surface region could act as an active site for corrosion formation due to its higher energy state possessed by shot peening.

Moreover, Figure 6.15 is very clear that the protective oxide film (passive film) formed on the surface after the corrosion test which is assessed

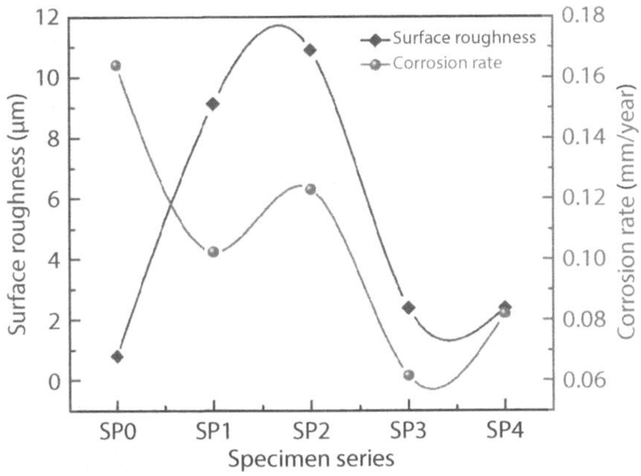


**Figure 6.14** Influence of surface roughness on corrosion rate.

by the presence of $O_2$ element. Since the presence of $O_2$ percentage is associated with the rate of corrosion, corrosion always occurs in the surface region of oxygen deficiency [41]. Plastic deformation in the surface region after shot peening caused to raise the Ra values and dislocation densities which are the prime factors in lowering the $O_2$ percentage and hindering the formation of oxide film. The passive film acts as a barrier between the metal surface and the corrosive environment NaCl, preventing direct contact and inhibiting corrosion. It is a protective layer, reducing the rate of metal dissolution and minimizing the likelihood of localized corrosion, such as pitting or crevice corrosion. This passive film also can offer resistance to certain chemical substances and environments and it can protect the metal surface from the deleterious effects of acids, alkalis, and other aggressive chemicals by preventing direct contact between the metal and the corrosive agents.

The $O_2$ percentage after the corrosion test is found to be higher in the SP3 specimen when compared to other specimens which favors in reduction of the corrosion rate and formation of dense passive film in the surface region. The SP3 specimen contains the lowest aluminum percentage in this work and it is an index to the forming less corrosion and dense passive film [42]. In contrast, in this investigation, it is observed that the dislocation density increased with increasing peening time shown detrimental to the formation of the protective oxide film. A similar type of behavior was seen in different studies and reported that the combination of lower surface roughness and low dislocation density along NC structure contributed to the formation of protective oxide film and improved corrosion resistance in the case of ultrasonically shot-peened Al alloy [24]. Therefore,

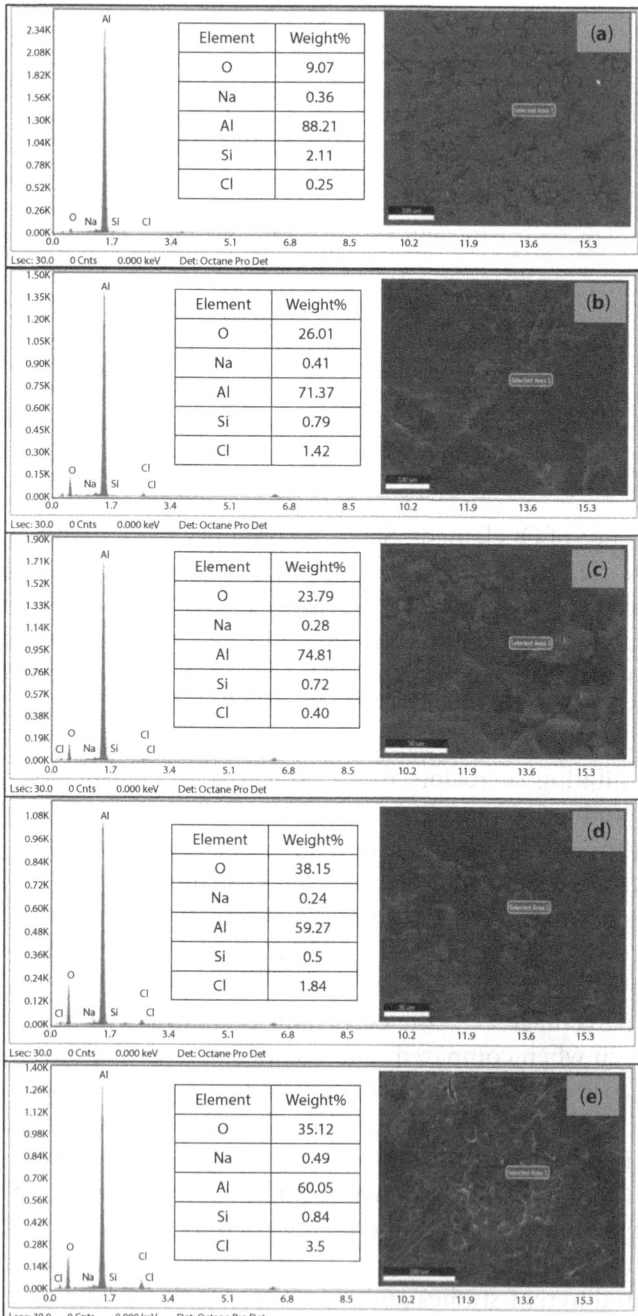


**Figure 6.15** EDAX results after immersion in 3.5 wt % NaCl solution: (a) SP0, (b) SP1, (c) SP2, (d) SP3, and (e) SP4.

the presence of a stable and dense passive film significantly extends the life span of MMC, as it provides long-term protection against corrosion. This is especially important in applications where the material is exposed to harsh environments or aggressive substances.

Moreover, crystallite refinement causes the formation of the passive film, resulting in a significant improvement in the corrosion resistance seen in the case of Ti-6Al-3Nb-2Zr-1Mo (Ti8o) alloy treated by laser shot peening and the corrosion tests conducted in 3.5% NaCl solution [43, 44]. Thus, the above observations are confirmed in the SP3 specimen after immersion in 3.5 wt % NaCl solution in the present investigation.

The developed compressive residual stress by shot peening shown can also have a significant impact on the corrosion behavior of a material. The specimen SP3 possessing compressive residual stress of 16 MPa can influence the propagation of cracks and corrosion-related damage.

The specimen SP3 with surface properties possessed by shot peening can hinder crack initiation and slow down crack propagation, thereby improving the material's resistance to corrosion-induced cracking. On the other hand, tensile residual stresses can promote crack initiation and accelerate crack growth, making the material more susceptible to corrosion damage seen in the specimen without shot peening. These tensile residual stresses seen in the case of SP0 can act as the driving force for crack initiation, particularly in the presence of corrosive agents. The combination of residual stress and corrosion can accelerate the rate of crack propagation, leading to the failure of the material. Thus, the corrosion resistance in the case of metallic materials is influenced by several factors such as crystallite size, hardness, dislocation density, surface roughness, and associated stresses like tensile and compressive residual stresses seen among specimens in this work. The optimum combination of several factors in the surface region of AA6061-$SiC_p$ MMC shows more influence and ultimately improves the corrosion resistance.

## 6.8 Conclusion

In this study, AA6061-$SiC_p$ MMC specimens were shot-peened for about 70- and 140-s duration and one set was subjected to grinding. The effect of surface modification on the corrosion behavior of the composite was investigated. The findings of this study on the influence of NC structure, roughness, residual stresses, and dislocation density developed by shot peening on the corrosion behavior of the composite are summarized as follows:

(1) The increment of surface roughness, dislocation density, and crystallite refinement occurred in the surface of MMC through plastic deformation and residual stresses induced by the effect of shot peening and maximum values of the above properties obtained with 140-s duration.
(2) The measured residual stresses by XRD are expressed in terms of DS rings and in the form of concentric diffraction circles presented. Features of DS rings clearly show the nature of residual stresses in MMC under the influence of shot-peening treatment.
(3) The value of residual stresses in the specimen before shot peening (SP0) is about +2 MPa, and after shot peening with 70 and 140 s, the residual stresses are changed to −16 and −23 MPa respectively. These changes are more evident in understanding the effect of shot peening, which transforms the tensile residual stresses into compressive residual stresses of MMC.
(4) Improvement of NC structure and hardness associated with FWHM broadening by work-hardening effect induced by shot peening.
(5) Corrosion resistance of the MMC increased with shot peening + grinding and high corrosion resistance exhibited by SP3 consisting of the dense protective passive film among all the specimens tested in NaCl solution.
(6) The deterioration of corrosion resistance in MMC specimens is mainly due to higher values of surface roughness, residual stresses, and dislocation densities in spite of the NC structure developed by shot peening.
(7) It confirms that the corrosion behavior of MMC treated by shot peening is not only governed by the NC structure obtained but also due to surface roughness, dislocation densities, and passive film developed in the surface of the MMC specimens.

## References

1. Jin, Y., Wang, K., Wang, W., Peng, P., Zhou, S., Huang, L., Yang, T., Qiao, K., Zhang, B., Cai, J., Microstructure and mechanical properties of AE42 rare earth-containing magnesium alloy prepared by friction stir processing. *Mater. Charact.*, 150, 52–61, 2019.

2. Abbas, A. and Huang, S.J., Investigation of severe plastic deformation effects on microstructure and mechanical properties of WS2/AZ91 magnesium metal matrix composites. *Mater. Sci. Eng.: A*, 780, 139211, 2020.
3. Qiao, X.G., Ying, T., Zheng, M.Y., Wei, E.D., Wu, K., Hu, X.S., Gan, W.M., Brokmeier, H.G., Golovin, I.S., Microstructure evolution and mechanical properties of nano-SiCp/AZ91 composite processed by extrusion and equal channel angular pressing (ECAP). *Mater. Charact.*, 121, 222–230, 2016.
4. Sun, J., Yang, Z., Han, J., Liu, H., Song, D., Jiang, J., Ma, A., High strength and ductility AZ91 magnesium alloy with multi-heterogenous microstructures prepared by high-temperature ECAP and short-time aging. *Mater. Sci. Eng.: A*, 734, 485–490, 2018.
5. Aydin, F., Sun, Y., Turan, M.E., The effect of TiB 2 content on wear and mechanical behavior of AZ91 magnesium matrix composites produced by powder metallurgy. *Powder Metall. Met. Ceram.*, 57, 9-10, 564–572, 2019.
6. Hutchings, I. and Shipway, P., *Tribology: friction and wear of engineering materials*, Butterworth-Heinemann, United Kingdom, 2017.
7. Kovacı, H., Bozkurt, Y.B., Yetim, A.F., Aslan, M., Çelik, A., The effect of surface plastic deformation produced by shot peening on corrosion behavior of a low-alloy steel. *Surf. Coat. Technol.*, 360, 78–86, 2019.
8. Mishra, R. and Balasubramaniam, R., Effect of nanocrystalline crystallite size on the electrochemical and corrosion behavior of nickel. *Corros. Sci.*, 46, 12, 3019–3029, 2004.
9. Wang, L., Lin, Y., Zeng, Z., Liu, W., Xue, Q., Hu, L., Zhang, J., Electrochemical corrosion behavior of nanocrystalline Co coatings explained by higher crystallite boundary density. *Electrochim. Acta*, 52, 13, 4342–4350, 2007.
10. Wang, T., Yu, J., Dong, B., Surface nanocrystallization induced by shot peening and its effect on corrosion resistance of 1Cr18Ni9Ti stainless steel. *Surf. Coat. Technol.*, 200, 16-17, 4777–4781, 2006.
11. Kumar, P., Mahobia, G.S., Mandal, S., Singh Vand Chattopadhyay, K., Enhanced corrosion resistance of the surface modified Ti-13Nb-13Zr alloy by ultrasonic shot peening. *Corros. Sci.*, 189, 109597, 2021.
12. Alharbi, N., Corrosion resistance of 3D printed SS316L post-processed by ultrasonic shot peening at optimum energy level. *Proc. Inst. Mech. Eng., Part B: J. Eng. Manuf.*, 237, 5, 745–57, 2023 Apr.
13. Pan, H., Tao, W., Zhang, B., Jiang, P., Wang, Z., Wu, W., Liu, L., Li, J., Wu, Z., Cai, Z., Effect of Shot Peening Strengths on Microstructure and Mechanical Properties of 316L Stainless Steel Prepared by 3D Printing. *Adv. Eng. Mater.*, 2201675, 2023 Feb 3.
14. Zine, P.U. and Joshi, P.K., Inducing compressive residual stress to minimize effect of crack generation by using controlled shot peening process. *Mater. Today: Proc.*, 72, 870–7, 2023 Jan 1.

15. Jose, B., Manoharan, M., Natarajan, A., Technology development for *in-situ* measurement of residual stress in arc welded joints of MDN 250 by portable Cosα X-ray diffraction method. *Proc. Inst. Mech. Eng., Part E: J. Process Mech. Eng.*, 236, 4, 1308–1318, 2022.
16. Krishna, K.G., Sivaprasad, K., Narayanan, T.S., Kumar, K.H., Localized corrosion of an ultrafine crystalliteed Al–4Zn–2Mg alloy produced by cryorolling. *Corros. Sci.*, 60, 82–89, 2012.
17. Yoo, J., Han, S., Nam, Y., Jeong, S., Effect of shot-peening on the passive film formation and corrosion of carbon steel in LiBr aqueous solution. *J. Mech. Sci. Technol.*, 34, 1–5, 2020.
18. Toloei, A., Stoilov, V., Northwood, D., Simultaneous effect of surface roughness and passivity on corrosion resistance of metals. *WIT Trans. Eng. Sci.*, 90, 355–367, 2015.
19. Bagherifard, S., Hickey, D.J., Fintová, S., Pastorek, F., Fernandez-Pariente, I., Bandini, M., Webster, T.J., Guagliano, M., Effects of nanofeatures induced by severe shot peening (SSP) on mechanical, corrosion and cyto-compatibility properties of magnesium alloy AZ31. *Acta Biomater.*, 66, 93–108, 2018.
20. Mhaede, M., Pastorek, F., Hadzima, B., Influence of shot peening on corrosion properties of biocompatible magnesium alloy AZ31 coated by dicalcium phosphate dihydrate (DCPD). *Mater. Sci. Eng.: C*, 39, 330–335, 2014.
21. Wu, Y., Kang, Y., Chen, Y., Effect of shot peening on the corrosion and wear resistance of 2205 duplex stainless steel. *J. Mater. Eng. Perform.*, 32, 7, 3186–201, 2023 Apr.
22. Shen, L., Wang, L., Wang, Y., Wang, C., Plasma nitriding of AISI 304 austenitic stainless steel with pre-shot peening. *Surf. Coat. Technol.*, 204, 20, 3222–3227, 2010.
23. Asquith, D.T., Yerokhin, A.L., Yates, J.R., Matthews, A., The effect of combined shot-peening and PEO treatment on the corrosion performance of 2024 Al alloy. *Thin Solid Films*, 516, 2-4, 417–421, 2007.
24. Pandey, V., Singh, J.K., Chattopadhyay, K., Srinivas, N.S., Singh, V., Influence of ultrasonic shot peening on corrosion behavior of 7075 aluminum alloy. *J. Alloys Compd.*, 723, 826–840, 2017.
25. Lainé, S.J., Knowles, K.M., Doorbar, P.J., Cutts, R.D., Rugg, D., Microstructural characterisation of metallic shot peened and laser shock peened Ti–6Al–4V. *Acta Mater.*, 123, 350–361, 2017.
26. Prevey, P.S., X-ray diffraction residual stress techniques, in: *Metals Handbook 10*, pp. 380–392, ASM Metals Park, OH, Materials Park Campus, Russell Township, Geauga County, Ohio, 1986.
27. Guo, J., Fu, H., Pan, B., Kang, R., Recent progress of residual stress measurement methods: A review. *Chin. J. Aeronaut. 2019*, 34, 2, 54–78, 2021.
28. Zakaria, H.M., Microstructural and corrosion behavior of Al/SiC metal matrix composites. *Ain Shams Eng. J.*, 5, 3, 831–838, 2014.

29. Zhao, J., Zhou, W., Tang, J., Jiang, T., Liu, H., Analytical and experimental study on the surface generation mechanism in shot peening. *Arch. Civ. Mech. Eng.*, 22, 3, 1–8, 2022.
30. Azar, V., Hashemi, B., Yazdi, M.R., The effect of shot peening on fatigue and corrosion behavior of 316L stainless steel in Ringer's solution. *Surf. Coat. Technol.*, 204, 21-22, 3546–3551, 2010.
31. Lai, H.H., Cheng, H.C., Lee, C.Y., Lin, C.M., Wu, W., Effect of shot peening time on δ/γ residual stress profiles of AISI 304 weld. *J. Mater. Process. Technol.*, 284, 116747, 2020.
32. Xu, Q., Cao, Y., Cai, J., Yu, J., Si, C., The influence of ultrasonic shot peening on the surface roughness, microstructure, and mechanical properties of TC2 thin-sheet. *J. Mater. Res. Technol.*, 15, 384–393, 2021.
33. Hou, L.F., Wei, Y.H., Liu, B.S., Xu, B.S., Microstructure evolution of AZ91D induced by high energy shot peening. *Trans. Nonferrous Met. Soc. China*, 18, 5, 1053–1057, 2008.
34. Luan, W., Jiang, C., Ji, V., Wang, H., Effect of shot peening on surface mechanical properties of TiB 2/Al composite. *J. Mater. Sci.*, 44, 10, 2454–2458, 2009.
35. Yolcu, C. and Kahraman, F., Effect of reinforcement particle amounts on dry sliding wear behavior of shot-peened SiC/A356 composites. *Mater. Test.*, 64, 4, 594–601, 2022.
36. Ye, H., Sun, X., Liu, Y., Rao, X.X., Gu, Q., Effect of ultrasonic surface rolling process on mechanical properties and corrosion resistance of AZ31B Mg alloy. *Surf. Coat. Technol.*, 372, 288–298, 2019.
37. Lee, H.S., Kim, D.S., Jung, J.S., Pyoun, Y.S., Shin, K., Influence of peening on the corrosion properties of AISI 304 stainless steel. *Corros. Sci.*, 51, 12, 2826–2830, 2009.
38. Conroy, K.G. and Breslin, C.B., The electrochemical deposition of polyaniline at pure aluminium: electrochemical activity and corrosion protection properties. *Electrochim. Acta*, 48, 6, 721–732, 2003.
39. Hao, Y.W., Deng, B., Zhong, C., Jiang, Y.M., Li, J., Effect of surface mechanical attrition treatment on corrosion behavior of 316 stainless steel. *J. Iron. Steel Res. Int.*, 16, 2, 68–72, 2009.
40. Laleh, M. and Kargar, F., Effect of surface nanocrystallization on the microstructural and corrosion characteristics of AZ91D magnesium alloy. *J. Alloys Compd.*, 509, 37, 9150–9156, 2011.
41. Avner, S.H., *Introduction to Physical Metallurgy*, 2nd Edition, McGraw Hill Publishing, New York, 1974.
42. Pérez, O.R., Garcia-Hinojoosa, J.A., Gomez, F.R., Mejia-Sintillo, S., Salinas-Bravo, V.M., Lopes-Sensenez, R., Gonzalez-Rodriguez, J.G., Garcia-Perez, C.A., Corrosion Behavior of A356/SiC Alloy Matrix Composites in 3.5% NaCl Solution. *Int. J. Electrochem. Sci.*, 14, 7426–7436, 2019.

43. Luo, G., Zhang, L., Xiong, Y., Zhang, B., Chen, X., Wu, Y., Wang, S., Cao, W., Microstructure Evolution and Improved Surface Properties of Ti-6Al-3Nb-2Zr-1Mo Alloy by Laser Shot Peening. *J. Mater. Eng. Perform.*, 31, 1–12, 2022.
44. Thanigaivelan, R., Saravanan, K.G., Makesh, M., Surface Modification Techniques for Improving the Material Performance, in: *Advanced Manufacturing Techniques for Engineering and Engineered Materials*, pp. 1–19, IGI Global, Hershey, Pennsylvania, USA, 2022.

7

# 3D Printing: Fundamentals, Applications, and Future Prospects

**Sonali Nag[1], Partha Pratim Borah[2] and Kankan Kishore Pathak[2*]**

[1]*Department of Mechanical Engineering, R.V.S College of Engineering and Technology, Jamshedpur, Jharkhand, India*
[2]*Department of Mechanical Engineering, Girijananda Chowdhury University, Guwahati, Assam, India*

## *Abstract*

Layer-based manufacturing technologies, also known as 3D printing, represent an emerging method in manufacturing. It allows transforming items designed in the digital setting into physical stuff using layered methods of manufacturing. In 3D printing technology, complex geometric parts with detailed internal features may be manufactured using a lesser amount of material without the requirement of extra production steps such as making line installation and design of the mold. Therefore, any component may be manufactured at once, which justifies the suitability of the 3D printing procedure for the making of biomedical devices as well as other devices frequently encountered in our day-to-day lives. In 3D printing, different materials such as metals, plastics, composites, organic, and biomaterials may be used. In most cases, the generation of 3D models relies on the utilization of CAD design software or the employment of 3D scanning technology. 3D printers make parts by regulating the placement and adhesion of consecutive layers of building material in 3D space, so 3D printing is also known as additive layered manufacturing. Today, 3D printers are considered one of the most sophisticated equipment for rapid prototyping (RP). This chapter examines the fundamental operational principles of 3D printers, along with the modern technologies associated with them and the various materials utilized in 3D printing technology. Additionally, the advantages, various challenges, applications, and future prospects of 3D printing are also addressed.

**Corresponding author*: kankan_me@gimt-guwahati.ac.in; kankankishore@gmail.com

R. Thanigaivelan, Pradeep Kumar Krishnan, Kamalakanta Muduli and Santosh Kumar Tamang (eds.) New Materials, Processing and Manufacturability: Fabrication and Processing of Advanced Materials, (123–138) © 2024 Scrivener Publishing LLC

***Keywords***: 3D printing, additive manufacturing, rapid prototyping

## 7.1 Introduction

The phrase "3D printing" encompasses several additive manufacturing methods wherein objects are built by layering plastic, metal, or other materials successively. Today, 3D printing has numerous uses in many different industries around the world [1]. This technology is commonly utilized in the large-scale fabrication of diverse-designed items for the agricultural, health care, automotive, railway, and aviation sectors [2]. From a computer-aided design (CAD) model, it is feasible to manufacture an object by applying successive layers of various materials. A pivotal element in a burgeoning industrial transformation characterized by digitalization, data integration, and connectivity is three-dimensional (3D) printing. Product designers enhance the adaptability of 3D printing in response to shifts in both demand and technological advancements.

In the realm of 3D printing, light or heat catalyzes the transformation of polymers, powders, or filaments into tangible objects. These methodologies encompass processes such as the laser solidification of liquid polymers (stereolithography, SLA), the laser fusion of metal powder (selective laser sintering, SLS), and the extrusion of molten plastic (fused deposition modeling, FDM) to fabricate solid entities. The integration of 3D printing necessitates sharp coordination and management efforts due to the complex blend of digital and physical technologies in play. 3D printing is different from the traditional manufacturing process in so many ways. 3D printing follows an additive approach, where it constructs solid objects by incrementally depositing material layer by layer. This reduces raw material waste, but energy consumption is higher compared to conventional processes. 3D printing transitions seamlessly from a digital model to a physical object, eradicating the necessity for tooling, as indicated by reference [3]. People live in a digital world now. 3D printing has become a part of this digitization. Generally, CAD software provides the design of objects that may be fabricated using 3D printers. Furthermore, by leveraging various computational technologies like artificial intelligence, machine learning, and robotics, 3D printing proves advantageous in delivering superior-quality products. Manufacturers use additive manufacturing to be agile in responding to developments in demand and innovation [4].

## 7.2 Materials for 3D Printing

The realm of 3D printing technology employs an array of materials, including ceramics, metals, and polymers, as well as their amalgamations in the form of hybrids, composites, or functionally graded materials (FGMs), as noted in reference [5]. These materials exhibit outstanding physical characteristics, making them suitable for 3D printing applications ranging from manufacturing human organs to aerospace components. Typical metallic options encompass aluminum alloys, cobalt-based alloys, nickel-based alloys, stainless steels, and titanium alloys [6].

Cobalt-based alloys boast exceptional attributes such as high specific stiffness, resilience, robust recovery capabilities, elongation, and adaptability to heat treatment [7]. In the aerospace sector, nickel-based alloys are the material of choice for crafting components. Objects fabricated through 3D printing with nickel-based alloys are suitable for use in hazardous environments, owing to their remarkable corrosion resistance. Additionally, 3D printing technology extends its capabilities to the utilization of titanium alloys [8], which exhibit exclusive qualities, including ductility, resistance to corrosion, oxidation resistance, and low density [9]. The versatility of 3D printing technologies encompasses the utilization of polymers for various applications. Polymers, particularly those in liquid form or with low melting points, are extensively adopted in the 3D printing industry due to their cost-effectiveness, lightweight nature, and ease of processing [10]. Notably, polymers play a pivotal role in the development of biomaterials and medical devices. In contemporary times, ceramics have gained prominence as a preferred material for 3D printing technology, primarily due to their commendable mechanical properties. The aerospace industry often employs composite materials such as carbon fibers, reinforced polymer composites, and glass fibers, owing to their exceptional characteristics, including high specific stiffness, strength, resistance to corrosion, and fatigue resistance. Furthermore, 3D printing technology extends its reach to the use of smart materials in the creation of soft robotic systems [11].

Almost every business might benefit from 3D printers. They are not only for prototypes, either. The printing of finished objects is a common use for 3D printers. The construction sector is currently printing complete homes using this cutting-edge printing technology. Due to their adaptability, accuracy, and speed, 3D printers constitute an exciting possibility for the coming years of production [12]. Educational institutions worldwide are integrating 3D printers into their classrooms, enhancing interactive learning experiences. This technology enables the production of

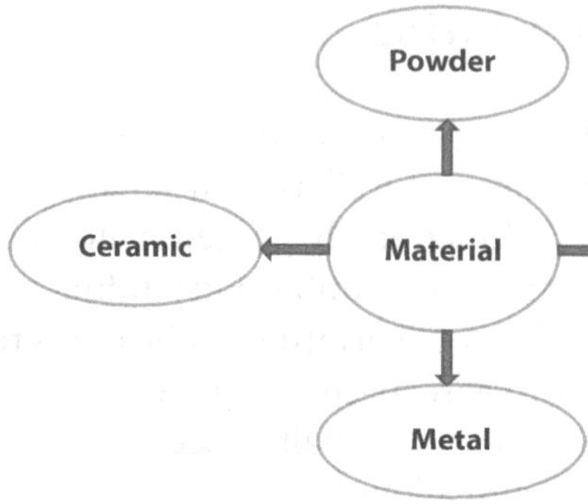


**Figure 7.1** Materials employed in 3D printing.

three-dimensional dinosaur bones and components for robotics projects, fostering hands-on learning opportunities for students. The adaptability/versatility of 3D printing technology makes it a game-changer in any sector. Plastics can be employed by 3D printing machines to create stiff items such as eyewear. To manufacture flexible products like phone cases and bike grips, they may also use a hybrid mixture of plastic and rubber powder. To produce highly resilient products for industry, certain 3D printers can additionally employ carbon fiber and metallic powders [13]. Figure 7.1 shows the chart with materials employed in 3D printing.

### 7.2.1 Stages Involved in the Process of 3D Printing

Figure 7.2 shows the flow diagram of different stages involved in 3D printing.

The various stages are as follows:

Modeling: The process of 3D printing starts with the digital formation of the three-dimensional product using various computer-aided software programs like SolidWorks and AutoCAD [14].

3D printing: After 3D modeling, begin the process of layer-upon-layer configuration of the whole process using a 3D slicer. Each subsequent layer is added atop the neighboring layer, a process that continues until all layers are formed, culminating in the creation of a three-dimensional object.

Finishing: Regardless of the adequacy of the 3D-printed product for a variety of uses, a larger version of the object is produced, enabling the

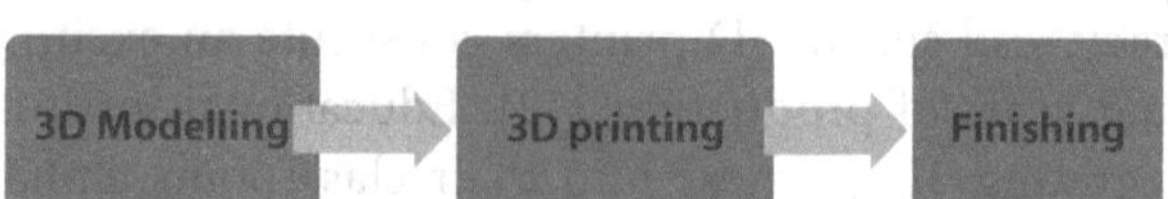


**Figure 7.2** Different stages involved in 3D printing.

elimination of extra material through a more exact and high-resolution method, thus improving precision [15].

### 7.2.2 3D Printing Technologies

Several layering methods and materials are particular to each type of 3D printing technology. The following fundamental procedures are used in 3D printing:

(i) Stereolithography: Using a laser beam to sculpt photopolymer resins, the SLA method creates 3D models [16]. A design is created on the polymer bed when the UV laser beam impacts the polymer layer; the design solidifies and advances one inch downward; a second sheet is then polymerized in the same manner; this procedure is repetitive until the entire 3D item is made [17].

(ii) Selective laser sintering (SLS): A laser beam and powder are utilized in this process. The 3D-modeled file is supplied into the framework, and an effective pulsed laser beam, like a $CO_2$ laser beam, is projected onto the powder bed [18]. Any material that has thermal stability and endurance, including polyamide, polystyrene, and polycarbonate, may be utilized to create powder beds. The 3D model is created stratum by stratum by melting and solidifying each of the layers of powder, which is subsequently combined to create the final design [19, 20].

(iii) Photopolymerization: Photopolymerization-based 3D printing techniques have become popular among polymer chemists, material scientists, and engineers [21]. Using UV light, photopolymers are cured. This process is referred to as photopolymerization. A resin material called a photopolymer solidifies when exposed to UV light. It makes the procedure faster than others since it aids in the continuous solidification of various layers at once. The exposed material becomes robust and long-lasting thanks to photopolymerization [22].

(iv) 3D slicing: This technique involves dissecting a design into multiple levels. It merely entails cutting a design into layers, which are then placed one on top of the other. A wide array of applications is available for the process of 3D slicing during printing, such as Cura Slicer, Slic3r, and Simply3D, as mentioned in reference [23].

(v) STL file format: Stereolithography is the primary use for STL files. When the printing procedure begins, the exterior structure of a component is described using the format of an STL file [24].

(vi) Multi-jet fusion: This technique is employed in industry to create 3D prototypes. The technique includes the mixing of a detailing agent and a fusing agent. The main substance used to create prototypes is a layer of nylon powder. With the use of a binding agent, also known as a fusing agent, a single layer of material is fused with another layer in a specific pattern. After fusing, the layers are subjected to thermal energy sources for improved adhesion. A detailing agent is then used to produce design features and smooth surfaces once the layers have fused [25].

(vii) Electron beam melting: This 3D printing technique is frequently employed in the creation of heavy metal components [26]. The metal elements are liquefied by a beam of electrons in a vacuum chamber, multiple layers are generated one at a time, and then those layers are hardened to create a 3D-printed object. Compared to other 3D printing processes, the finished object does not require thermal treatment to solidify the subsequent layers [27, 28].

(viii) Fused deposition model (FDM): This technique for 3D printing is currently the most popular. Small-scale final goods and 3D prototypes are both produced with FDM. The procedure uses polyacetic acid as the core material, which is a thermoplastic [29]. By heating the thermoplastic material and extruding it into the layers using extrusion nozzles, a 3D item is printed in layers [30]. The movement of the liquifier head and extrusion nozzles in the X and Y coordinates aligns with the predefined instructions provided to the printer, adapting to the specific design of the 3D object. Each layer that is created is fused with the layer behind it, which becomes harder over time. The FDM is efficient and produces robust, beautifully finished 3D objects [31].

(ix) Laminated object manufacturing (LOM): LOM uses laminated sheets that have been covered in an adhesive substance. Depending on the specifications of the three-dimensional model, the layers can be made of paper or plastic [32, 33]. Under specific conditions of pressure and temperature, all of the laminated sheets are cemented together. The laminated sheets are then sliced using a laser or similar cutting-edge equipment into the appropriate 3D shape [34, 35].

Under precise temperature and pressure conditions, sintering is the merging of particles into a single solid mass without melting. This process is known as direct metal laser sintering (DMLS). Only metal powders can be utilized for DMLS. At an industrial scale, DMLS is frequently utilized for the manufacturing of metallic components and prototypes [36].

(x) 3D printers that use poly-jet technology project photopolymers onto the design bed; subsequently, they are solidified through exposure to UV light. The entire 3D model is created using an additive approach that adds layers one at a time [37]. The utilization of two or more materials for a single prototype or product is the most astounding aspect of poly-jet printing. When utilizing poly-jet printers, further processing is not required since the 3D-modeled object is usable immediately after production. While pricey and quick, poly-jet 3D printing has higher design accuracy than other 3D printing methods [38].

### 7.2.3 Applications of 3D Printing

Numerous uses exist for 3D printing. It makes it possible to produce high-end goods faster and more effectively for less money. General idea of the application of the 3D printing may be understood from the Figure 7.3. Through additive manufacturing, which conventional production methods cannot replicate, manufacturers and engineers gain the capability to fabricate intricate designs to a greater extent [39]. Figure 7.3 shows the chart of application of 3D printing. Applications include the following:

- Rapid prototyping: 3D printing is utilized for rapid 3D structure prototyping. Small-scale production: This aspect of 3D printing has made itself available on a commercial scale. It is employed for the final products in industries as well [40].
- Aerospace sector: 3D printing technology is used in the aerospace industry to create enhanced, complex geometries and frivolous parts that may demand less energy and resources. Nickel-based alloys have gained prominence in the aerospace industry due to their exceptional tensile properties, resistance to oxidation and corrosion, and impressive damage tolerance [41].

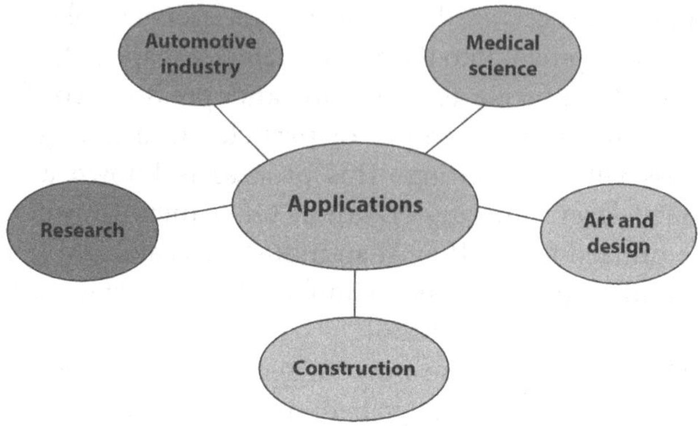


**Figure 7.3** Applications of 3D printing.

- Automotive industry: The fabrication of components for the car industry employs 3D printing. Additionally, 3D printing technology may reduce material waste and consumption, prices, and time; as a consequence, it allows for the rapid testing of innovative designs. Due to their low cost, small assembly parts composed of powdered metals are also 3D printed for mass production [42, 43].
- Medical equipment: 3D printing has made a mark in the realm of biomechanical engineering as well, from the small-scale production of braces and splints to the prototype of artificial limbs. Printing 3D skin, bones, organs, and tissues for medication and pharmaceutical research has become possible because of 3D printing technology. For teaching reasons, 3D-printed models of body parts and systems are available [44].
- Research and development: Basically, any form of 3D model may be produced using 3D printing steps, and these models are then used for research [45].
- Manufacturing tools: This involves developing and producing tools required for the manufacturing of tools, such as dies and molds, and it also uses 3D printers [46].
- Toys and games: Legos and other small toys composed of plastic suitable for manufacturing have entered the market thanks to 3D printing [47].
- Jewelry: Limited-edition jewelry is currently quite popular, even though the production cost is not as high as the tags claim. Women purchase it for their own delight, though 3D-fabricated items are also exclusive [48].

- Architectural designs: For examination and approval of everything that includes construction and development, the architectural business generates scalable 3D models of the buildings and bridges [49].

### 7.2.4 Advantages of 3D Printing

Figure 7.4 shows the pictorial representation of advantages of 3D printing. Here are a few benefits related to 3D printing that have undoubtedly improved and eased our lives:

(i) Speed: CAD models enable the swift production of intricate designs within a matter of hours, while 3D printing further facilitates rapid validation and the generation of design ideas.
(ii) Cost-effectiveness: 3D printing stands out as a highly cost-effective process, particularly well-suited for small-scale production. It demands minimal machinery and a reduced workforce for part manufacturing. This efficiency arises from the method's ability to build components from scratch, which results in significantly less waste material compared to traditional machining techniques.
(iii) Flexibility: The utilization of 3D printing allows for the integration of multiple materials into a single item, providing the flexibility to combine and synchronize a wide array of colors, textures, and even mechanical properties. All users, even those with little CAD skills, can alter designs using 3D printing to create one-of-a-kind, personalized new parts.

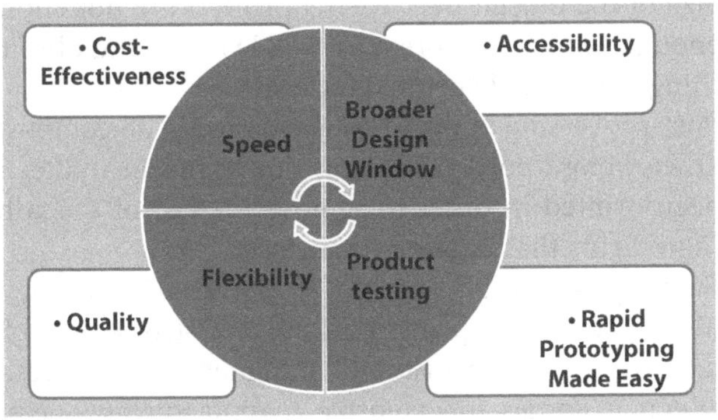


**Figure 7.4** Advantages of 3D printing.

(iv) Product testing: The physical prototype may be examined, and if any problems are found, the CAD file can be modified, allowing for the printing of a new version the following day.
(v) Quality: Because 3D printing enables step-by-step assembly of the product or a component of it, it ensures that the design will be enhanced and that the parts or products will be of higher quality.
(vi) Accessibility: Compared to traditional manufacturing configurations, 3D printing technologies are more accessible and can therefore be utilized by a greater number of people.
(vii) Broader design window: The process of 3D printing makes it simple to produce intricate components and products; traditional methods of production had many restrictions when it came to complex and elaborate designs.
(viii) Rapid prototyping simplified: You may finish your prototype in a matter of weeks or days because of 3D printing [50].

### 7.2.5 Disadvantages of 3D Printing

Figure 7.5 shows the pictorial representation of disadvantages of 3D printing. Some limitations in 3D printing are discussed as follows:

(i) Material restrictions: 3D printing may be utilized to test a diversity of materials during prototyping for design and development, but in the end, only materials with highly precise properties may be used for the final product and its mass manufacturing.
(ii) Size of the object: Because 3D printers are not enormous enough to manufacture massive 3D forms and objects, the things that are 3D printed have smaller sizes.
(iii) Post-processing of the 3D object: The 3D object must cool down before post-processing can begin on it after it has been printed in order to remove the last of the adhesive ingredients that were on its surface.
(iv) Fragility of the 3D structures: Some 3D printing techniques create 3D prints that are not strong enough and may crack under pressure or at higher temperatures.
(v) When it comes to extensive manufacturing operations, there is typically an escalation in costs associated with

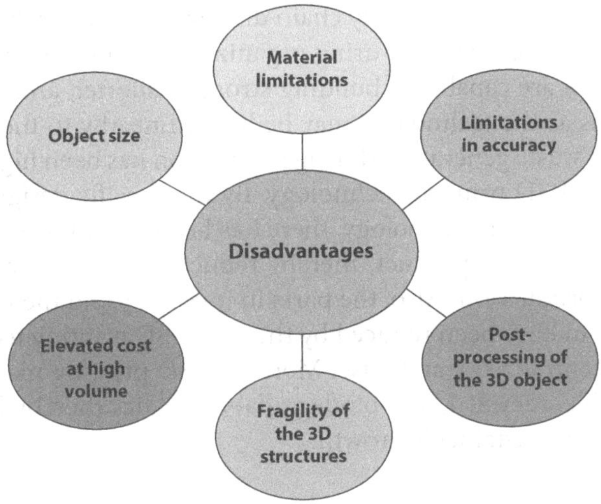


**Figure 7.5** Disadvantages of 3D printing.

using 3D printing, as the impact of economies of scale is relatively diminished.

(vi) Accuracy limitations: The accuracy of a printed component relies on the specific machine and method employed in the production process. The precision of 3D-printed items is not always guaranteed [51].

## 7.3 Future Prospects of 3D Printing Technology

In the contemporary context, a rapid growth in the 3D printing technology is observed. The rise in demand for specialized materials in order to fulfill the requirements of end parts will keep advancing the types and variety of available alternatives. The ability to effectively manage a broader range of advanced materials will play a pivotal role in the development of future printers, particularly those tailored for industrial applications. Thus, businesses benefit from additive manufacturing methods, especially in such areas where earlier it was not possible. Moreover, as manufacturers are advancing toward industry 5.0, fully automated production, post-production, and integrated usability are the major trends these days. In the modern era, industries are adopting 3D printing technology to complete their various needs which leads to a more holistic production environment. Employing the technology of additive manufacturing in industries minimizes the problem of supply chain disruptions, thereby boosting supply chain resilience. Manufacturing organizations adopting the 3D printing technology are capable of building stronger, shorter, and more robust supply chains as the technology may be located nearby to the consumer's location. The waste generation during production has been highly reduced by the usage of 3D printing technology. By the specific design of a product using 3D printing technology, there has been a significant decrease in the weight of the end product, thereby reducing the amount of material required for production. Also, the parts in inventory and the waste associated with them have been reduced by the use of 3D printing technology as part of a decentralized warehouse. Moreover, 3D printing materials make use of reusable, recycled, and biodegradable plastics thereby leading to a forward drive in sustainable growth.

## 7.4 Conclusions

3D printers have a versatile role in each industry and it has become an important part of a new industrial revolution. The development of 3D printing as a pioneering technology has opened up a wide range of possibilities and uses. It has transformed over time from a niche idea to a familiar manufacturing method, upending established production processes and altering the design, production, and innovation. Ceramics, metallic, polymers, combination of polymers in the form of hybrid, composites, and FGMs are certain variety of materials employed in the process of 3D printing. Nickel-based alloy 3D-printed components are employed in hazardous conditions due to their exceptional corrosion resistance.

The capacity of 3D printing to make detailed designs and complicated geometries that were not practically possible in the earlier years is one of its main advantages. This has created new opportunities for customization, quick prototyping, and effective manufacturing techniques in industries like consumer products, aerospace, automotive, healthcare, and architecture. Sectors like healthcare have been profoundly impacted by the ability to produce highly customized and personalized items, allowing for the more accurate and comfortable fabrication of prostheses and patient-specific implants. Furthermore, by lowering energy consumption and wasteful material usage, 3D printing has established itself to be ecologically approachable.

Despite the immense potential of 3D printing, there are still certain difficulties associated with it. 3D printers are only able to produce smaller objects because the 3D printers are not large enough to produce large 3D objects.

Also, some 3D structures are not sturdy enough and may break down at high temperatures and pressure. Moreover, 3D-printed objects are not cost-efficient if produced at high volume and there is a limitation in the accuracy of any 3D-printed part.

Looking forward, 3D printing has a promising future. Automation, machine learning, and advancements in materials science are all expected to further push the limits of this technology. The technology has the ability to democratize production, strengthen local economies, and transform supply chains as it becomes more developed and widely available. Moreover, 3D printing materials make use of reusable, recycled, and biodegradable plastics which strengthens sustainable growth. In the years to come, 3D printing is likely to keep reshaping sectors, fostering innovation, and altering the nature of manufacturing and consumable items.

## References

1. Ghaffar, S.H., Corker, J., Fan, M., Additive manufacturing technology and its implementation in construction as an eco-innovative solution. *Autom. Constr.*, 93, 1–11, 2018.
2. Buswell, R.A., Soar, R.C., Gibb, A.G., Thorpe, A., Freeform construction: Mega-scale rapid manufacturing for construction. *Autom. Constr.*, 16, 2, 224–231, 2007.
3. Shahrubudin, N., Lee, T.C., Ramlan, R.J.P.M., An overview on 3D printing technology: Technological, materials, and applications. *Proc. Manuf.*, 35, 1286–1296, 2019.
4. Anastasiou, A., Tsirmpas, C., Rompas, A., Giokas, K., Koutsouris, D., 3D printing: Basic concepts mathematics and technologies, in: *13th IEEE International Conference on BioInformatics and BioEngineering*, pp. 1–4, 2013, November.
5. Ngo, T.D., Kashani, A., Imbalzano, G., Nguyen, K.T., Hui, D., Additive manufacturing (3D printing): A review of materials, methods, applications and challenges. *Composites, Part B*, 143, 172–196, 2018.
6. Stansbury, J.W. and Idacavage, M.J., 3D printing with polymers: Challenges among expanding options and opportunities. *Dent. Mater.*, 32, 1, 54–64, 2016.
7. Chattoraj, P., Debnath, K., Manmadhachary, A., A Review paper on 3D-Printing Aspects and Various Processes Used in the 3D-Printing. *Int. J. Mech. Eng.*, 7, 3, 0974–5823, (March), 2022, ISSN:0974-5823.
8. Gibson, I., Rosen, D.W., Stucker, B., Khorasani, M., Rosen, D., Stucker, B., Khorasani, M., *Additive manufacturing technologies*, vol. 17, pp. 160–186, Springer, Cham, Switzerland, 2021.

9. Jandyal, A., Chaturvedi, I., Wazir, I., Raina, A., Haq, M.I.U., 3D printing–A review of processes, materials and applications in industry 4.0. *Sustain. Oper. Comput.*, 3, 33–42, 2022.
10. Tan, L.J., Zhu, W., Zhou, K., Recent progress on polymer materials for additive manufacturing. *Adv. Funct. Mater.*, 30, 43, 2003062, 2020.
11. Goodridge, R.D., Tuck, C.J., Hague, R.J.M., Laser sintering of poly- amides and other polymers. *Prog. Mater Sci.*, 57, 2, 229–267, 2012.
12. Rajan, K.M., Sahoo, A.K., Routara, B.C., Panda, A., Kumar, R., A review on various approaches of 3D printing of Ti-Alloy. *Mater. Today: Proc.*, 62, 3865–3868, 2022.
13. Chatham, C.A. and Washington II, A.L., A framework for forming thermoset polymer networks during laser powder bed fusion additive manufacturing. *Addit. Manuf.*, 72, 103620, 2023.
14. Gokhare, V.G., Raut, D.N., Shinde, D.K., A review paper on 3D-printing aspects and various processes used in the 3D-printing. *Int. J. Eng. Res. Technol.*, 6, 06, 953–958, 2017.
15. Raul, V.A. and Wakchaure, K., A Review Paper On 3d-printing Aspects And Various Processes Used In The 3d-printing. *Int. J. Innov. Eng. Res. Technol.*, 1–6, 2018.
16. Melchels, F.P., Feijen, J., Grijpma, D.W., A review on stereolithography and its applications in biomedical engineering. *Biomaterials*, 31, 24, 6121–6130, 2010.
17. Corbel, S., Dufaud, O., Roques-Carmes, T., Materials for stereolithography, in: *Stereolithography: Materials, processes and applications*, pp. 141–159, Springer US, Boston, MA, 2011.
18. Kumar, S., Selective laser sintering: a qualitative and objective approach. *JOM*, 55, 43–47, 2003.
19. Kruth, J.P., Wang, X., Laoui, T., Froyen, L., Lasers and materials in selective laser sintering. *Assem. Autom.*, 23, 4, 357–371, 2003.
20. Yap, C.Y., Chua, C.K., Dong, Z.L., Liu, Z.H., Zhang, D.Q., Loh, L.E., Sing, S.L., Review of selective laser melting: Materials and applications. *Appl. Phys. Rev.*, 2, 041101, 2015.
21. Bagheri, A. and Jin, J., Photopolymerization in 3D printing. *ACS Appl. Polym. Mater.*, 1, 4, 593–611, 2019.
22. Bao, Y., Recent trends in advanced photoinitiators for vat photopolymerization 3D printing. *Macromol. Rapid Commun.*, 43, 14, 2200202, 2022.
23. Ding, D., Pan, Z., Cuiuri, D., Li, H., van Duin, S., Advanced design for additive manufacturing: 3D slicing and 2D path planning, in: *New trends in 3d printing*, 1-23, 2016.
24. Hu, J., Study on STL-based slicing process for 3D printing, in: *2017 International solid freeform fabrication symposium*, University of Texas at Austin, 2017.

25. Kim, H., Zhao, Y., Zhao, L., Process-level modeling and simulation for HP's Multi Jet Fusion 3D printing technology, in: *2016 1st international workshop on cyber-physical production systems (CPPS)*, IEEE, pp. 1–4, 2016, April.
26. Murr, L.E., Gaytan, S.M., Ramirez, D.A., Martinez, E., Hernandez, J., Amato, K.N., Shindo, P.W., Medina, F.R., Wicker, R.B., Metal fabrication by additive manufacturing using laser and electron beam melting technologies. *J. Mater. Sci. Technol.*, 28, 1, 1–14, 2012.
27. Zhang, L.C., Liu, Y., Li, S., Hao, Y., Additive manufacturing of titanium alloys by electron beam melting: A review. *Adv. Eng. Mater.*, 20, 5, 1700842, 2018.
28. Körner, C., Additive manufacturing of metallic components by selective electron beam melting—A review. *Int. Mater. Rev.*, 61, 5, 361–377, 2016.
29. Medellin-Castillo, H.I. and Zaragoza-Siqueiros, J., Design and manufacturing strategies for fused deposition modelling in additive manufacturing: A review. *Chin. J. Mech. Eng.*, 32, 53, 1, 1–16, 2019.
30. Daminabo, S.C., Goel, S., Grammatikos, S.A., Nezhad, H.Y., Thakur, V.K., Fused deposition modeling-based additive manufacturing (3D printing): techniques for polymer material systems. *Mater. Today Chem.*, 16, 100248, 2020.
31. Ning, F., Cong, W., Qiu, J., Wei, J., Wang, S., Additive manufacturing of carbon fiber reinforced thermoplastic composites using fused deposition modeling. *Composites, Part B*, 80, 369–378, 2015.
32. Dermeik, B. and Travitzky, N., Laminated object manufacturing of ceramic-based materials. *Adv. Eng. Mater.*, 22, 9, 2000256, 2020.
33. Mekonnen, B.G., Bright, G., Walker, A., A study on state of the art technology of laminated object manufacturing (LOM), in: *CAD/CAM, Robotics and Factories of the Future: Proceedings of the 28th International Conference on CARs & FoF 2016*, Springer India, New Delhi, pp. 207–216, 2016, January.
34. Griffey, J., The types of 3-D printing. *Libr. Technol. Rep.*, 50, 5, 8–12, 2014.
35. Bhushan, B. and Caspers, M., An overview of additive manufacturing (3D printing) for microfabrication. *Microsyst. Technol.*, 23, 1117–1124, 2017.
36. Venkatesh, K.V. and Nandini, V.V., Direct metal laser sintering: a digitised metal casting technology. *J. Indian Prosthodont. Soc.*, 13, 4, 389–392, 2013.
37. Muthuram, N., Madhav, P.S., Vasan, D.K., Mohan, M.E., Prajeeth, G., A review of recent literatures in poly jet printing process. *Mater. Today: Proc.*, 68, 1906–1920, 2022.
38. Aslani, K.E., Vakouftsi, F., Kechagias, J.D., Mastorakis, N.E., Surface roughness optimization of poly-jet 3D printing using Grey Taguchi method, in: *2019 international conference on control, artificial intelligence, robotics & optimization (ICCAIRO)*, IEEE, pp. 213–218, 2019, December.
39. Berman, B., 3-D printing: The new industrial revolution. *Bus. Horiz.*, 55, 2, 155–162, 2012.
40. Macdonald, E., Salas, R., Espalin, D., Perez, M., Aguilera, E., Muse, D., Wicker, R.B., 3D printing for the rapid prototyping of structural electronics. *IEEE Access*, 2, 234–242, 2014.

41. Joshi, S.C. and Sheikh, A.A., 3D printing in aerospace and its longterm sustainability. *Virtual Phys. Prototyping*, 10, 4, 175–185, 2015.
42. Lim, C.W.J., Le, K.Q., Lu, Q., Wong, C.H., An overview of 3-D printing in manufacturing, aerospace, and automotive industries. *IEEE Potentials*, 35, 4, 18–22, 2016.
43. Elakkad, A.S., 3D technology in the automotive industry. *Int. J. Eng. Tech. Res.*, 8, 11, 110–122, 2019.
44. Chana-Rodríguez, F., Mañanes, R.P., Rojo-Manaute, J., Gil, P., Martínez-Gómiz, J.M., Vaquero-Martín, J., 3D surgical printing and pre contoured plates for acetabular fractures. *Injury*, 47, 11, 2507–2511, 2016.
45. Javaid, M., Haleem, A., Singh, R.P., Suman, R., 3D printing applications for healthcare research and development. *Glob. Health J.*, 6, 4, 217–226, 2022.
46. Thakar, C.M., Parkhe, S.S., Jain, A., Phasinam, K., Murugesan, G., Ventayen, R.J.M., 3d Printing: Basic principles and applications. *Mater. Today: Proc.*, 51, 842–849, 2022.
47. Petersen, E.E., Kidd, R.W., Pearce, J.M., Impact of DIY home manufacturing with 3D printing on the toy and game market. *Technologies*, 5, 3, 45, 2017.
48. Yap, Y.L. and Yeong, W.Y., Additive manufacture of fashion and jewellery products: a mini review: This paper provides an insight into the future of 3D printing industries for fashion and jewellery products. *Virtual Phys. Prototyping*, 9, 3, 195–201, 2014.
49. Žujović, M., Obradović, R., Rakonjac, I., Milošević, J., 3D printing technologies in architectural design and construction: a systematic literature review. *Buildings*, 12, 9, 1319, 2022.
50. Pîrjan, A. and Petroşanu, D.M., The impact of 3D printing technology on the society and economy. *J. Inf. Syst. Oper. Manage.*, 7, 2, 360–370, 2013.
51. Tserovski, S., Georgieva, S., Simeonov, R., Bigdeli, A., Röttinger, H., Kinov, P., Advantages and disadvantages of 3D printing for pre- operative planning of revision hip surgery. *J. Surg. Case Rep.*, 2019, 7, rjz214, 2019.

# 8

# Cutting Zone Temperature and Cutting Force in 3D-Milling Operations Using ABAQUS

**Jagadeesha T.[1], R. Thanigaivelan[2*], Avinash Malladi[3] and Natrayan L.[4]**

[1]*Department of Mechanical Engineering, National Institute of Technology NIT, Calicut, Kerala, India*

[2]*AKT Memorial College of Engineering and Technology, Kallakurichi, Tamil Nadu, India*

[3]*Department of Mechatronics Engineering, Faculty of Science and Technology, IcfaiTech, Icfai Foundation for Higher Education, Hyderabad, Telangana, India*

[4]*Department of Mechanical Engineering, Saveetha School of Engineering, SIMATS, Chennai, Tamil Nadu, India*

***Abstract***

The 3D finite element method (FEM) is a widely used numerical method for solving problems in physics, engineering, and other fields. It provides a powerful tool for solving complex problems that cannot be easily solved using analytical methods. An attempt has been made to study the complex milling process of aluminum using the 3D FEM software ABAQUS. This integrated model helps in the prediction of cutting zone temperature, cutting forces, burr formation, and chip flow along the rake face in a contouring end-milling operation. Multiple simulations are done to check the effect of process parameters which is difficult to attain physically through experiments. The 3D CAD model of the milling cutter is modified in SOLIDWORKS according to the required geometry, which is one of the important aspects of the study. The force distribution in three directions and chip flow characteristics in various cutting conditions are studied. These obtained 3D model simulation results are beneficial in predicting the influence of various machining parameters prior to the final machining, which reduces the time for conducting multiple experiments.

**Corresponding author*: tvelan10@gmail.com

R. Thanigaivelan, Pradeep Kumar Krishnan, Kamalakanta Muduli and Santosh Kumar Tamang (eds.) New Materials, Processing and Manufacturability: Fabrication and Processing of Advanced Materials, (139–156) © 2024 Scrivener Publishing LLC

***Keywords*:** End-milling, cutting force, chip formation, 3D FEM

## 8.1 Introduction

The milling process is a common machining process in all manufacturing industries to create parts and complex shapes from different materials. The process starts by securing the workpiece in a vice or fixture on the milling machine table. The rotating cutter will remove material from the workpiece. The machine's cutting tool, usually a carbide end mill, is then used to remove material from the block by making a series of passes across the surface. The cutting tool can be of different shapes and sizes, including end mills, ball nose cutters, and face mills. Because of its robustness and versatility, it has been used in various applications to create complex geometries and structures even though the latest machining process is available.

For the simulation of the machining process, finite element analysis (FEA) is considered a powerful tool for the study of the milling process in materials. The 3D finite element method (FEM) can accurately model the behavior of material that varies in three dimensions. Automobile industries, medical industries, aerospace industries, etc., use this application of FEM for analyzing stress distribution, plastic deformation, and strain variations in various solid objects. The figure representing the milling process is shown in Figure 8.1.

Process parameters refer to the variables that affect the behavior of the system during the manufacturing process. These parameters can include

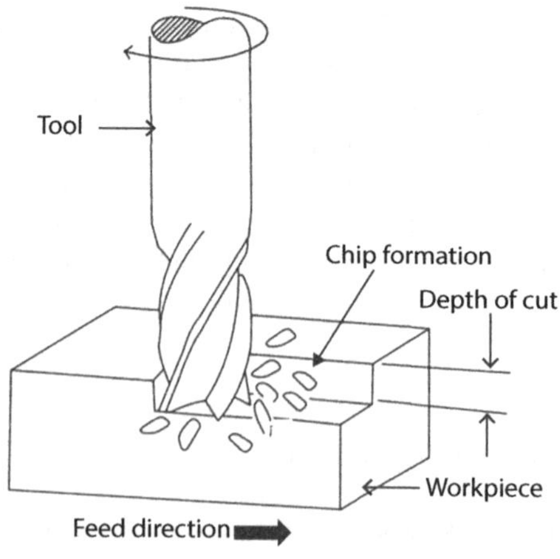


**Figure 8.1** Milling process using end-milling cutter.

feed, speed, depth of cut, and mesh size. FEM can be used to model the physical behavior of the system under various process parameters and simulate the response of the system. By using FEM, we are able to optimize the parameters to attain the most economical and efficient way to machine the workpiece. Few authors did 2D FEM simulation using high-speed cutting of 200,000 rpm and analyzed the burr formation of titanium alloy Ti6Al4V. To analyze the effect of the cutting tool on-chip flow and burr formation, the 3D FEM method gives an additional advantage over 2D FEM analysis.

In this study, aluminum is taken as the workpiece material by taking advantage of its high machinability. Aluminum which is lightweight, least dense, and ductile in nature makes it more machinable and forms an uncomplicated chip. The current work is focused on investigating cutting force and stress distribution on the workpiece.

## 8.2 Literature Review

3D FEM simulations were used extensively in the machining processes to analyze and verify the best optimum parameters. Since milling is a complex process, 2D simulation cannot be done. Therefore, more effective 3D simulation of milling has been studied by many researchers in the past decade. Wu and Zhang proposed a 3D finite element model of contour end-milling of titanium alloy Ti6Al4V. The simulation was done in ABAQUS software. The Johnson-Cook equation is taken for the material constitutive model and the damage constitutive equation is taken for the chip separation criterion. Furthermore, they found the advantages and capability of simulation of complex milling process using Ti6Al4V for the optimum selection of parameters and to improve the efficiency of the process [1]. In another study, Sun *et al.* conducted FEM analysis in aluminum alloy 7075-T and proposed an equation relating the feed engagement and cutting forces along with chip formation. Radial depth of cut is taken as an important parameter and simulations are done at different values. DEFORM software is used for simulation purposes. They concluded that the milling force increases gradually with an increase in the radial depth of the cut [2].

Researchers proposed an integrated FEM model for the milling process to analyze the milling force, temperature distribution, burr formation, chip flow prediction, etc. Tool deflection is considered by taking the tool as a deformable part. Through simulation, they could accurately observe the burr formation height from 1 to 5 μm depending on the milling settings chosen. Furthermore, the simulation results were compared with physical experiments and a 4% deviation was observed. Tang *et al.* studied the effect of cutting parameters in the orthogonal cutting of hard-to-machine material D2 steel. The dry hard machining process is chosen with stress and temperature as the most important parameters. They compared the experiment results and previously done simulation results by iterative procedure to get the calibrated properties. They found that the cutting temperature has a direct relationship with the residual stresses. The cutting temperature first shows a downside graph then it increases with the depth of the cut. Also, shear angles show a drastic increase at the lower depth of the cut then gradually increase followed by a slight decrease at the higher depth of the cut [3].

## 8.3 Objectives

The objective of this simulation work is to investigate the variation of stress and cutting forces on an aluminum workpiece using a TiAlN-coated tungsten carbide tool with two flutes. Simulations were done by changing the cutting conditions and studying various aspects of the machining process to analyze the force distribution and von Mises stress. The constants in the Johnson-Cook model were procured after referring to multiple literatures for the machining of aluminum. This simulation work was carried out on ABAQUS 2020 CAE FEM software.

## 8.4 Methodology

After thoroughly studying the machining process, a flow of procedures was built to meet the objectives described above. Parameters like density, melting temperature, elastic modulus, and thermal conductivity are obtained from ASTM standards for aluminum [3]. The fracture properties of aluminum were given as input to ABAQUS to calculate the plastic strain of damage [4].

A 3D model of the workpiece and cutting tool was built using SolidWorks software and the created parts were imported to ABAQUS for FEA [5]. The tool and workpiece were divided into small finite elements using the meshing tool. The mesh at the edges of the tool and the top surface of the workpiece were kept fine or decreased in size or increased in number [6]. This was done for the sole purpose of improving the results and accuracy of our FEA model. To attain the objectives, the material properties of aluminum were taken from ASTM for aluminum.

The objective of the simulation work shown in Figure 8.2 is to study the variation of stress and cutting forces on aluminum workpieces. The objective of the simulation work is to study the variation of stress and cutting forces on aluminum workpieces. A number of simulations are done by varying the cutting conditions, and an assessment of various aspects of the machining process is done to analyze the force distribution of von Mises stress. The parameters of Johnson-cook failure values were procured after studying various literature and using those values for modeling aluminum. This is achieved by the use of ABAQUS FEM software.

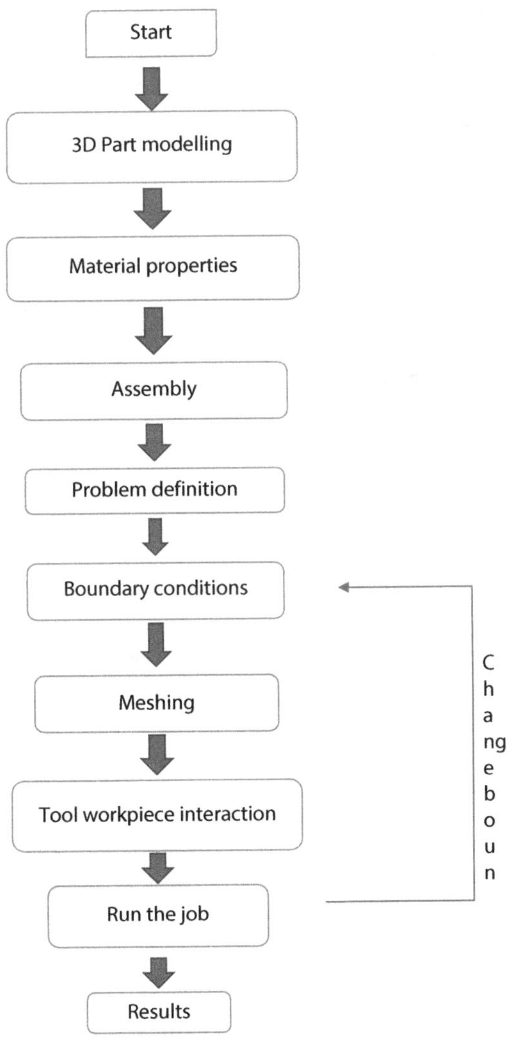


**Figure 8.2** Flow diagram of simulation work.

# 8.5 Simulation

## 8.5.1 Geometry

A three-dimensional geometry of the workpiece was created using the part manager feature in the part module of ABAQUS as shown in Figure 8.3. Also, the 3D CAD model of the two-fluted TiAlN tungsten carbide–coated (8-mm diameter) end-milling cutter is imported into the part module. The dimensions of the tool are scaled up to the required ratio with the workpiece.

These two-part models are combined to the required positions in the assembly module and the required depth of the cut is given. Assuming symmetric geometry along the center of the workpiece helps to reduce the computational time and reduce the file size.

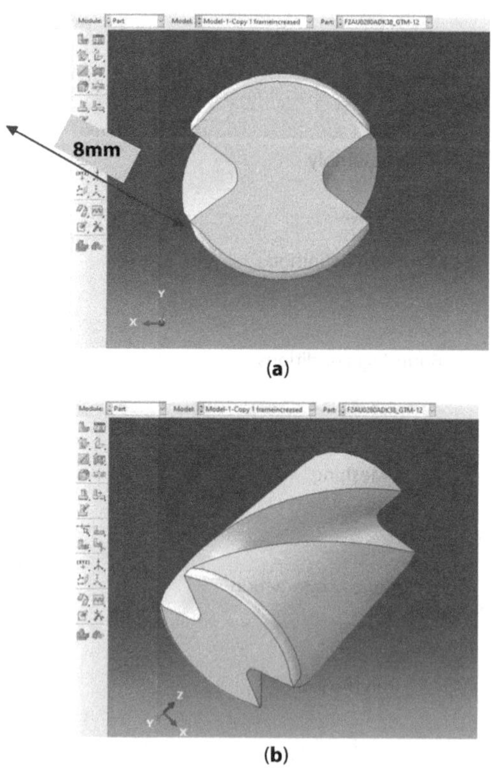


**Figure 8.3** Three-dimensional geometry of tool model: (a) Front view and (b) isometric view.

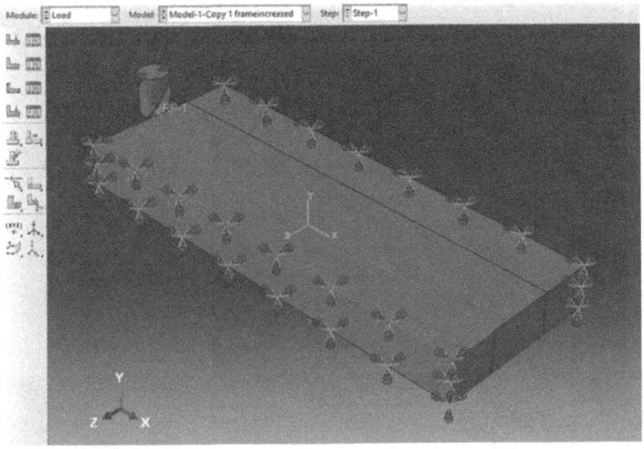

**Figure 8.4** Boundary condition.

### 8.5.2 Boundary Conditions

The two parallel faces of the workpiece are fixed by the ENCASTRE boundary condition in the load module as shown in Figure 8.4. In the same load module, linear and angular velocities are specified in the boundary condition manager. Two sets of values are taken for the parameters feed and speed, and it is specified as V1 and VR2, respectively.

The given Table 8.1 below shows different combinations of feed and speed values used in different trials of simulation.

### 8.5.3 Meshing Details

The workpiece is divided into three partitions to give fine meshing along the tool path as shown in Figure 8.5. By using seed edges from the mesh module, the number of elements along each partition edge is given to obtain fine meshing along the tool path direction, and the remaining area is given

**Table 8.1** Values of feed and speed used in simulation.

| Parameters | Set 1 | Set 2 | Set 3 | Set 4 |
|---|---|---|---|---|
| V1 (mm/s) | 0.2 | 0.4 | 0.2 | 0.4 |
| VR2 (rad/s) | 500 | 1000 | 500 | 1000 |

The milling depth is given as 0.1 mm.

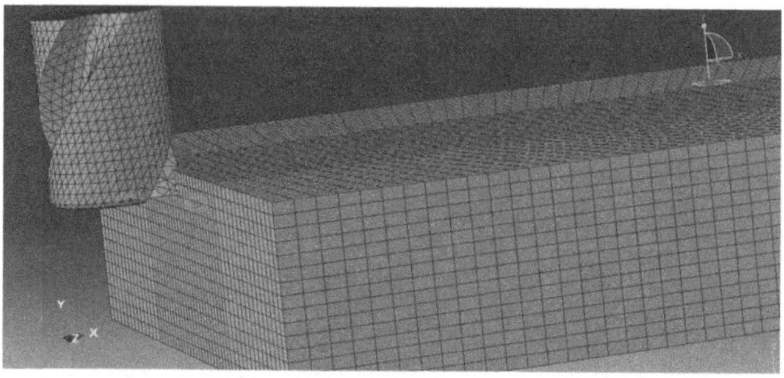
(a)

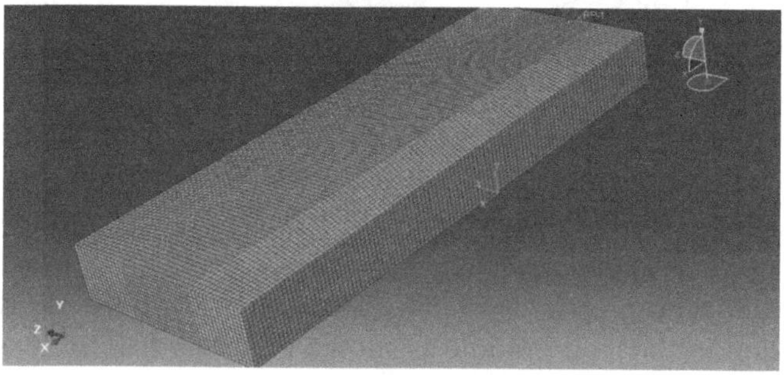
(b)

**Figure 8.5** (a) Meshing of tool and workpiece and (b) fine meshing of inner machining region.

course meshing. Fine meshing is given for obtaining accurate results; however, this leads to an increase in computational time and a further increase in step time. The meshing details are given in Table 8.2 below.

The tetrahedral mesh was given for the coated carbide cutting tool, and the hexagonal mesh was applied for the workpiece material. A finer mesh is provided at the workpiece surface where the tool comes in contact, which will give closer result values to actual results. Tetra mesh in FEM analysis is stiff but accurate in capturing geometry and also takes less time to generate. Hexamesh shows good result accuracy in FEM analysis, but its

**Table 8.2** Instance table of meshing.

| Instance name | Number of elements | Element type |
|---|---|---|
| Workpiece | 70,305 | C3D8R |
| Tool | 12,708 | C3D4 |

accuracy is affected due to the fact that it cannot capture geometry accurately and is also time-consuming in modeling (it takes more than hours). Other cutting parameters under consideration, a cutting depth of 1 mm, and a cutting width of 8 mm are kept constant.

### 8.5.4 Step Information

Since time-dependent study is done, step information should be provided in the step module, which is one of the important modules in ABAQUS that enables users to study changes happening over a set period of time. The step file generated by the software is shown in Table 8.3. Since milling is a complex process, the analysis step has been taken as a dynamic explicit procedure [7].

## 8.6 Milling Conditions

The milling conditions provided for the simulation are listed in Table 8.4 mentioned below.

Various material models are used for the simulation of the machining process. Among that Johnson-Cook material model is commonly used to describe the materials under high strength, high-velocity impacts, or explosive loading, which can cause significant plastic deformation and damage. The Johnson-Cook material model used for this study is given by [8]

$$\sigma = (A + B\varepsilon^n)\left(1 + C\ln\frac{\dot{\varepsilon}}{\dot{\varepsilon}_0}\right)\left[1 - \left(\frac{T - T_r}{T_m - T_r}\right)^m\right] \tag{8.1}$$

Where,

ε - Equivalent plastic strain

**Table 8.3** Step module information.

| Name | Procedure | Step increment time | Total time | Target time increment |
|---|---|---|---|---|
| Step 1 | DYNAMIC, EXPLICIT | 2.000 | 2.000 | 1e−05 |

**Table 8.4** Material properties of the workpiece.

| Material properties of aluminum | Values |
|---|---|
| Hardness (HRC) | 31.00 |
| Density (kg/m$^3$) | 4431.00 |
| Thermal conductivity (W/mK) | 8.00 |
| Elastic modulus (MPa) | 123 |
| Linear thermal expansion ($10^{-6}$/°C) | 7.450 |

$\dot{\varepsilon}$ & $\dot{\varepsilon}_0$ - Equivalent and reference plastic strain rates
T - Material's cutting zone temperature
Tm - Melting temperature
Tr - Room temperature
n - Hardening strain index
m - Thermal softening index
A, B, and C - Material's yield strength, strain, and strain rate sensitivities, respectively.

Material constants for the Johnson-Cook model of aluminum are given below in Table 8.5.

The friction force that occurs between two surfaces that are in contact and slide against each other is described using the Coulomb friction model [8]. In this FEM model, the frictional force is assumed to be proportional to the normal force that presses the surfaces together and the coefficient of friction (μ), which is a dimensionless constant that characterizes the frictional properties of the two surfaces. The friction model is given by

**Table 8.5** Material constants of aluminum.

| Johnson-Cook parameters | Values |
|---|---|
| A | 1000.00 |
| B | 780.00 |
| n | 0.469 |
| c | 0.03300 |
| m | 1.0200 |

$$\tau = \mu\sigma_n \tag{8.2}$$

where $\tau$ = frictional stress at the rake face of the tool
$\sigma_n$ = the normal stress.

## 8.7 Result and Discussion

### 8.7.1 Effect of Cutting Parameters on Force Distribution

In the milling process, cutting forces are crucial and have a big impact on both the tool and workpiece. Therefore, it is crucial to take cutting forces and their effect into account while milling for a number of reasons. Some of them include tool wear, surface quality, dimensional accuracy, machine stability, and energy consumption. The force distribution in the milling process of aluminum at different cutting times by changing parameter sets is shown in Figure 8.6.

In the milling process, cutting forces are crucial and have a big impact on both the tool and workpiece. Therefore, it is crucial to take cutting forces and their effect into account while milling for a number of reasons. For four different sets of cutting speed and feed, we obtained four different force v/s time graphs. The forces in three different directions (Fx, Fy, and Fz) are represented in the graph. The maximum force was obtained for set 2 and set 3 at a time instant of 1.5 seconds along the *x* direction. Table 8.6 given below shows the maximum forces obtained in three different directions for varying parameters.

Design of experiment (DOE) analysis of force variation with respect to feeding and speed done in Minitab 19.1 (64-bit) version [9]. To determine the impact of speed and feed on force variation, nine sets of parameters (V1 = 0.2, 0.3,0.4 mm/s and VR1= 500, 750, 1000 rad/s) are taken and the main effect plot is studied in Figure 8.7. The factor that shows a significant difference between low and high levels of a factor is considered the factor that has an effect on the response variable. If there is an interaction effect between the two factors, then the impact of each factor on the response variable is dependent on the level of the other factor.

The significant factor is obtained as feed and Minitab's DOE tools can be used to optimize the process by adjusting the levels of factor to achieve the desired force variation [10]. The given below equation is a multiple regression equation that relates the force with four independent variables,

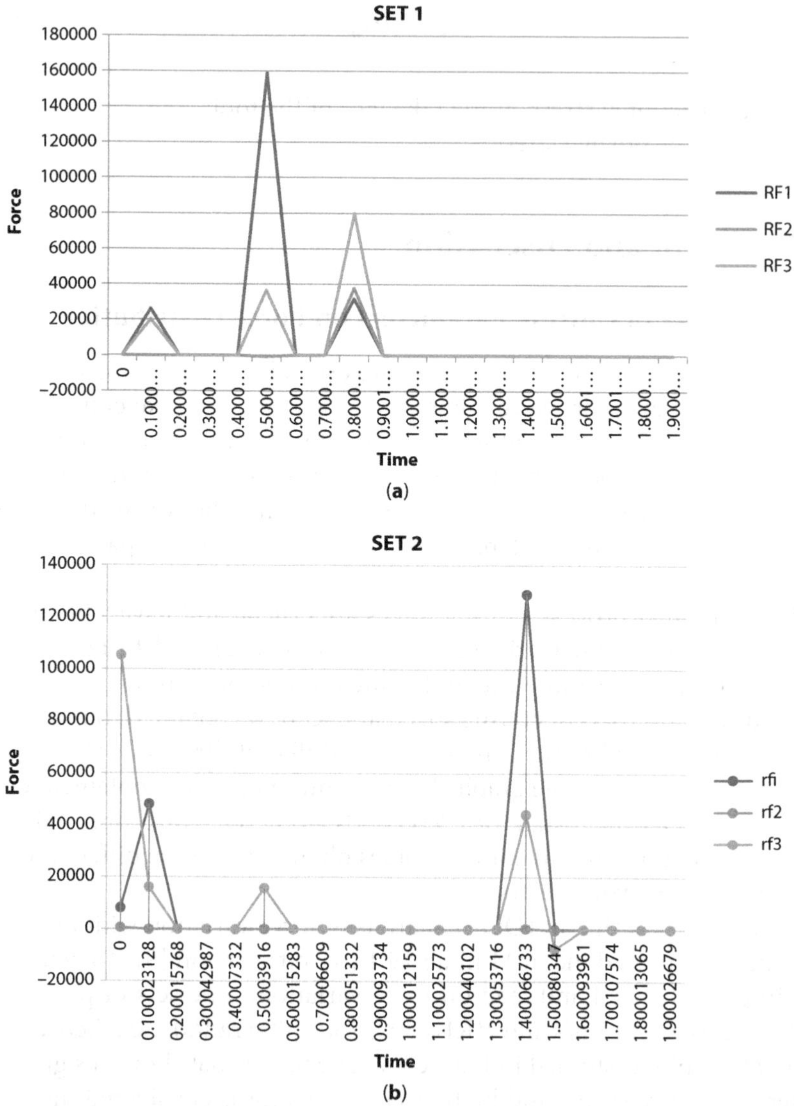


**Figure 8.6** (a) Force distribution for set 1, (b) force distribution for set 2. *(Continued)*

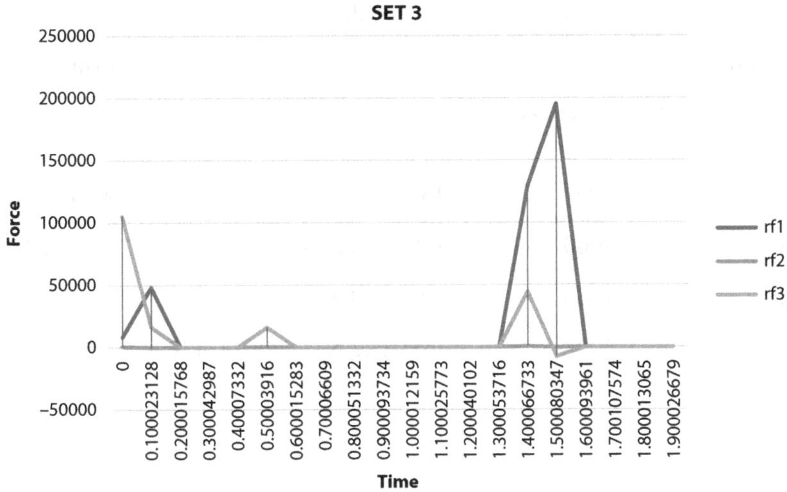


(c)

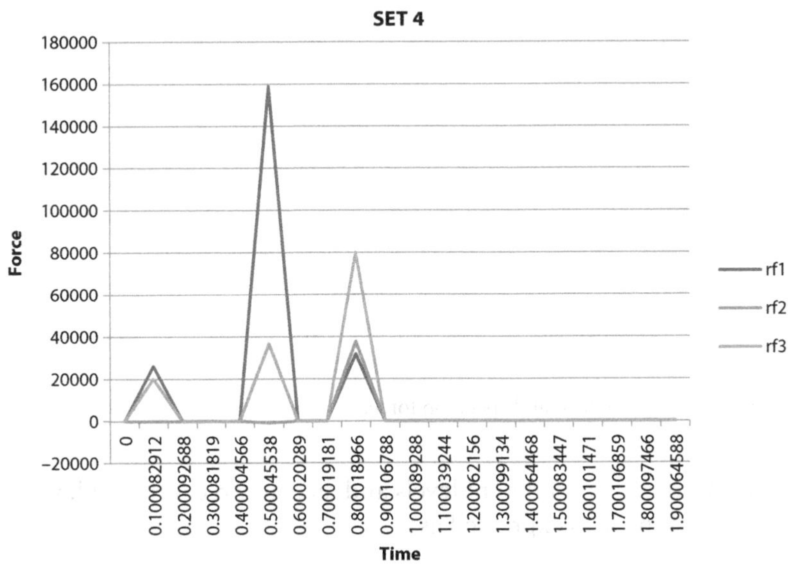


(d)

**Figure 8.6 (Continued)** (c) force distribution for set 3, and (d) force distribution for set 4.

**Table 8.6** Maximum Force values in three directions.

| Parameter | | Fx (max) | Fy (max) | Fz (max) |
|---|---|---|---|---|
| Speed (mm/s) [V1] | Feed (rad/s) [VR2] | | | |
| 0.2 | 500 | 158825.8 N | 785.2474 N | 80050.87 N |
| 0.4 | 1000 | 128804.3 N | 548.2158 N | 7557.262 N |
| 0.2 | 1000 | 195215.8 N | 548.2158 N | 7557.262 N |
| 0.4 | 500 | 158825.7 N | 785.2474 N | 80050.87 N |

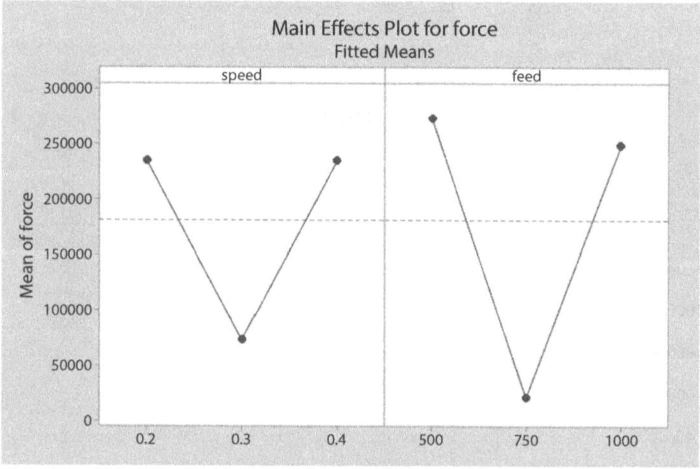


**Figure 8.7** Influence of feed and speed on force.

namely, speed at three different levels and feed at three different levels, and their interactions. It is expressed as:

**Force** =180999 + 53873 speed_0.2 - 107746 speed_0.3 + 53873 speed_0.4 + 92209 feed_500- 159727 feed_750 + 67518 feed_1000 + 42667 speed*feed_0.2500 - 65961 speed*feed_0.2750 + 23294 speed*feed_0.2 1000 - 85334 speed*feed_0.3 500 + 131922 speed*feed_0.3 750 - 46588 speed*feed_0.3 1000 + 42667 speed*feed_0.4 500 - 65961 speed*feed_0.4 750+ 23294 speed*feed_0.4 1000

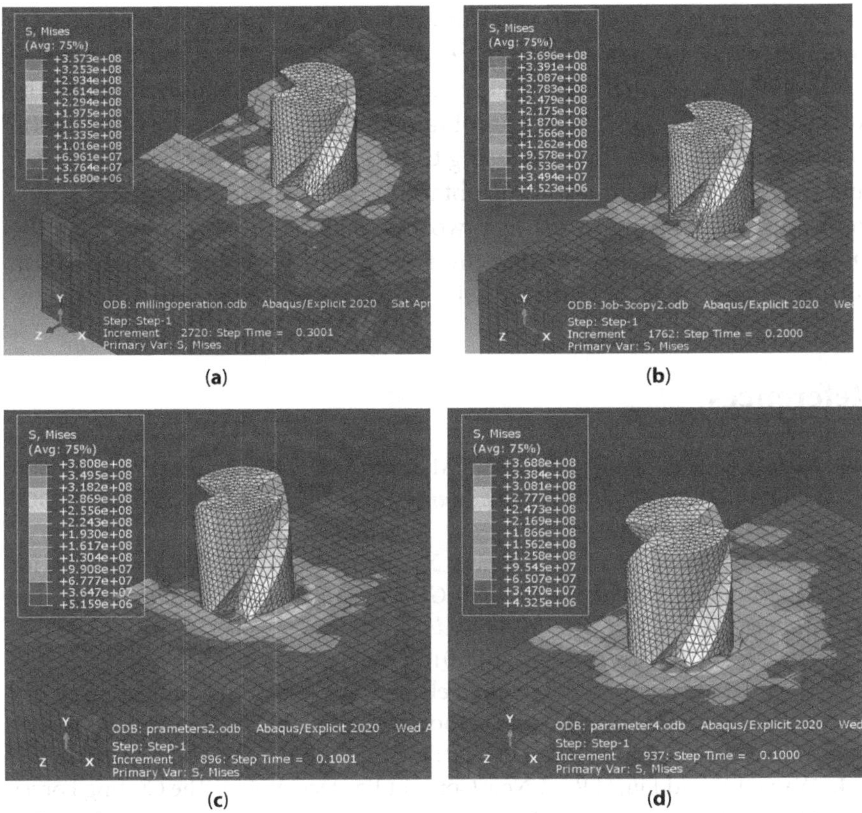


**Figure 8.8** von Misses stress: (a) set 1, (b) set 2, (c) set 3, and (d) set 4.

### 8.7.2 Effect of Cutting Parameters on Stress Distribution

Understanding the von Mises stress distribution in the milling zone is crucial for optimizing cutting parameters and ensuring the integrity of the workpiece and the cutting tool [11]. It is the measure of maximum shear stress that the material can withstand without yielding. FEA simulation is done with four sets of parameters. For each set of parameters, the stress variation along the surface of the workpiece is shown in Figure 8.8.

## 8.8 Conclusion

A 3D FEA of the milling process was simulated and the results were presented. From Table 8.2, it was observed that the maximum cutting force is along the feed direction which is in the *x* direction. It was found that when the feed rate is increased, the axial and radial forces tend to increase as well. This is because a higher feed rate leads to more material being removed per unit of time, which requires the cutting tool to exert more force on the workpiece. The force-time graph obtained shows that the force variation follows an increasing and decreasing trend and a no-force zone in between successive peaks. The reason for obtaining this trend in the graph is due to intermittent cutting by using a two-fluted end-milling cutter. From the DOE analysis, it was found that the feed variable has more influence on cutting force than speed provided the depth of cut is kept constant.

## References


1. Wu, H.B. and Zhang, S.J., 3D FEM simulation of milling process for titanium alloy Ti6Al4V. *Int. J. Adv. Manuf. Technol.*, 71, 1319–1326, 2014, DOI 10.1007/s00170-013-5546-0.
2. Sun, Q., Cheng, X., Liu, Y., Yang, X., Li, Y., Modeling and Simulation for Micromilling Mechanisms. *13th Global Congress on Manufacturing and Management, GCMM 2016, Procedia Engineering*, vol. 174, pp. 760 – 766, 2017, DOI:10.1016/j.proeng.2017.01.219.
3. Tang, L., Huang, J., Xie, L., Finite element modeling and simulation in dry hard orthogonal cutting AISI D2 tool steel with CBN cutting tool. *Int. J. Adv. Manuf. Technol.*, 53, 1167–1181, 2011, DOI 10.1007/s00170-010-2901-2.
4. Wang, J.Y. and Liu, C.R., A New Concept For Decoupling The Cutting Forces Due To Tool Flank Wear And Chip Formation In Hard Turning. *Mach. Sci. Technol. Int. J.*, 2, 1, 77–90, 1998, doi.org/10.1080/10940349808945660.
5. Özel, T., & Liu, X., Investigations on mechanics-based process planning of micro-end milling in machining mold cavities. *Mater. Manuf. Processes*, 24, 12, 1274–1281, 2009.
6. Zong, W.J., Li, Z.Q., Zhang, L., Liang, Y.C., Sun, T., An, C.H., Zhang, J.F., Zhou, L., Wang, J., Finite element simulation of diamond tool geometries affecting the 3D surface topography in fly cutting of KDP crystals. *Int. J. Adv. Manuf. Technol.*, 68, 1927–1936, 2013, DOI 10.1007/s00170-013-4804-5.
7. Jung, D.W., Study of dynamic explicit analysis in sheet metal forming processes using faster punch velocity and mass scaling scheme. *J. Mater. Eng. Perform.*, *7*, 479–490, 1998.
8. Khare, S., Kumar, K., Choudhary, S. *et al.*, Determination of Johnson–Cook Material Parameters for Armour Plate Using DIC and FEM. *Met. Mater. Int.*, 27, 4984–4995, 2021, https://doi.org/10.1007/s12540-020-00895-3.
9. Maniraj, S. and Thanigaivelan, R., Optimization of electrochemical micromachining process parameters for machining of AMCs with different% compositions of GGBS using Taguchi and TOPSIS methods. *Trans. Indian Inst. Met.*, *72*, 3057–3066, 2019.

10. Mythili, T. and Thanigaivelan, R., Optimization of wire EDM process parameters on Al6061/$Al_2O_3$ composite and its surface integrity studies. *Bull. Pol. Acad. Sci.: Tech. Sci.*, 68, 6, 1403–1412, 2020.
11. Suraidah, S., Ridzuwan, M., Asmelash, M., Azhar, A., Mulubrhan, F., End milling finite element method for cutting force prediction and material removal analysis. *IOP Conf. Ser.: Mater. Sci. Eng.*, 788, 1, 012020, 2020, April, IOP Publishing.

9

# A Rare Ferroelectric Material: Fresnoite ($Ba_2TiSi_2O_8$)

**Shiv K. Barbar**

*Materials Science Laboratory, Department of Physics, Jai Narain Vyas University, Jodhpur, Rajasthan, India*

***Abstract***

The solid-state reaction method was employed to synthesize the ceramic samples of fresnoite (barium titanium silicate mineral) doped with calcium ($Ca^{2+}$) and having general formula $Ba_{2-x}Ca_xTiSi_2O_8$ (BCTS), utilizing high purity oxides and carbonates. X-ray diffraction technique and scanning electron microscopic (SEM) technique were used to explore the phase pure crystallization and structural characteristics of compounds. Furthermore, the analysis of obtained PXRD data was done via Rietveld refinement using the space group *P4bm*. The lattice constants and other structural parameters computed using the Rietveld refinement are consistent with the preceding study thus far, as evidenced by a good agreement between observed and computed interplanar d-spacing. The Raman spectra of tetragonal $Ba_{2-x}Ca_xTiSi_2O_8$ ceramics were measured at 300 K, revealing distinct modes. Detailed investigations were conducted to analyze the changes in Raman modes corresponding to varying concentrations of $Ca^{2+}$ substitution. The surface morphology and grain bonding in the synthesized compounds were performed using scanning electron microscopy. To determine the activation energies of the compounds, their dc electrical conductivities were measured. The pristine compound exhibited significantly high activation energy (4.74 eV), which decreased as the doping concentrations increased. The dielectric constant and dissipation factor with varying frequency and temperature were studied and it has been established that $Ba_{2-x}Ca_xTiSi_2O_8$ is insulating at room temperature (300 K). The impact of Ca doping on heat flow, specific heat, and other thermal parameters was investigated using modulated differential scanning calorimetry (MDSC).

*Email*: barbar_shivkumar@yahoo.co.in

R. Thanigaivelan, Pradeep Kumar Krishnan, Kamalakanta Muduli and Santosh Kumar Tamang (eds.) New Materials, Processing and Manufacturability: Fabrication and Processing of Advanced Materials, (157–186) 

***Keywords*:** Ferroelectrics, ceramics, X-ray diffraction, scanning electron microscopy, Raman spectroscopy, electrical properties, MDSC

## 9.1 Introduction

During the geological survey of the sanbornite deposits in eastern Fresno County, California, a variety of newly discovered silicate minerals containing barium have been identified. One of these is the mineral fresnoite ($2BaO\text{-}TiO_2\text{-}2SiO_2$), which has the general formula $Ba_2TiSi_2O_8$ [1]. Fresnoite is a rare mineral that exhibits a polar tetragonal structure characterized by the space group $C^2_{4V}$ *-P4bm* [2, 3]. This material has good piezo, pyroelectric, optical, and other nonlinear characteristics suitable for many practical devices because of its noncentrosymmetric crystal structure [4–11]. A theoretical investigation using first principle calculation measured the SHG (second harmonic generation) properties of $Ba_2TiSi_2O_8$ and $Sr_2TiSi_2O_8$ and found that it is useful for second-order nonlinear optical materials [12]. Because of the electro-optic effect and function of frequency shifting, these materials can be used in the area of photonics [13]. Yan *et al.* suggested that $SiO_2$ is very economical and effective for the improvement of the breakdown strength of materials. Since titanium can act as a network former ($Ti^{4+}$) and as a network modifier ($Ti^{3+}$), it can play a significant role in the crystallization of $Ba_2TiSi_2O_8$ [14]. The breakdown strength of $Ba_{0.4}Sr_{0.6}TiO_3\text{-}(BaO\text{-}TiO_2\text{-}SiO_2)$ glasses was measured at around 16.51 kV/mm [15]. The thermophysical properties have been measured on titanosilicates $Ba_2TiSi_2O_8$ and $Sr_2TiSi_2O_8$ synthesized via sol-gel auto-combustion and complex polymerization methods [16]. $Ba_2TiSi_2O_8$ single crystal can serve as a substrate material for surface acoustic wave (SAW) devices, thanks to its significant electromechanical coupling factor (k = 0.28) in piezoelectric devices [17, 18]. It is interesting to observe that the dielectric permittivity data for a single crystal exhibited a large peak at about 433 K, yet the peak is missing from the dissipation factor curve. In contrast, the bulk samples did not exhibit the dielectric anomaly at 433 K that was seen in the single-crystal cases [5]. In addition, a minor reproducible dielectric anomaly at 805 K has also been found in the 0.1 to 100 kHz range [19]. It has been observed that a commensurate-incommensurate phase transition may be connected to an elastic anomaly seen at 435 K [20]. Nevertheless, the symmetry remained unaltered, with only minor variations in atomic position, as evidenced by the single-crystal XRD patterns obtained at 297 K and 573 K [10]. Moreover, the differential scanning calorimetry (DSC) analysis conducted between 300 K and 875 K demonstrated a significant and consistent thermal irregularity around 433 K, spanning approximately 60 K. This anomaly was accompanied by an enthalpy change of 60(10) J $mol^{-1}$ and an associated entropy change of 0.19137 J $mol^{-1}$ $K^{-1}$. Another anomaly is also obtained at 810(5) K with an enthalpy change of $\Delta H$ = 160 J $mol^{-1}$ and entropy change of $\Delta S$ = 0.2 J $mol^{-1}$ $K^{-1}$ [19]. Zhao *et al.* observed the ultraviolet cutoff edge at around 240 nm for a single crystal of $Ba_2TiSi_2O_8$ grown by the LiF-$H_3BO_3$ flux system. It has also been observed that fresnoite reveals blue-green fluorescence on excitation at 254-nm wavelength [13]. On the other hand, a nanocrystallized glass with optical transparency displays blue photoluminescence SHG. In contrast, the fresnoite glass ceramics doped with $Sm^{3+}$ exhibit visible upconversion luminescence [21, 22]. Apart from this, optical isotropy in polycrystalline $Ba_2TiSi_2O_8$ has also been investigated using grain size–dependent dielectric models [23]. The optical performance of fresnoite revealed that, below 433 K, the material experiences a structural transition, shifting from a 4-mm point group to a square millimeter point group [24]. To evaluate the transition metal coordination ($Ti^{4+}$) in glassy fresnoite, electron energy loss spectroscopy (EELS) was utilized [25]. In recent years, electrical properties and crystallization kinetics of glassy materials were also investigated [26–29]. In-depth, research has been done on and comparisons with single-crystal data made about the crystallography of fresnoite glass ceramic. This study discovered that the expansion and contraction of lattice parameters were caused by thermal stress during the cooling process and crystallization-induced stress during the crystallization process [30]. The thermal glass stability against crystallization was confirmed by thermal investigation on ZnO–doped 30BaO-50$TiO_2$-20$SiO_2$ glass system synthesized via the melt-quenching method [31].

It has been observed from the existing literature that single crystal, ceramics, glass ceramics, and thin film of fresnoite and its analog have been prepared by employing different preparation techniques such as the Czochralski method [5, 6, 10, 13, 32–44], flux growth [13, 45], solid-state reaction [19, 46–52], sol-gel route [53, 54], hydrothermal method [55], surface crystallization [56–63], electrochemically induced nucleation (EiN) [64–66], sputtering [67, 68], and pulsed laser deposition [69, 70]. Apart from this, fresnoite has a lower dielectric constant than other ferroelectrics. Hence, it can be used for high-frequency applications. Because of its good piezoelectric properties, it can be used as an airbag sensor in automotive technology and in easy production of biomorphs. The SHG properties make it useful for optical switching and modulation and light control optical devices [71–74]. This silicate material is also useful for nuclear waste storage. Inspired by its huge application area, we have synthesized it in pristine and substituted form and tried to emphasize anomalous physical behavior by adopting various experimental techniques. This chapter focuses on the synthesis and characterization of pristine as well as calcium-doped fresnoite ceramics ($Ba_2TiSi_2O_8$).

## 9.2 Experimental

### 9.2.1 Materials and Methods

The polycrystalline ceramics of calcium-doped fresnoite of chemical composition $Ba_{2-x}Ca_xTiSi_2O_8$ (x = 0.0, 0.2, 0.4, 0.8, and 1.0) were synthesized by solid-state diffusion method using $BaCO_3$, $CaCO_3$, $TiO_2$, and $SiO_2$ oxides as precursors having high purity (99.9%). Stoichiometric quantities of these constituents were thoroughly mixed by using mortar-pestle and

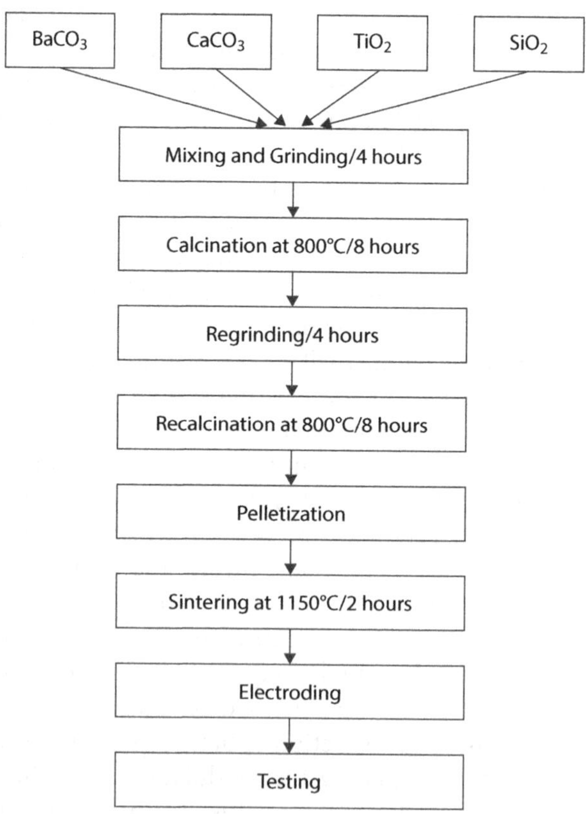


**Figure 9.1** Flow chart diagram of synthesis of Ca-modified $Ba_2TiSi_2O_8$ ceramics.

calcined at 800°C and 900°C for 8 hours in a silica crucible in a high-temperature chamber furnace in air. The mixing and calcination were repeated to get a homogeneous fine powder. Furthermore, the resulting mixture was pelletized by applying 5 tonnes/cm$^2$ pressure using a hydraulic press. The cylindrical pellets of pristine compound ($Ba_2TiSi_2O_8$) heated at 1300°C for ½ an hour in a platinum crucible were completely melted. The melted powder was cooled at the rate of 1°C/min until reaching 1100°C, followed by a faster cooling rate until it reached room temperature. Subsequently, this melted sample was scratched from the crucible and ground for 1 hour. The resultant powder was again pelletized and sintered at 1100°C for 45 min. Both sides of these pellets were painted with silver paste for electrical measurement. The detailed preparation of pure compound is given in our paper [50]. The sintering of Ca-doped samples was done at 1150°C for 2 hours [75]. A flow chart diagram for the synthesis of calcium-substituted fresnoite ($Ba_{2-x}Ca_xTiSi_2O_8$) ceramics is shown in Figure 9.1.

## 9.3 Characterization Details

### 9.3.1 Powder X-Ray Diffraction

RT powder X-ray diffraction (XRD) data of synthesized compounds were recorded in 2θ range of 10° to 90° with a scanning rate of 2° per minute with CuKα target of wavelength 1.5405 Å, 45 kV, and 40 mA on a Rigaku X-ray diffractometer. Utilizing the standard silicon sample that came with the equipment, the instrument was calibrated. FullProf package software was used to get precise crystallographic parameters for pristine as well as Ca-doped compounds.

### 9.3.2 Scanning Electron Microscope

Surface morphology and crystallinity of the synthesized $Ba_{2-x}Ca_xTiSi_2O_8$ (x = 0.0, 0.2, 0.4, 0.8, and 1.0) have been checked by JEOL, JSM-5600 Scanning Electron Microscopy instrument in the backscattered mode at different magnifications.

### 9.3.3 Raman Spectroscopy

RT Raman spectra were measured in the range of 300 to 1100 cm$^{-1}$, utilizing backscattering geometry; the measurements were conducted by the Jobin-Yvon Horiba Labram system (HR800) which comprises a single spectrograph with 0.25 m of focal length, 1800 grooves $mm^{-1}$ holographic grating filter, and a Peltier-cooled CCD detector featuring 1024 x 256 pixels of 26 μm. The spectrum was stimulated by 632.8 nm light (1.95 eV) from an air-cooled He-Ne laser of 19 mW power. To prevent sample degradation, the laser power at the sample was consistently maintained at 5 mW, except during the laser power dependence experiments. White light illumination was utilized to illuminate the microscope stage after each spectrum was recorded.

### 9.3.4 DC Conductivity and Dielectric

Using a lab-made two-probe setup, the dc electrical resistivity measurement was performed for the examination of conductivity with varying temperatures (300 K to 800 K). The measurement of dielectric constant (ε') and dissipation factor (ε") with varying frequency (5 Hz to 10 MHz) and temperatures (RT to 673 K) in an air atmosphere was conducted on a Hawlett-Packard 4192A impedance analyzer.

### 9.3.5 Modulated Differential Scanning Calorimetry

A TA instrument MDSC Model 2910 was used to examine the specific heat and heat flow from 400K to 840K in an inert nitrogen environment at a 5 K/min heating rate and 0.75 K per 60-second modulation rate.

## 9.4 Result and Discussion

### 9.4.1 X-Ray Diffraction Analysis

The powder X-ray diffraction diagrams of $Ba_{2-x}Ca_xTiSi_2O_8$ (x = 0.0, 0.2, 0.4, 0.8, and 1.0) are shown in Figure 9.2. The PXRD diagrams indicated a close match between the Bragg reflections of the pristine compound ($BaTiSi_2O_8$) and the JCPDS Card No. #84-0924 of fresnoite. However, during the Rietveld analysis, two minor impurity peaks of low intensity were identified at 2θ = 26.37° and 31.49°. The XRD figure does not exhibit any foreign peak when the $Ba^{2+}$ ion site is doped with various concentrations of $Ca^{2+}$ ion, with the exception of samples where x = 0.2 and 1.0, where an impurity peak may be detected between 2θ = 20° to 40°. Rietveld refinement analysis of XRD data has confirmed the phase purity of synthesized samples. The X-ray diffraction figure reveals that the peak positions for $Ba_{2-x}Ca_xTiSi_2O_8$ (x = 0.2 and 0.4) remain unchanged; however, XRD peak positions of

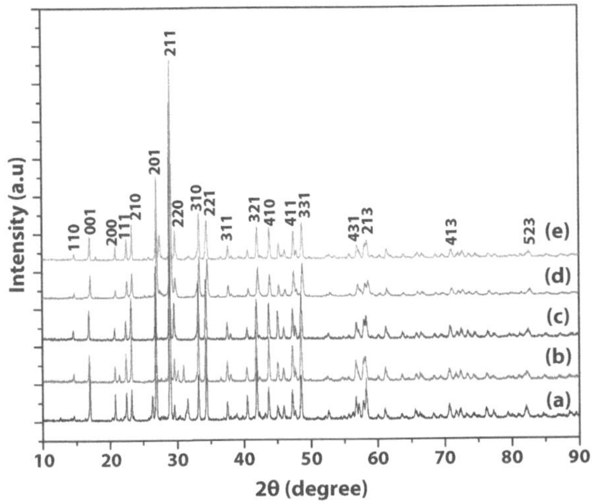


**Figure 9.2** X-ray diffraction diagrams of $Ba_{2-x}Ca_xTiSi_2O_8$ recorded at RT (a) 0.0, (b) x = 0.2, (c) x = 0.4, (d) x = 0.8, and (e) x = 1.

x = 0.8 and 1.0 compounds have a little shift in peak positions. As a result, it can be seen that each sample of the $Ba_{2-x}Ca_xTiSi_2O_8$ system exhibits the same crystal structure.

Figures 9.3(a to e) displays the Rietveld refinement performed on the RT PXRD data of $Ba_{2-x}Ca_xTiSi_2O_8$ (x = 0.0, 0.2, 0.4, 0.8, and 1) ceramics. The FullProf program [76] was utilized for this purpose. During the refinement process, the lattice constants were adjusted, while the positions of atoms, isotropic thermal parameters, and occupancy were maintained at the values provided earlier [3]. The Rietveld refinement of the $Ba_{2-x}Ca_xTiSi_2O_8$ pattern reaffirmed the presence of the *P4bm* space group of tetragonal structure, which aligns with the findings of the previous report [10]. The Rietveld analysis confirms a strong matching between the calculated and observed patterns, which is further supported by the minimum difference pattern between calculated and observed patterns. The pristine compound (x = 0.0) pattern exhibits two minor peaks of low intensity that were not included in the current fitting. The first impurity peak at 2θ = 26.37° is likely attributed to unreacted $SiO_2$, while the second peak at 2θ = 31.49° corresponds to the highest intensity peak of $BaTiO_3$, matching with JCPDS: #05-0626. No additional impurity peaks were detected in this compound. Furthermore, a closer inspection of the Rietveld refined patterns revealed minor impurity peaks with very low intensity between 2θ = 20° to 40° in the doped compounds with x = 0.2. Specifically, the peaks at 2θ = 21.51° and 30.19° are likely due to unreacted $SiO_2$ (JCPDS: #85-0621

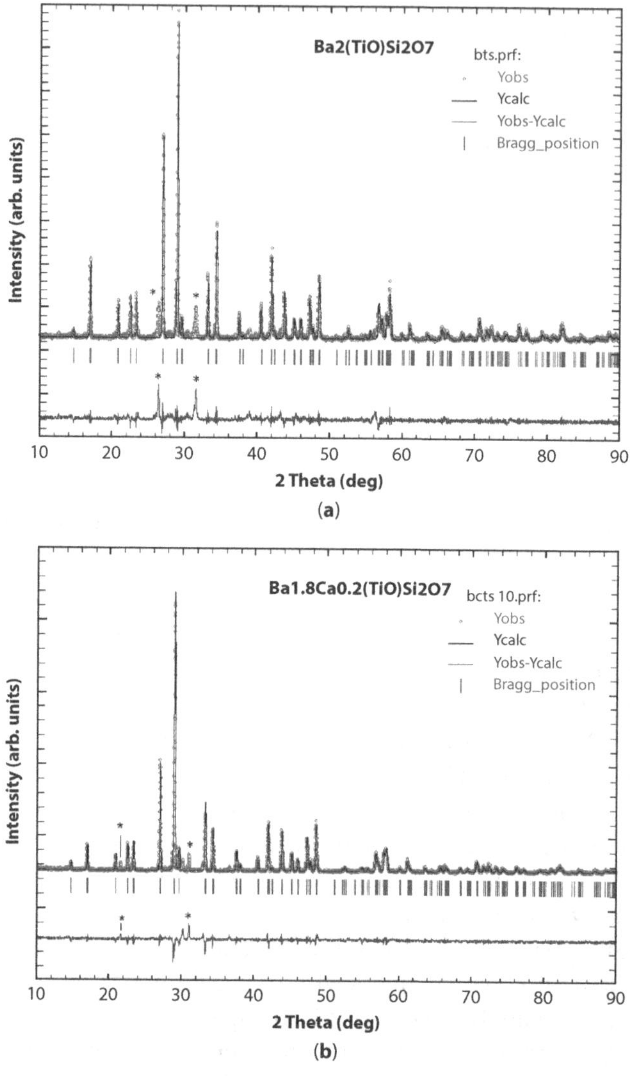


**Figure 9.3** (a) Rietveld refined PXRD pattern of $Ba_{2-x}Ca_xTiSi_2O_8$ (x = 0.0) ceramic. (b) Rietveld refined PXRD pattern of $Ba_{2-x}Ca_xTiSi_2O_8$ (x = 0.2) ceramic.

and #88-2483). Additionally, the peaks at 2θ = 30.98° and 36.58° are due to the formation of the $BaTiO_3$ phase (JCPDS: #05-0626) and unreacted $TiO_2$ (JCPDF: #89-6975). For x = 1.0, two low-intensity extra peaks at 2θ = 17.9° and 24.47° were found and identified. These two peaks were attributed to

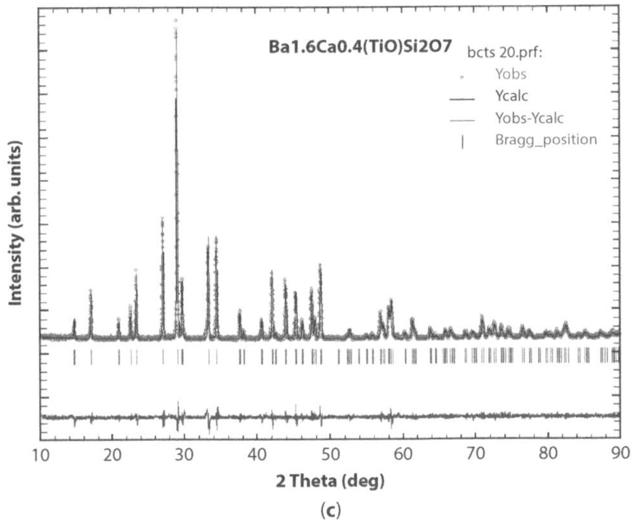


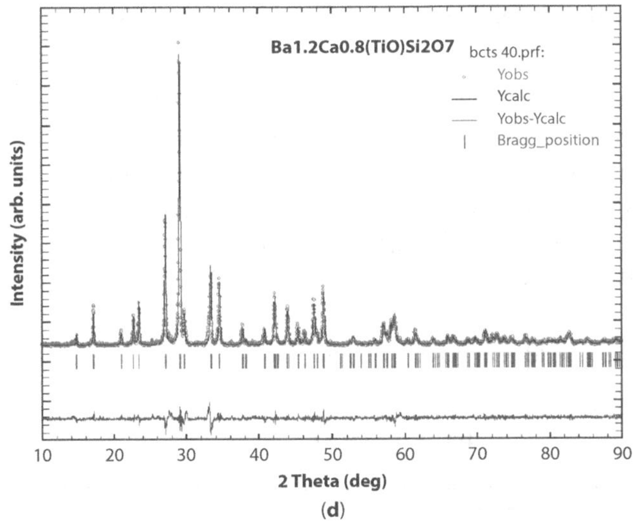


**Figure 9.3** (c) Rietveld refined PXRD pattern of $Ba_{2-x}Ca_xTiSi_2O_8$ (x = 0.4) ceramic. (d) Rietveld refined PXRD pattern of $Ba_{2-x}Ca_xTiSi_2O_8$ (x = 0.8) ceramic.

$SiO_2$ and $TiO_2$. Importantly, all impurity peaks within the 2θ range of 20° to 40° in all X-ray diffraction patterns fall within the 5% error limit compared to the pure compound. As a result, these impurities neither affect the structure nor physical properties of the synthesized ceramics. A summary of the

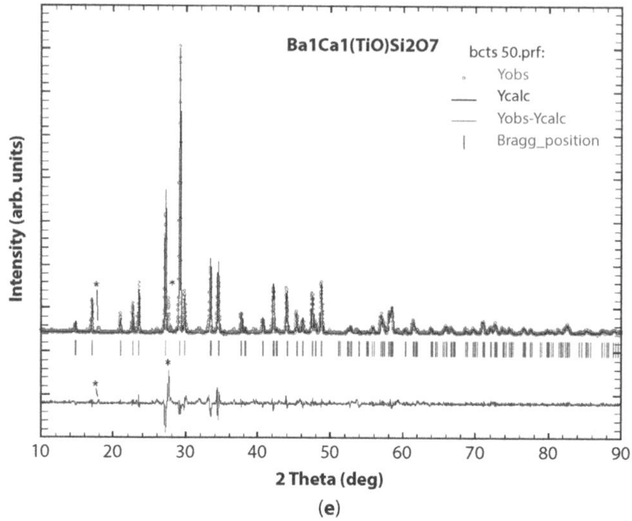


**Figure 9.3** (e) Rietveld refined PXRD pattern of $Ba_{2-x}Ca_xTiSi_2O_8$ (x = 1).

detailed analysis is presented in Table 9.1. Table 9.2 demonstrates a favorable agreement between the d-values obtained from the observed PXRD diagrams and those calculated based on values of Miller indices and cell parameters for all compositions (x = 0.0, 0.2, 0.4, 0.8, and 1.0). Analysis of the table reveals that the change in interplanar spacing is negligible when the $Ca^{2+}$ ion concentration increases up to x = 0.2. However, a slight change

**Table 9.1** Rietveld refined crystallographic parameters of $Ba_{2-x}Ca_xTiSi_2O_8$ ceramics.

| | $Ba_{2-x}Ca_xTiSi_2O_8$ | | | | |
|---|---|---|---|---|---|
| **Parameters** | **x = 0.0** | **x = 0.2** | **x = 0.4** | **x = 0.8** | **x = 1** |
| Lattice constant | | | | | |
| a (Å) | 8.5420 | 8.5210 | 8.5000 | 8.4950 | 8.4930 |
| c (Å) | 5.2200 | 5.2250 | 5.2060 | 5.1890 | 5.2040 |
| Cell volume (Å$^3$) | 381.08 | 379.40 | 376.11 | 374.42 | 375.47 |
| $R_{Bragg}$ | 2.6 | 1.8 | 1.5 | 1.6 | 2.3 |
| $R_f$ | 2.4 | 1.3 | 1.0 | 1.4 | 2.0 |
| $\chi^2$ | 6.44 | 7.03 | 4.85 | 3.47 | 6.52 |

**Table 9.2** Interplanar d-spacing and corresponding Miller indices of $Ba_{2-x}Ca_xTiSi_2O_8$ ceramics.

| x = 0.0 | | x = 0.2 | | x = 0.4 | | x = 0.8 | | x = 1 | | |
|---|---|---|---|---|---|---|---|---|---|---|
| $d_{obs}$ | $d_{cal}$ | $d_{obs}$ | $d_{cal}$ | $d_{obs}$ | $d_{cal}$ | $d_{obs}$ | $d_{cal}$ | $d_{obs}$ | $d_{cal}$ | hkl |
| 5.22 | 5.22 | 5.22 | 5.22 | 5.20 | 5.20 | 5.19 | 5.18 | 5.20 | 5.20 | 001 |
| 3.82 | 3.82 | 3.81 | 3.81 | 3.80 | 3.80 | 3.80 | 3.79 | 3.80 | 3.79 | 210 |
| 3.30 | 3.30 | 3.30 | 3.30 | 3.29 | 3.29 | 3.28 | 3.28 | 3.29 | 3.29 | 201 |
| 3.08 | 3.08 | 3.08 | 3.07 | 3.07 | 3.07 | 3.06 | 3.06 | 3.07 | 3.06 | 211 |
| 2.70 | 2.70 | 2.69 | 2.69 | 2.68 | 2.68 | 2.68 | 2.68 | 2.68 | 2.68 | 310 |
| 2.61 | 2.61 | 2.61 | 2.61 | 2.60 | 2.60 | 2.60 | 2.59 | 2.60 | 2.60 | 221 |
| 2.15 | 2.15 | 2.15 | 2.15 | 2.14 | 2.14 | 2.14 | 2.14 | 2.14 | 2.14 | 321 |
| 2.07 | 2.07 | 2.06 | 2.06 | 2.06 | 2.06 | 2.06 | 2.06 | 2.06 | 2.06 | 410 |
| 1.92 | 1.92 | 1.92 | 1.92 | 1.91 | 1.91 | 1.91 | 1.91 | 1.91 | 1.91 | 411 |
| 1.87 | 1.87 | 1.87 | 1.87 | 1.87 | 1.87 | 1.86 | 1.86 | 1.86 | 1.86 | 331 |
| 1.58 | 1.58 | 1.58 | 1.58 | 1.57 | 1.57 | 1.57 | 1.57 | 1.57 | 1.57 | 213 |

in interplanar spacing is observed with further increments in the $Ca^{2+}$ concentration from x = 0.4, 0.8, and 1.0. This substantial change in interplanar spacing due to the increasing $Ca^{2+}$ concentration may be due to the difference in ionic radii of $Ca^{2+}$ and $Ba^{2+}$. Since the ionic radii of the $Ca^{2+}$ ion (0.99Å) is smaller compared to the ionic radii of the $Ba^{2+}$ ion (1.35Å), the interplanar spacing of $Ca^{2+}$ doped compounds decreases as the $Ca^{2+}$ ion concentration increases. Notably, the value of tetragonality (0.61) of all the compounds remains nearly constant (Table 9.1). This indicates that all the samples are crystallized in the same structure [75].

Table 9.3 presents the bond lengths (interatomic distances) calculated using the Bond_Str program. These values exhibit a strong agreement with the previously reported values [3], albeit with increased precision in the present study. Our structural results align well with the earlier findings [3, 10, 30] regarding the pristine compound. The analysis of single-crystal data demonstrates that the bond length Ba-O ranges from 2.65 to 3.34Å [10]. However, the present structural determination on fresnoite ceramic reveals that the Ba-O interatomic distances vary between 2.13 and 3.01 Å.

**Table 9.3** Interatomic distances for $Ba_{2-x}Ca_xTiSi_2O_8$ ceramics.

| Bond type | x = 0.0 | x = 0.2 | x = 0.4 | x = 0.8 | x = 1 |
|---|---|---|---|---|---|
| Ba-O1 | 2.8752 | 2.8284 | 2.7940 | 2.7268 | 2.7532 |
| Ba-O2 | 2.6753 | 2.7965 | 2.6760 | 2.7290 | 2.7184 |
| Ba-O3 | 3.0389 | 2.9991 | 3.1014 | 2.9543 | 2.9535 |
| Ti-O3 | 1.9439 | 2.0452 | 1.8863 | 1.8737 | 1.7868 |
| Ti-O4 | 1.2899 | 1.8650 | 2.6028 | 2.5944 | 2.6024 |
| Si-Si | 2.0492 | 2.0780 | 2.0420 | 2.0400 | 2.0490 |
| Si-O1 | 1.5854 | 1.6688 | 1.6989 | 1.6501 | 1.6447 |
| Si-O2 | 2.6269 | 1.5337 | 1.9006 | 1.7979 | 1.8157 |
| Si-O3 | 1.4847 | 1.5705 | 1.4928 | 1.5910 | 1.6866 |
| O1-O2 | 3.1138 | 2.6221 | 2.9980 | 2.9367 | 2.8724 |
| O1-O3 | 2.5205 | 2.5754 | 2.5411 | 2.5733 | 2.6748 |
| O2-O3 | 2.9447 | 2.5440 | 2.6494 | 2.6969 | 2.8106 |
| O3-O4 | 2.1313 | 2.1294 | 2.6106 | 3.0922 | 3.1257 |
| Ca-O1 | - | 2.8284 | 2.7940 | 2.7268 | 2.7532 |
| Ca-O2 | - | 2.7965 | 2.6760 | 2.7290 | 2.7184 |
| Ca-O3 | - | 2.9991 | 3.1014 | 2.9543 | 2.9535 |

This slight discrepancy in bond lengths may be attributed to the differences between the ceramic and single-crystal forms of the samples. Additionally, it is noteworthy that the Ba-O distances fall within the range of 2.22 to 2.90 Å, which is shorter than the ionic bond distance of 2.94 Å, affirming the existence of covalent characteristics in these Ba-O bonds. As for the Ti-O bond lengths, the longer Ti-O3 bond (1.94Å) and the shorter Ti-O4 bond (1.29Å) exhibit good agreement with earlier reports [10]. Table 9.3 further displays the Si-O interatomic distances, where Si-O1, Si-O2, and Si-O3 bonds correspond to the pyrosilicate group. Notably, considerable changes in the Ba-O, Ti-O, and Si-O lengths are observed with the introduction of $Ca^{2+}$ ions, which can be due to the distinct ionic radii of $Ca^{2+}$ and $Ba^{2+}$ ions (Table 9.3) [75, 77].

### 9.4.2 Morphological Analysis

The density of the compounds was determined using the conventional technique of volume and weight measurements, resulting in values ranging from approximately 85% to 95% of the theoretical values. To examine the morphology of these ceramics, SEM (scanning electron microscopy) was employed. The SEM photographs of all the samples are presented in Figures 9.4(a to d). These micrographs reveal a nearly uniform and spherical distribution of grains across the entire surface. The sizes of grains, measured from these micrographs, range from 1 to 2 μm (Table 9.4). Importantly, the average particle size found from the SEM pictures demonstrates good agreement with the size calculated using the Scherrer formula in X-ray diffraction, where $t = 0.89\lambda/\beta\cos\theta$, with $\beta$ representing the FWHM and $\lambda$ denoting the wavelength of X-ray [75].

### 9.4.3 Raman Spectroscopic Analysis

Figure 9.5(a to e) displays the room temperature Raman spectrum of $Ba_{2-x}Ca_xTiSi_2O_8$ (x = 0.0, 0.2, 0.4, 0.8, and 1) measured in the range of wave number from 300 to 1100 $cm^{-1}$. The observed bands for pristine fresnoite compound are notably consistent with the spectrum already reported [36, 47]. Fresnoite, characterized by a [$TiO_5$] pyramid and [$Si_2O_7$] layered structure, exhibits distinctive vibrational characteristics influenced by its molecular groups. We have examined the Raman modes of the pristine fresnoite ceramic and compared them to those from earlier investigations [39, 42], and the corresponding mode positions are presented in Table 9.5. This table also demonstrates the shift in mode positions resulting from the doping of various concentrations of $Ca^{2+}$ ions in place of $Ba^{2+}$ ions. From factor group theory, it has been determined that a unit cell of the *P4bm* space group structure, consisting of 26 atoms, gives rise to 47 active modes, namely, $11A_1$, $6B_1$, $10B_2$, and 20E. These modes encompass acoustic modes, rotatory modes of $TiO_5$ and $Si_2O_7$ groups, translatory modes of Ba, and stretching and bending modes of $TiO_5$ and $Si_2O_7$ groups.

Due to its classification as a polar (noncentrosymmetric) material, fresnoite exhibits both Raman and IR-active modes. Both modes can experience separation, known as TO-LO splitting, under the influence of long-range electrostatic forces. In the case of fresnoite, this splitting can be obtained with the representation of $A_1$ and E modes [42]. The prominent modes at 872 and 858 $cm^{-1}$ can be attributed to Ti and Si ions and are linked to a combined mode involving the vibration of $\upsilon(SiO_3)$ and the stretching of the bond between Ti and O (Ti-O). Another mode at

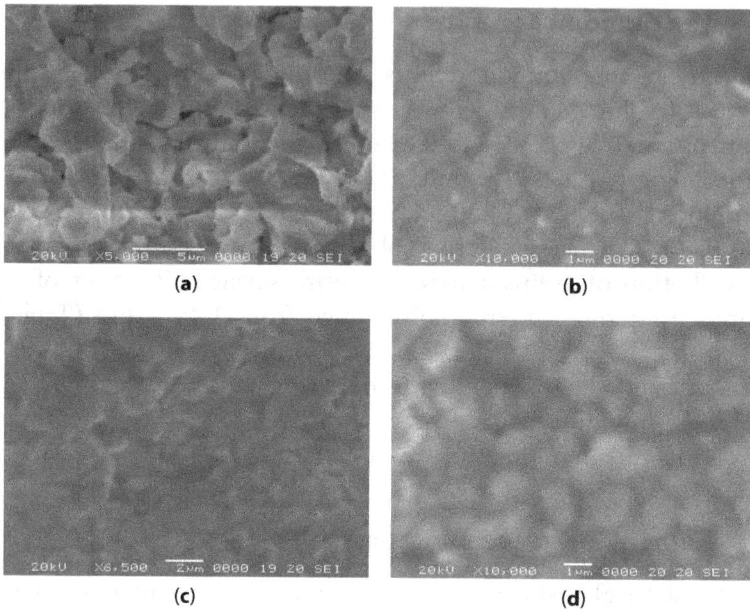


**Figure 9.4** (a) SEM micrograph of $Ba_{2-x}Ca_xTiSi_2O_8$ (x = 0.0) ceramic. (b) SEM micrographs of $Ba_{2-x}Ca_xTiSi_2O_8$ (x = 0.2) ceramic. (c) SEM micrographs of $Ba_{2-x}Ca_xTiSi_2O_8$ (x = 0.4) ceramic. (d) SEM micrographs of $Ba_{2-x}Ca_xTiSi_2O_8$ (x = 1) ceramic.

**Table 9.4** Physical parameters of $Ba_{2-x}Ca_xTiSi_2O_8$ ceramics.

| | $Ba_{2-x}Ca_xTiSi_2O_8$ | | | | |
|---|---|---|---|---|---|
| | **x = 0.0** | **x = 0.2** | **x = 0.4** | **x = 0.8** | **x = 1** |
| Mean grain size (μm) | ~2 | ~1.5 | 1.25 | ~1 | ~1 |
| Activation energy (eV) | 4.74 | 1.26 | 1.23 | 0.84 | 0.75 |

870 $cm^{-1}$ relates to the Ti-O* vibrations within the $TiO_5$ polyhedral, with O* representing the apical oxygen in the polyhedral. The mode around 663$cm^{-1}$ solely involves Si ions and arises from the υ(Si-O-Si) vibration in the [$Si_2O_7$] group, whether 597 $cm^{-1}$ mode is exclusively concerned with Ti and corresponds to the υ($TiO_5$) vibration. Ti-O and Si-O vibrations are significantly mixed as a result of the stretching vibrations shifting toward the higher side of wave numbers due to the higher covalency of Ti-O bonds. Consequently, the modes at higher wave numbers, specifically at 1060, 990,

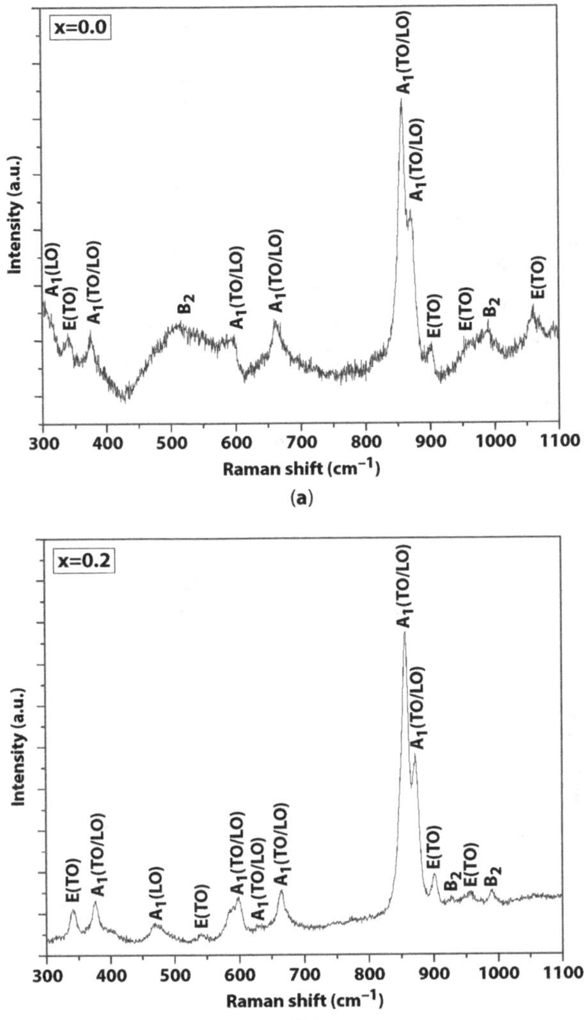


**Figure 9.5** (a) RT Raman spectra of $Ba_{2-x}Ca_xTiSi_2O_8$ (x = 0.0) ceramic. (b) RT Raman spectra of $Ba_{2-x}Ca_xTiSi_2O_8$ (x = 0.2) ceramic.

and 900 $cm^{-1}$ in fresnoite's spectrum, cannot be attributed to the υ(Si-O-Si) bond, but rather to the υ(Si-O) vibration mode. Notably, the bands at 375 and 339 $cm^{-1}$ correspond to the bending and translational modes of the $TiO_5$ and $Si_2O_7$ groups, respectively. Furthermore, a band at 515 $cm^{-1}$ is identified as the δ(Si-O) deformation mode, which represents an inactive Raman mode.

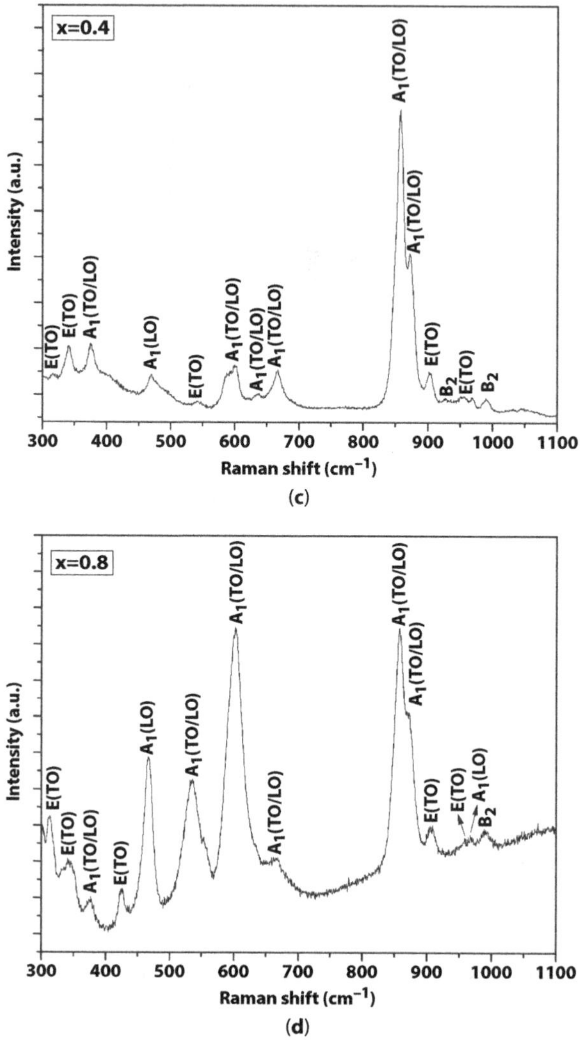


**Figure 9.5** (c) RT Raman spectra of $Ba_{2-x}Ca_xTiSi_2O_8$ (x = 0.4) ceramic. (d) RT Raman spectra of $Ba_{2-x}Ca_xTiSi_2O_8$ (x = 0.8) ceramic.

Upon substitution of $Ca^{2+}$ ions on the $Ba^{2+}$ site, the $A_1$(TO/LO) mode at 603 $cm^{-1}$ exhibits a slight shift toward higher wavenumbers, accompanied by the unexpected appearance of certain modes. As the frequency of Raman modes depends on ionic masses and force constants, the band position shift can be attributed to the mass difference between $Ba^{2+}$ and $Ca^{2+}$ ions. With higher concentrations of $Ca^{2+}$ ion substitution, specifically for x = 0.8 and 1, significant changes in intensity are observed for

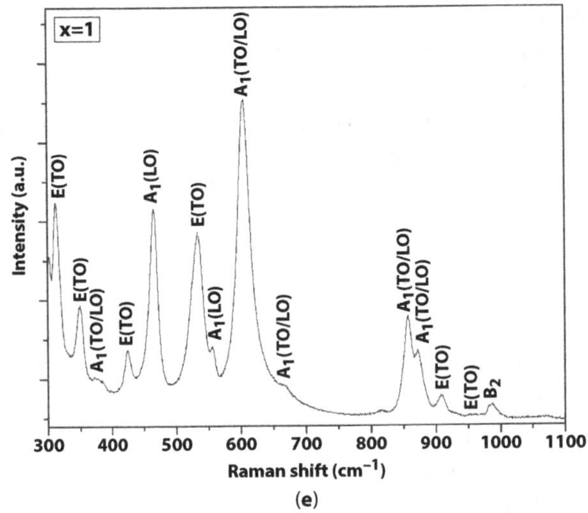


**Figure 9.5** (e) RT Raman spectra of $Ba_{2-x}Ca_xTiSi_2O_8$ (x = 1) ceramic.

**Table 9.5** Raman modes of $Ba_{2-x}Ca_xTiSi_2O_8$ ceramics.

| Assigned modes | Positions (cm⁻¹) of modes | | | | | | |
|---|---|---|---|---|---|---|---|
| | x = 0.0 | Ref. [42] | Ref. [39] | x = 0.2 | x = 0.4 | x = 0.8 | x = 1 |
| A-mode | 304 | - | - | - | - | - | - |
| | 375 | 376 | 376 | 376 | 376 | 376 | 375 |
| | - | 474 | 476 | 468 | 468 | 468 | 468 |
| | 597 | 602 | 603 | 598 | 600 | 602 | 605 |
| | - | 626 | 628 | 627 | 636 | - | - |
| | 662 | 664 | 664 | 665 | 666 | 667 | 669 |
| | 858 | 858 | 858 | 858 | 858 | 858 | 858 |
| | 871 | 873 | 871 | 872 | 872 | 873 | 873 |
| | - | 964 | 965 | - | - | 965 | - |
| E-mode | - | 317 | 318 | - | 316 | 314 | 312 |
| | 338 | 340 | 339 | 341 | 341 | 341 | 348 |
| | - | 425 | - | - | - | 425 | 425 |
| | - | 539 | 540 | 540 | 541 | 535 | 534 |
| | 902 | 900 | 902 | 902 | 903 | 905 | 908 |
| | 952 | 952 | - | 957 | 957 | 958 | 958 |
| | 1060 | 1065 | - | - | - | - | - |
| B-mode | 515 | - | - | - | - | - | - |
| | - | 927 | - | 925 | 927 | - | - |
| | 990 | 993 | - | 990 | 990 | 990 | 990 |

certain modes at both lower and higher wave numbers. This suggests that increasing Ca concentrations play a crucial role in reducing strong modes at higher wave numbers and enhancing weak modes at lower wave numbers. This change in band intensity can be attributed to molecule relaxation and bond stretching, respectively. The presence of new modes in Ca-doped samples can be attributed to the disorder in bonds of fresnoite [78].

### 9.4.4 Electrical Analysis

The dc electrical conductivity ($\sigma_{dc}$) of $Ba_{2-x}Ca_xTiSi_2O_8$ ceramics was assessed by measuring the steady-state current values at various temperatures, and the results were plotted in Figure 9.6 for an elevated temperature range of 643 K to 753 K. Since the resistivity is found to be extremely high, exhibiting almost no temperature dependence from room temperature (RT) to 643 K, the resistivity of these samples was plotted from 643 K to 753 K. Beyond 643 K, the conductivity increases by increasing the temperature up to studied temperature limit. This increasing trend in conductivity can be attributed to thermally activated processes and mathematically explained by the Arrhenius equation:

$$\sigma = \sigma o \exp(-Ea/kT)$$

where σo is a pre-exponential factor, σ is the conductivity, k is the Boltzmann constant ($k = 8.6 \times 10^{-5}$ eV/K), Ea is the activation energy, and T is representing temperature.

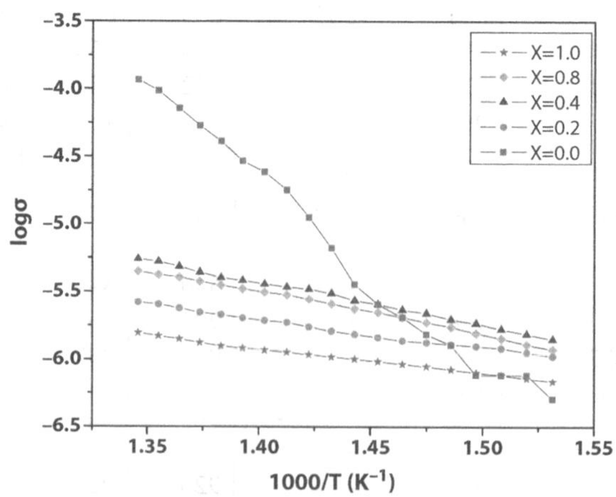


**Figure 9.6** log σ vs. $10^3/T$ curves of $Ba_{2-X}Ca_XTiSi_2O_8$ ceramics.

The activation energy of all samples was determined by analyzing the logσ versus 1/T curve and the results are presented in Table 9.4. It is widely recognized that the formation of oxide materials requires elevated temperatures during the processing stage. At temperatures exceeding 1000°C, it is expected that the oxygen ions present in the sample will out from the lattice, resulting in the creation of oxygen vacancies. Upon cooling, oxygen reenters the lattice; however, it cannot fully compensate for the oxygen loss. This phenomenon is observed in titanates and other compounds with layered structures, including the ones studied in this research [79, 80].

At low temperatures, the mobility of oxygen ions is limited but increases as the temperature rises. Consequently, pristine fresnoite compounds exhibit high conductivity and greater value of activation energy at high temperatures. When $Ca^{2+}$ ions replace $Ba^{2+}$ ions, the $Ca^{2+}$ ions may effectively act as acceptor dopants, resulting in an increased number of oxygen vacancies. The formation of charge complexes by these oxygen vacancies is responsible for longer relaxation times. At high temperatures, these charge complexes may be unstable, resulting in an increased number of oxygen vacancies and high conductivity. Further $Ca^{2+}$ ion concentration leads to more oxygen ion vacancies, which increases conductivity and lowers activation energy as shown in Table 9.4.

The greater activation energy value seen in the pristine fresnoite and its decreasing trend with $Ca^{2+}$ substitution can also be elucidated on the basis of bond angles, bond lengths, and the nature of bonds within the sheets. The sheets in fresnoite consist of $TiO_5$ tetragonal pyramids and $SiO_4$ tetrahedra, which are interconnected by barium-oxygen bonds. The bonding within the sheets is predominantly covalent, while the bonding between the intersheet layers is ionic in nature. This structural arrangement is consistent with previous studies [3, 30].

To investigate the dielectric response in $Ba_{2-x}Ca_xTiSi_2O_8$ (x = 0.0, 0.2, 0.4, 0.8, and 1), the variation of the dielectric constant (ε') was plotted as a function of frequency at RT shown in Figure 9.7. The graph depicted that the dielectric constant exhibits a rapid decrease at lower frequencies and becomes almost constant at higher frequencies, showing a relaxation behavior. This type of behavior can be readily explained by the Maxwell-Wagner-Sillars phenomena and Koop theory's interfacial polarization. Dipoles can follow an applied ac field at low frequencies because space charge and orientation polarizations control the dielectric response. However, at high frequencies, charge carriers reverse their orientation and decrease polarization, making it possible for dipoles to follow an applied ac field. Consequently, the dielectric constant reduces with increasing

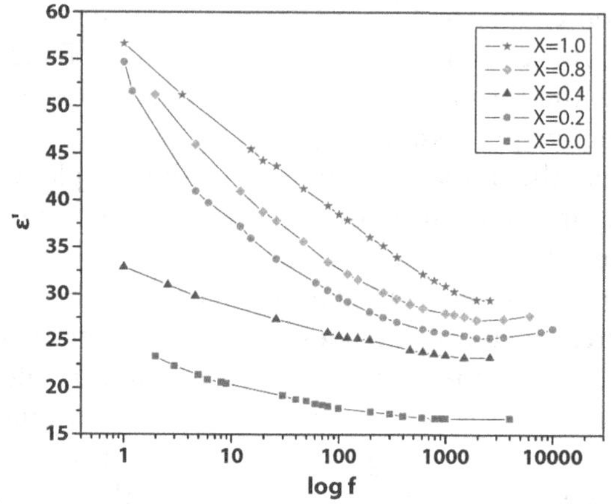


**Figure 9.7** Dielectric constant vs. log f curves of $Ba_{2-x}Ca_xTiSi_2O_8$ ceramics.

frequency and eventually reaches a constant value, indicating a relaxation behavior, at higher frequencies.

Figures 9.8 and 9.9 show the temperature-dependent dielectric constant (ε') characteristics at two fixed frequencies of 100 kHz and 1 MHz. The figure exhibits that the dielectric constant of pristine compound (x = 0.0) remains relatively constant up to 573 K, after which it increases with rising temperature. Furthermore, it is evident from the plot that the dielectric constant exhibits a greater enhancement with higher concentrations

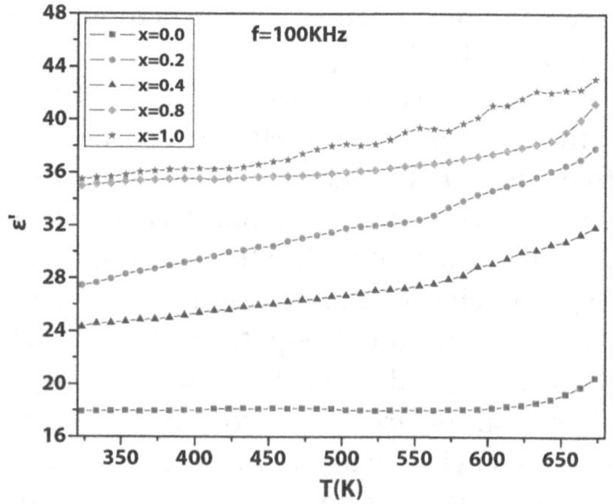


**Figure 9.8** Dielectric constant vs. temperature curves of $Ba_{2-x}Ca_xTiSi_2O_8$ ceramics at 100 kHz.

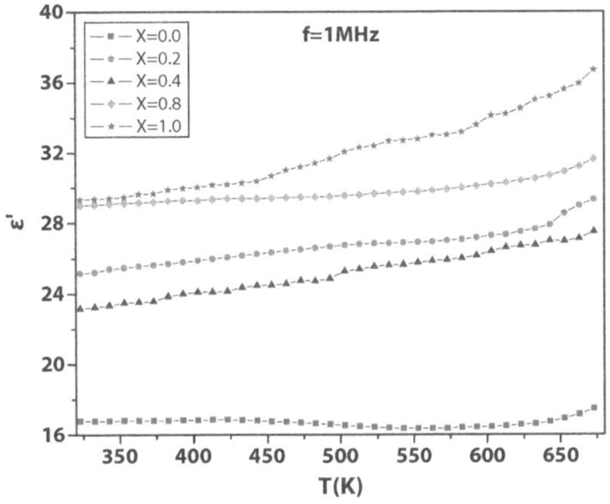


**Figure 9.9** Dielectric constant vs. temperature curves of $Ba_{2-x}Ca_xTiSi_2O_8$ ceramics at 1 MHz.

of $Ca^{2+}$ ions. However, no significant anomalies in the dielectric behavior have been observed within the measured temperature range. Moreover, the dissipation factor (ε") versus temperature graph at the same fixed frequencies is shown in Figures 9.10 and 9.11 and also exhibits the same trend as seen in ε' versus temperature plot. The dissipation factor of the pristine compound exhibits a nearly constant behavior up to 523K, with a slight increase beyond 523K. The dissipation factor shows a progressive increase

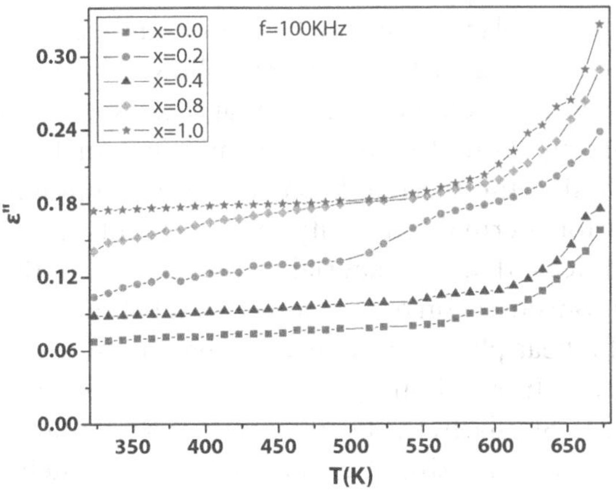


**Figure 9.10** Temperature dependence of dissipation factor for $Ba_{2-x}Ca_xTiSi_2O_8$ ceramics at 100 kHz.

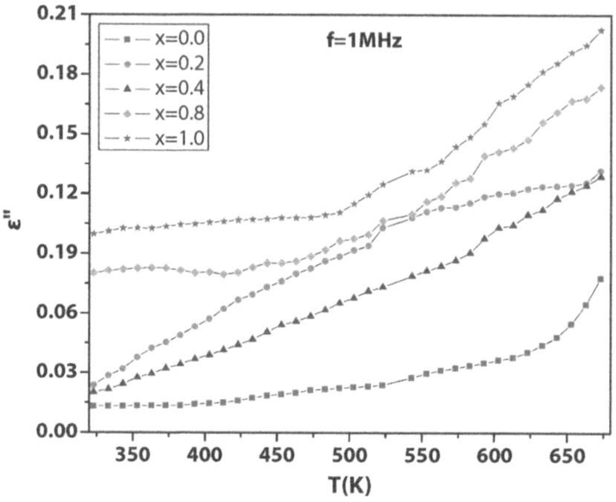


**Figure 9.11** Temperature dependence of dissipation factor for $Ba_{2-x}Ca_xTiSi_2O_8$ ceramics at 1 MHz.

with higher concentrations of x; however dielectric anomalies have not been detected in any synthesized sample [75].

### 9.4.5 Thermal Analysis

To investigate the thermal anomalies in $Ba_{2-x}Ca_xTiSi_2O_8$ (x = 0.0, 0.2, 0.4, 0.8, and 1), the specific heat and heat flow curves were plotted in the temperature range of 400 K to 840 K. Both the curves are presented in Figures 9.12(a to e). Figure 9.12(a) clearly displays that the specific heat graph of the pristine compound (x = 0.0) exhibits two broad humps around 480 K and 645 K. Similar to this, the heat flow curve displays a broad and shallow dip around the same temperatures followed by a modest endothermic peak at 798 K. It should be noted that at these specific points, the heat exchange occurring within the material is minimal, making it challenging to accurately predict the thermal behavior without employing more precise equipment for thermal sensitivity measurements. Nevertheless, these observations indicate that the material exhibits excellent thermal stability.

When a low concentration of $Ca^{2+}$ ions was substituted for $Ba^{2+}$ ions (x = 0.2), the specific heat plot represents a similar trend as observed for the pristine sample, with broad humps observed at 485 K and 685 K (as seen in Figure 9.12(b)). Similarly, the heat flow curve depicts a shallow dip close to these temperatures, making it challenging to accurately determine the heat exchange magnitude at these two points. As the $Ca^{2+}$ concentration is further increased from x = 0.2 to x = 0.4 (Figure 9.12(c)), both specific heat and heat flow curves continue to display a similar trend, with broad humps and dips occurring at approximately the same temperatures of 490 K and 680 K, respectively. Furthermore, for x = 0.8, both curves exhibit an identical pattern to the three previous compositions, featuring broad humps and shallow valleys at approximately 490 K and 700 K, respectively (see Figure 9.12(d)).

When 50% of the $Ba^{2+}$ ions were replaced by $Ca^{2+}$ ions (x = 1), both plots exhibited a broad hump and shallow dip at 485 K and 690 K (see Figure 9.12(e)). The specific heat and heat flow protocols also showed a

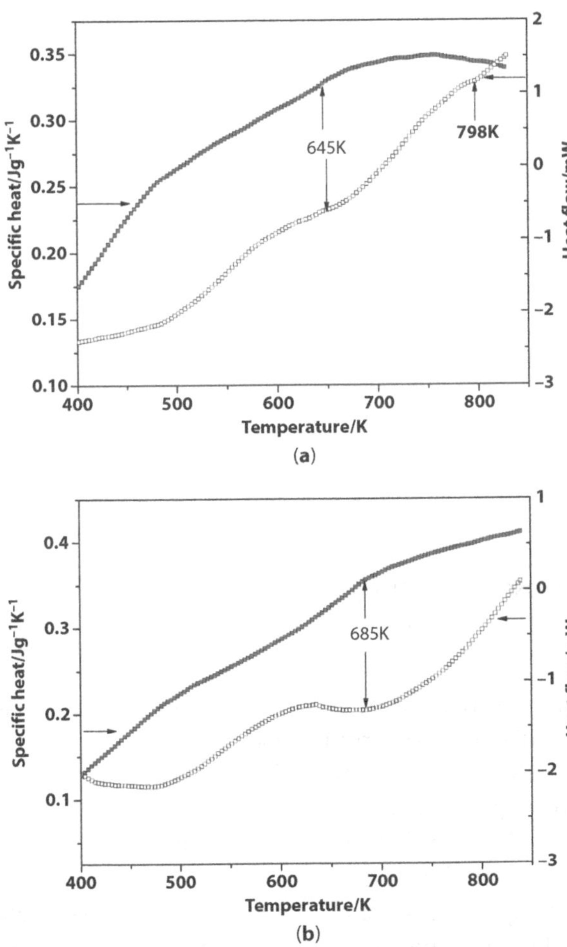


**Figure 9.12** (a): MDSC curves of $Ba_{2-x}Ca_xTiSi_2O_8$ (x = 0.0) ceramic. (b): MDSC curves of $Ba_{2-x}Ca_xTiSi_2O_8$ (x = 0.2) ceramic.

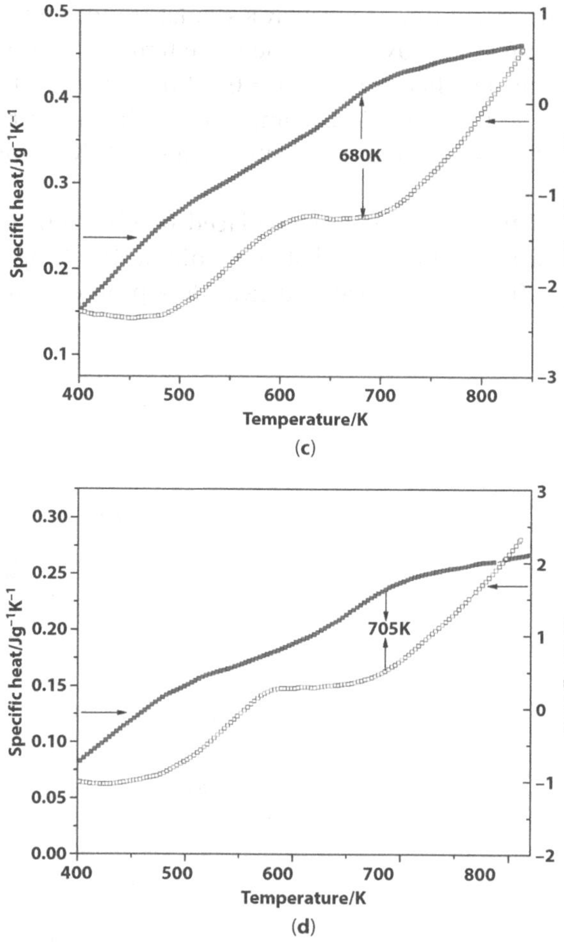


**Figure 9.12** (c): MDSC curves of $Ba_{2-x}Ca_xTiSi_2O_8$ (x = 0.4) ceramic. (d): MDSC curves of $Ba_{2-x}Ca_xTiSi_2O_8$ (x = 0.8) ceramic.

clear exo- and endothermic peak at 770 K. The specific heat amount of this peak is calculated to be around 0.45 J $g^{-1}$ $K^{-1}$. The thermal anomaly observed at 770 K could potentially be attributed to a structural transformation, although further confirmation would be required through high-temperature XRD and other precise techniques.

The investigation of these DSC curves reveals a sluggish thermal behavior, posing challenges in accurately determining the heat exchange occurring at the shallow dips and broad humps. Additionally, it seems that the

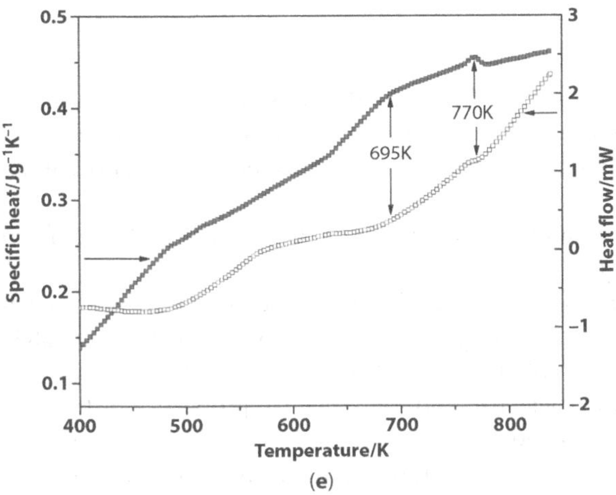


**Figure 9.12** (e): MDSC curves of $Ba_{2-x}Ca_xTiSi_2O_8$ (x = 1) ceramic.

materials exhibit excellent thermal stability, rendering them highly valuable for high-temperature functional devices [81].

## 9.5 Conclusion

X-ray diffraction analysis confirmed that $Ba_{2-x}Ca_xTiSi_2O_8$ (x = 0.0, 0.2, 0.4, 0.8, and 1) samples synthesized via conventional solid-state reaction method are crystallized in phase pure tetragonal symmetry with noncentrosymmetric space group *P4bm*. Furthermore, it is noted that a reasonably good fit to the Rietveld refined XRD patterns of Ca-doped fresnoite is suggestive of a dominant tetragonal structure. The calculated bond lengths exhibit good consistency with previous findings. Morphological results affirm the favorable time and temperature conditions of synthesis. Raman spectroscopy reveals clear and sharp optical phonon modes in both the pure and Ca-substituted compounds, with some modes shifting toward higher frequency upon Ba to Ca substitution. However, due to their low intensity, some modes appear indistinct and cannot be accurately distinguished. The low dielectric constant and high resistivity contribute to their favorable dielectric and ferroelectric properties, making them suitable for various practical devices. The thermal studies of pristine and doped fresnoite exhibit that thermal stability renders them suitable for high-temperature functional devices.

## References


1. Alfors, J.T., Stinson, M.C., Matthews, R.A., Pabst, A., Seven new barium minerals from eastern Fresno County, California. *Am. Mineral.*, 50, 314, 1965.
2. Moore, P.B. and Louisnathan, J., Fresnoite: unusual titanium coordination. *Science*, 156, 1361, 1967.
3. Moore, P.B. and Louisnathan, S.J., The crystal structure of fresnoite, $Ba_2(TiO)Si_2O_7$. *Z. fur Krist.*, 130, 438, 1969.
4. Abrahams, S.C., Structurally based prediction of ferroelectricity in inorganic materials with point group 6mm. *Acta Crystallogr. B.*, 44, 585, 1988.
5. Halliyal, A., Bhalla, A.S., Markgraf, S.A., Cross, L.E., Newnham, R.E., Unusual pyroelectric and piezoelectric properties of fresnoite ($Ba_2TiSi_2O_8$) single crystal and polar glass-ceramics. *Ferroelectrics*, 62, 27, 1985.
6. Kimura, M., Fujino, Y., Kawamura, T., New piezoelectric crystal: Synthetic Fresnoite ($Ba_2TiSi_2O_8$). *Appl. Phys. Lett.*, 29, 227, 1976.
7. Kimura, M., Elastic and piezoelectric properties of $Ba_2TiSi_2O_8$. *J. Appl. Phys.*, 48, 2850, 1977.
8. Haussühl, S., Eckstein, J., Recker, K., Wallrafen, F., Growth and physical properties of fresnoite $Ba_2TiSi_2O_8$. *J. Cryst. Growth*, 40, 200, 1977.
9. Bechthold, P.S., Haüssuhl, S., Michael, E., Eckstein, J., Recker, K., Wallrafen, F., Second harmonic generation in fresnoite, $Ba_2TiSi_2O_8$. *Phys. Lett. A*, 65, 453, 1978.
10. Markgraf, S.A., Halliyal, A., Bhalla, A.S., Newnham, R.E., Prewitt, C.T., X-ray structure refinement and pyroelectric investigation of fresnoite, $Ba_2TiSi_2O_8$. *Ferroelectrics*, 62, 17, 1985.
11. Takahashi, Y., Kitamura, K., Benino, Y., Fujiwara, T., Komatsu, T., Second-order optical nonlinear and luminescent properties of $Ba_2TiSi_2O_8$ nanocrystallized glass. *Appl. Phys. Lett.*, 86, 091110, 2005.
12. Song, N., Momida, H., Oguchi, T., Kim, B.G., Polar phase transitions and physical properties in fresnoite $A_2TiSi_2O_8$ (A= Ba, Sr) by first principles calculations. *J. Solid State Chem., 242*, 13, 2016.
13. Zhao, W., Zhang, F., Liu, J., Hao, B., Pan, S., Zhang, F., Liu, L., Flux crystal growth of $Ba_2TiOSi_2O_7$. *J. Cryst. Growth, 413*, 46, 2015.
14. Yan, Y., Liu, L., Ning, C., Yang, Y., Xia, C., Zou, Y., Liu, G., Improved electrical properties of $SiO_2$-added $BaTiO_3$ ceramics by microwave sintering. *Mater. Lett., 165*, 135, 2016.
15. Wu, T., Pu, Y., Zong, T., Gao, P., Microstructures and dielectric properties of Ba0.4Sr0. 6TiO3 ceramics with $BaO–TiO_2–SiO_2$ glass–ceramics addition. *J. Alloys Compd., 584*, 461, 2014.
16. Sahu, M., Mukherjee, S., Keskar, M., Krishnan, K., Dash, S., Tomar, B.S., Investigations of thermophysico-chemical properties of $Ba_2TiSi_2O_8$ and $Sr_2TiSi_2O_8$. *Thermochim. Acta, 663*, 215, 2018.
17. Yamauchi, H., Surface-acoustic-wave characteristics on Fresnoite ($Ba_2TiSi_2O_8$) single crystal. *J. Appl. Phys.*, 49, 6162, 1978.

18. Melngailis, J., Vetelno, J.F., Jhunjunwala, A., Reed, T.B., Fahey, R.E., Stein, E., Surface acoustic wave properties of fresnoite, $Ba_2Si_2TiO_8$. *Appl. Phys. Lett.*, 32, 203, 1978.
19. Foster, M.C., Arbogast, D.J., Nielson, R.M., Photinos, P., Abrahams, S.C., Fresnoite: A new ferroelectric mineral. *J. Appl. Phys.*, 85, 2299, 1999.
20. Chang, Z.P. and Bhalla, A.S., Elastic anomaly in fresnoite ($Ba_2TiSi_2O_8$) single crystal. *Mater. Lett.*, 8, 418, 1989.
21. Takahashi, Y., Kitamura, K., Inoue, S., Benino, Y., Fujiwara, T., Komatsu, T., Effect of Heat-Treatment Temperature on Optical Properties of $Ba_2TiSi_2O_8$ Nanocrystallized Glasses. *J. Ceram. Soc. Jpn.*, 113, 419, 2005.
22. Zhu, B., Zhang, S., Lin, G., Zhou, S., Qiu, J., Enhanced Multiphoton Absorption Induced Luminescence in Transparent $Sm^{3+}$-doped $Ba_2TiSi_2O_8$ Glass-ceramics. *J. Phys. Chem. C*, 111, 17118, 2007.
23. Mayerhöfer, T.G., Shen, Z., Keding, R., Musfeldt, J.L., Optical isotropy in polycrystalline $Ba_2TiSi_2O_8$: Testing the limits of a well-established concept. *Phys. Rev. B*, 71, 184116, 2005.
24. Asahi, T., Osaka, T., Kobayashi, J., Abrahams, S.C., Nanamatsu, S., Kimura, M., Optical study on a phase transition of fresnoite $Ba_2Si_2TiO_8$. *Phys. Rev. B*, 63, 094104, 2001.
25. Höche, T., Grodzicki, M., Heyroth, F., Van Aken, P.A., Assessment of transition-metal coordination in glasses by electron energy-loss spectroscopy. *Phys. Rev. B*, 72, 205111, 2005.
26. Takahashi, Y., Osada, M., Masai, H., Fujiwara, T., Softer region at boundary of supercooled liquid–crystal in glassy Fresnoite. *Appl. Phys. Lett.*, 94, 241909, 2009.
27. Rangarajan, B., Shrout, T., Lanagan, M., Crystallization kinetics and dielectric properties of Fresnoite BaO-$TiO_2$-$SiO_2$ Glass ceramics. *J. Am. Ceram. Soc.*, 92, 2642, 2009.
28. Park, T., Davis, M.J., Vullo, P., Nenoff, T.M., Krumhansl, J.L., Navrotsky, A., Thermochemistry and Aqueous Durability of Ternary Glass Forming Ba-Titanosilicates: Fresnoite ($Ba_2TiSi_2O_8$) and Ba-Titanite ($BaTiSiO_5$). *J. Am. Ceram. Soc.*, 92, 2053, 2009.
29. Rangarajan, B., Bharadwaja, S.S.N., Furman, E., Shrout, T., Lanagan, M., Impedance Spectroscopy Studies of Fresnoite in BaO-$TiO_2$-$SiO_2$ system. *J. Am. Ceram. Soc.*, 93, 522, 2010.
30. Ochi, Y., Fresnoite crystal structure in glass-ceramics. *Mater. Res. Bull.*, 41, 740, 2006.
31. Lin, S., Han, J., Xu, K., Lu, P., Effects of ZnO on the stability of the $SiO_2$-BaO-$TiO_2$ system. *J. Non-Cryst. Solids*, 499, 8, 2018.
32. Markgraf, S.A., Randall, C.A., Bhalla, A.S., Reeder, R.J., Incommensurate phase in $Ba_2TiSi_2O_8$. *Solid States Commun.*, 75, 821, 1990.
33. Höche, T., Neumann, W., Esmaeilzadeh, S., Uecker, R., Lentzen, M., Rüssel, C., The crystal structure of $Sr_2TiSi_2O_8$. *J. Solid State Chem.*, 166, 15, 2002.

34. Eckstein, J., Recker, K., Wallrafen, F., Züchtung, von, Fresnoit, $Ba_2TiSi_2O_8$. *Naturwissenschaften*, 63, 435, 1976.
35. Shen, C., Zhang, H., Cong, H., Yu, H., Wang, J., Zhang, S., Investigations on the thermal and piezoelectric properties of fresnoite $Ba_2TiSi_2O_8$ single crystals. *J. Appl. Phys.*, 116, 044106, 2014.
36. Blasse, G., Fresnoite ($Ba_2TiSi_2O_8$): a luminescent compound with a long decay time. *J. Inorg. Nucl. Chem.*, 41, 639, 1979.
37. Iijima, K., Marumo, F., Kimura, M., Kawamura, T., Structure of a ferroelastic crystal $Ba_2TiGe_2O_8$ and its thermal phase transition. *J. Chem. Soc. Jpn.*, 10, 1557, 1981.
38. Markgraf, S.A. and Bhalla, A.S., Low-temperature phase transition in $Ba_2TiGe_2O_8$. *Phase Trans.*, 18, 55, 1989.
39. Mayerhöfer, T.G. and Dunken, H.H., Single-crystal IR spectroscopic investigation on fresnoite, Sr-fresnoite and Ge-fresnoite. *Vib. Spectrosc.*, *25*, 185–95, 2001.
40. Kimura, M., Doi, K., Nanamatsu, S., Kawamura, T., A new piezoelectric crystal: $Ba_2Ge_2SiO_8$. *Appl. Phys. Lett.*, 23, 531, 1973.
41. Halliyal, A., Bhalla, A.S., Cross, L.E., Phase transitions, dielectric, piezoelectric and pyroelectric properties of barium titanium germanate $Ba_2TiGe_2O_8$ single crystals. *Ferroelectrics*, 62, 3, 1985.
42. Markgraf, S.A., Sharma, S.K., Bhalla, A.S., Raman study of fresnoite-type materials: polarized single crystal, crystalline powders, and glasses. *J. Mater. Res.*, 8, 635, 1993.
43. Hoche, T., Heyroth, F., Keding, R., Sales, B., Mandrus, D., Electron energy-loss near-edge structure spectroscopy of titanium and vanadium in fresnoite framework structures. *Microscopy and Microanal.* 9, 74, 2003.
44. Shen, C., Zhang, H., Wang, D., Wang, J., Boughton, R.I., Optical properties in the Fresnoite $Ba_2TiSi_2O_8$ single crystal. *Crystals*, 7, 53, 2017.
45. Robbins, C.R., Synthesis and growth of fresnoite ($Ba_2TiSi_2O_8$) from a $TiO_2$ flux and its relation to the system $BaTiO_3$-$SiO_2$. *J. Res. Natl. Bur. Stand A: Phys. Chem.*, 74, 229, 1970.
46. Blasse, G., Fluorescence of compounds with fresnoite ($Ba_2TiSi_2O_8$) structure. *J. Inorg. Nucl. Chem.*, 30, 2283, 1968.
47. Gabelica-Robert, M. and Tarte, P., Vibrational spectrum of fresnoite ($Ba_2TiOSi_2O_7$) and isostructuralcompounds. *Phys. Chem. Minerals*, 7, 26, 1981.
48. Markgraf, S.A., Sharma, S.K., Bhalla, A.S., Raman study of glasses of $Ba_2TiSi_2O_8$ and $Ba_2TiGe_2O_8$. *J. Am. Ceram. Soc.*, 75, 2630, 1992.
49. Gopalakrishnan, J., Ramesha, K., Rangan, K.K., Pandey, S., In search of inorganic nonlinear optical materials for second harmonic generation. *J. Solid State Chem.*, 148, 75, 1999.
50. Roy, M., Barbar, S.K., Dave, P., Jangid, S., Bala, I., X-ray, scanning electron microscopy and electrical properties of synthetic fresnoite ($Ba_2TiSi_2O_8$) ceramics. *Appl. Phys. A*, 100, 1191, 2010.

51. Wong, C.L., Ferraris, C., White, T.J., Microstructural features and domain formation in $(BaSr)_2TiSi_2O_8$ fresnoites. *J. Solid State Chem.*, 184, 1768, 2011.
52. Schmid, H., Genequand, P., Tippmann, H., Pouilly, G., Guédu, H., Pyroelectricity and related properties in the fresnoitepseudobinary system $Ba_2TiGe_2O_8$-$Ba_2TiSi_2O_8$. *J. Mater. Sci.*, 13, 2257, 1978.
53. Shen, J., Zhou, J., Cui, X., Li, L., Synthesis and dielectric properties of $Ba_2TiSi_2O_8$ glass-ceramics from the sol–gel process. *J. Electroceram.*, 21, 565, 2008.
54. Dai, W., Zhu, M., Hou, Y.D., Wang, H., Yan, H., Shao, M. *et al.*, Preparation and characterization of $Ba_2TiSi_2O_8$ ferroelectric films produced by sol–gel method. *Mater. Lett.*, 58, 2927, 2004.
55. Mann, M., Abbott, E.E., Kolis, J.W., Hydrothermal crystal growth of fresnoite. *J. Cryst. Growth*, 312, 3395, 2010.
56. Takahashi, Y., Benino, Y., Fujiwara, T., Komatsu, T., Large second-order optical nonlinearities of fresnoite-type crystals in transparent surface-crystallized glasses. *J. Appl. Phys.*, 95, 3503, 2004.
57. Yamazaki, Y., Takahashi, Y., Ihara, R., Fujiwara, T., Surface crystallization of Fresnoite-type crystallized glasses with large thickness. *J. Ceram. Soc. Jpn.*, 119, 757, 2011.
58. Halliyal, A., Bhalla, A.S., Newnham, R.E., Cross, L.E., $Ba_2TiGe_2O_8$ and $Ba_2TiSi_2O_8$ pyroelectric glass-ceramics. *J. Mater. Sci.*, 16, 1023, 1981.
59. Schneider, M., Richter, W., Keding, R., Rüssel, C., XPS investigations on coordination and valency of Ti in fresnoite glasses and glass ceramics. *J. Non-Cryst. Solids*, 226, 273, 2013.
61. Jung, S.K., Cho, W.-S., Lee, C.-H., Structure and photoluminescence of $Ba_2TiSi_2O_8$ glass ceramics prepared by aerodynamic levitation. *J. Ceram. Process. Res.*, 16, 411, 2015.
62. Zhu, J., Zhang, Y., Song, X., Zhang, Q., Yang, D., Chen., Y., Dependence of crystallization behavior on particle size in barium strontium titanate glass-ceramics. *J. Am. Ceram. Soc.*, 98, 2340, 2015.
63. Halliyal, A., Bhalla, A.S., Markgraf, S.A., Cross, L.E., Newnham, R.E., Dielectric, piezoelectric and pyroelectric properties of $Sr_2TiSi_2O_8$ polar glass-ceramic: a new polar material. *J. Mater. Sci.*, 20, 3745, 1985.
64. Höche, T., Hergt, R., Keding, R., Rüssel, C., Microstructural characterization of grain-oriented glass-ceramics in the system $Ba_2TiSi_2O_8$ - $SiO_2$. *J. Mater. Sci.*, 34, 195, 1999.
65. Mayerhöfer, T.G., Dunken, H.H., Keding, R., Rüssel, C., Determination of the crystallographic orientation of oriented fresnoite glass ceramics by infrared reflectance spectroscopy. *Phys. Chem. Glasses*, 42, 353, 2001.
66. Wisniewski, W., Nagel, M., Rüssel, C., Macroscopic glass-permeated single-crystals of fresnoite. *CrystEngComm.*, 17, 5019, 2015.
67. Yamauchi, H., White, R.J., Ayukawa, M., Murray, T.C., Robinson, J.W., The structure of thin films sputter deposited from a $Ba_2Si_2TiO_8$ ceramic target. *J. Mater. Res.*, 3, 105, 1988.

68. Zhu, M.K., Yang, Y.H., Li, X.H., Wang, B., Wang, H., Yan, H., Preparation of $Ba_2Si_2TiO_8$ thin films by magnetron sputtering. *Microelectron. Eng.*, 66, 745, 2003.
69. Müller, A., Lorenz, M., Brachwitz, K., Lenzner, J., Mittwoch, K., Skorupa, W. *et al.*, Fresnoite thin films grown by pulsed laser deposition: photoluminescence and laser crystallization. *CrystEngComm.*, 13, 6377, 2011.
70. Lorenz, M., de Pablos-Martin, A., Patzig, C., Stötzel, M., Brachlitz, K., Hochmuth, H. *et al.*, Highly textured fresnoite thin films synthesized in situ by pulsed laser deposition with $CO_2$ laser direct heating. *J. Phys. D: Appl. Phys.*, 47, 034013, 2014.
71. Shen, C., Wang, D., Zhang, H., Wang, J., Boughton, R.I., Fresnoite $Ba_2TiSi_2O_8$: A novel stimulated Raman scattering active crystal. *Appl. Phys. Express*, 9, 122402, 2016.
72. Takahashi, Y., Saitoh, K., Benino, Y., Fujiwara, T., Komatsu, T., Formation of $Ba_2TiGe_2O_8$ phase in $BaO–TiO_2–GeO_2$ glasses and their optical nonlinearities. *J. Non-Cryst. Solids*, 345, 412, 2016.
73. dePablos-Martín, A., Ebert, M., Patzig, C., Krause, M., Dyrba, M., Miclea, P., Höche, T., Laser welding of sapphire wafers using a thin-film fresnoite glass solder. *Microsyst. Technol.*, 21, 1035, 2015.
74. dePablos-Martín, A., Lorenz, M., Grundmann, M., Höche, T., Laser welding of fused silica glass with sapphire using a non-stoichiometric, fresnoitic $Ba_2TiSi_2O_8{\cdot}3SiO_2$ thin film as an absorber. *Opt. Laser Technol.*, 92, 85, 2017.
75. Barbar, S.K. and Roy, M., Synthesis, Structural, and Electrical Properties of Ca-Modified $Ba_2TiSi_2O_8$ Ceramics. *J. Am. Ceram. Soc.*, 94, 843, 2011.
76. Rodriguez-Carvajal, J. and , Recent Developments of in Commission on Powder Diffraction. *IUCr. Newsletter*, 26, 12, 2001.
77. Barbar, S.K. and Roy, M., Synthesis of $Ba_2TiOSi_2O_7$ & $Ba_2TiOSi_{1.8}Ge_{0.2}O_7$ ferroelectric ceramics and the structure determination by Rietveld analysis. *Ceram. Int.*, 37, 2695, 2011.
78. Barbar, S.K. and Roy, M., Structural and vibrational studies of Ca- doped $Ba_2TiSi_2O_8$ ceramics. *J. Mol. Str.*, 1024, 132, 2012.
79. Newnham, R.E., Wolfe, E.W., Dorrian, J.F., Structural basis of ferroelectricity in the bismuth titanate family. *Mater. Res. Bull.*, 6, 1029, 1971.
80. Rama Sastry, P.S., Kumar, G.S., Bhimashankaram, T., Prasad, G., Pyroelectricity in calcium substituted sodium bismuth titanate layer structured ferroelectric ceramics. *Bull. Mater. Sci.*, 22, 59, 1999.
81. Barbar, S.K. and Roy, M., Structural, electrical and thermal properties in Ca-doped fresnoite ceramics. *J. Therm. Anal. Calorim.*, 111, 341, 2013.

10

# Research Progress and Developments in GTAW Process Using Visual Sensing and Weld Penetration Estimation

M. Soundarrajan[1]*, R. Prakash[2], D. Deepa[3], R. Thanigaivelan[4] and P. Suresh[1]

[1]*Department of Mechanical Engineering, Muthayammal Engineering College, Rasipuram, Namakkal (Dt), Tamil Nadu, India*
[2]*Department of Mechanical Engineering, AVS College of Technology, Salem, Tamil Nadu, India*
[3]*Department of Mechanical Engineering, Sri Shakthi Institute of Engineering & Technology, Coimbatore, Tamil Nadu, India*
[4]*Department of Mechanical Engineering, AKT Memorial College of Engineering and Technology, A.K.T Nagar, Neelamangalam, Kallakurichi (Dt), Tamil Nadu, India*

***Abstract***
The study of weld pool geometry is crucial as it elucidates the weld strength and mechanical properties such as surface roughness and characteristics of the weldment in the gas tungsten arc welding (GTAW) process. The output parameters such as weld penetration, structure of solidified weld, and gas porosity are essential to be visualized. The measurement of the geometry of the weld pool in real time plays a vital role in minimizing the distortion caused in the region of weldment. Therefore, different vision-based sensing technologies, algorithms, and artificial intelligence technologies are employed to visualize the convexity, top, and back width of the weld pool on the region of the weldment. This chapter explores various technologies implied to analyze the characteristics of the weld pool based on various visual sensing technologies. Also, this chapter quantifies the research of the latest implements in advanced GTAW and its monitoring techniques to academicians and industrialists.

***Keywords:*** GTAW, artificial intelligence, weld pool geometry, sensors

**Corresponding author*: soundarrajan05@gmail.com

R. Thanigaivelan, Pradeep Kumar Krishnan, Kamalakanta Muduli and Santosh Kumar Tamang (eds.) New Materials, Processing and Manufacturability: Fabrication and Processing of Advanced Materials, (187–206) © 2024 Scrivener Publishing LLC

## 10.1 Introduction

The GTAW method employs inert gasses as shielding gas to avoid the oxidation of the molten weld which generates porosity in the weldment. Because of moisture in the air, weldment gets oxidized and diminishes the strength of the weldment. Also, shielding gasses limits the unwanted reactions over the weldment [1]. The gasses such as argon (Ar), helium (He), and carbon dioxide ($CO_2$) are the most widely utilized in the GTAW process due to which are nonreactive throughout the process. Deep groove weld and flat position weld are easily done with the help of pure Ar gas which is used in welding aluminum and other nonferrous metals [2]. GTAW is also called tungsten inert gas (TIG) welding [3]. The pure Ar gasses are used in the welding process for shielding purposes which causes loss of nitrogen and the pictorial representation of the GTAW process is presented in Figure 10.1.

## 10.2 GTAW Process

Manually done welding process lags in weldment eminence due to uncontrolled supervision. The weld quality counts on the performance of the skilled labor. The excellence of welding depends on the high degree of control which decides the mechanical and microstructural characteristics of the weldment. The welding quality improves if the feed rate of welding is uniform. The extended heat cycle during the repositioning of the skilled labor causes uneven welding. Therefore, the corrosion properties of the weldment are affected and become more susceptible to the corrosion attack. Hence, an automated welding process may be adopted which ensures greater stability of the welds [4].

Commonly, welding is considered a very complex process as it involves multiple input parameters to obtain a single output. It is clearly understood that simply changing one of the input variables will not affect the result of the process. Changing the multiple inputs might bring a change in the output process. Welding is commonly intended as a nonlinear, progressive process. The results vary due to the discrete disturbances or continuous disturbances caused during the welding process. Discrete disturbances include filler changing time, deviation in the output due to oxidation, etc. This kind of discrete disturbance is caused due to the lack of skill of the labor. It can be avoided easily by regular practice of welding. However, continuous disturbances are caused due to variations in the input parameters

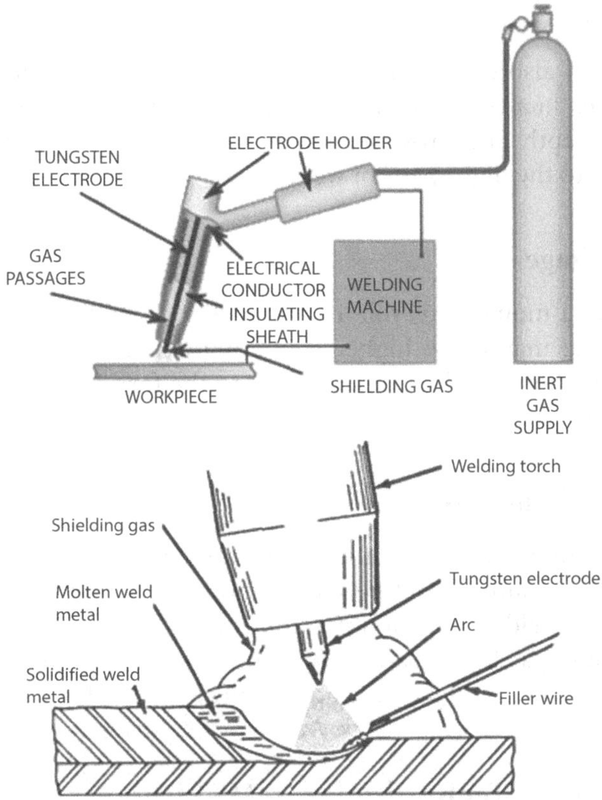


**Figure 10.1** Schematic sketch of the GTAW process [4].

which lead to excessive delay in the welding process [5]. There are different types of input variables such as voltage, current and polarity, welding pace, flow rate, type of shielding gas, Arc pulsing, and Arc length. The output parameters are properties of weldment and HAZ (heat-affected zone). These input and output parameters are grouped into two categories (controllable and fixed parameters). Controllable parameters include welding current, gas flow rate, welding voltage, and welding speed. Fixed parameters include welding gun size and electrode angle.

## 10.2.1 Current

Welding current, a predominant variable, governs the quality of weld. If the welding current is too low, the filler material sticks to the parent metal's surface on welding. If the welding current is very high, it might cause a spatter and damage the workpiece.

### 10.2.2 Speed

Welding speed also determines the quality of the weld. Welding at high speeds creates increased spattering and undercutting in the base metal. Penetration depth and cross-sectional area of the solidified weld also decrease due to the high speed of welding.

### 10.2.3 Voltage

Voltage plays a momentous role in the quality control of the weldment. Higher voltage may cause higher sputtering and causes an increase in HAZ. In the other case, low voltage leads to the formation of porosity in the weld area.

### 10.2.4 Shielding Gas

The most commonly used shielding gas is argon in the GTAW process as it is heavier than helium. Therefore, a lower flow rate is enough for making a shield over the weldment region if argon is used as the shielding gas over helium. Helium gas is comparatively lighter and escapes the weld zone if the flow rate is lower and its illustration is presented in Figure 10.2 [6].

## 10.3 Sensing Technologies in Welding

Nowadays, preprogrammed parameters are used in the GTAW process as highly engineered designs are coming into practice. Also, this automated process has cons which lack to reimburse the turbulence in the welding process. Henceforth, many sensing technologies are obligatory to ensure the quality of weld produced due to the automated welding process. Weld infiltration is the foremost factor that affects the weld quality. Therefore, various sensing techniques are employed in the automated process to

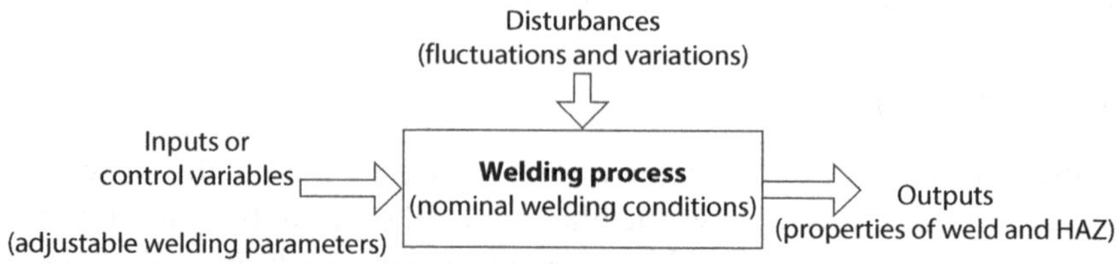


**Figure 10.2** Welding process parameters [6].

ensure high-quality weld penetration. One such technique is using an infrared sensor. This kind of sensing technique is worthwhile when monitoring is needed in inaccessible areas such as ship hulls, pipelines, and other inaccessible sites.

### 10.3.1 Sensor-Based Technology

In these kinds of sensing techniques, cameras and custom-built interfaces are used to send the signals of the temperature distribution to the computer to process the feedback on the weld quality. Using the sensing techniques, temperature distributions are calculated for plate thickness variation [7, 8].

### 10.3.2 Acoustic Emission Technology

Investigating the hot and cold cracking that forms during and after the welding process, which influences the weldment characteristics, is also essential. Another major factor is cracking which plays a main role in devastating the weld quality in a welding process. It is divided into two categories: hot cracking and cold cracking (cracking that formed after solidification) (cracking caused during the welding process). Because of the intense heat produced in the welding procedure, both physical and chemical characteristics of the HAZ change significantly. Cracking significantly reduces the fracture toughness. The main parts of the AE system are a transducer, preamplifier, attenuator, and cathode ray oscilloscope.

**Transducer**
The transducer is a device that creates the signals to analyze the propagation of cracks in the weld bead.

**Preamplifier**
This is a device used for amplification of the input signals before sending it to the cathode ray oscilloscope.

**Cathode Ray Oscilloscope**
This device is used to convert the input frequency or signals to a graphical representation which indicates the change of frequency with respect to time.

**Attenuator**
The attenuator is a device that reduces the amplified signals to a range of measurements in order to avoid damaging the component.

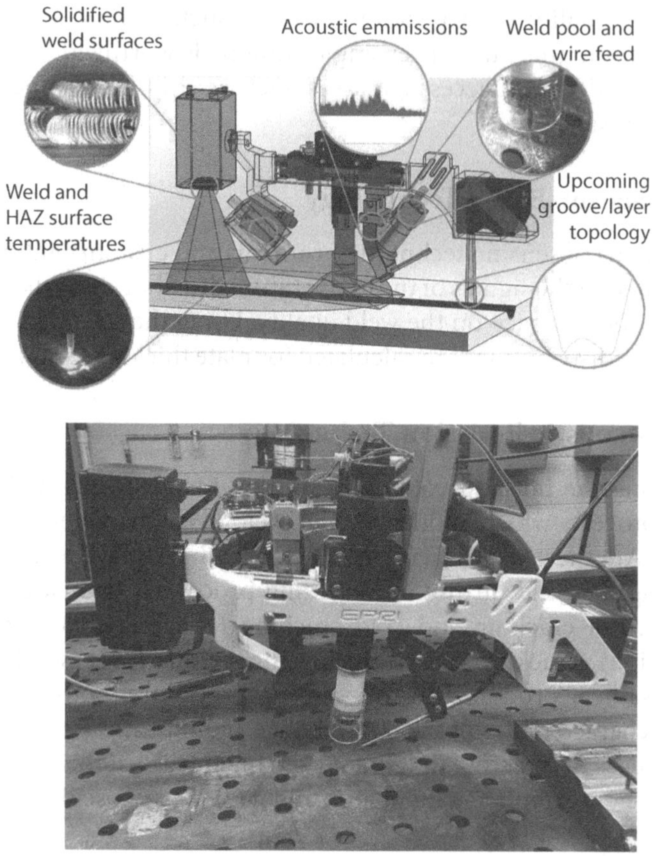


**Figure 10.3** Experimental setup: Sensor-based welding [9].

The method experimental setup for the sensor-based welding is shown in Figure 10.3.

The signals received from the transducer indicate a variation in the frequency for the micro prorogation of cracks and major cracks in the weld bead. This enables us to change the input parameters accordingly to get the desired quality of output [9].

### 10.3.3 Several Sensor Information Fusion Techniques

Several sensor fusion technologies utilize multiple sensors (sound sensor and weld arc sensor) to acquire the variations in the signals to investigate the arc length and sound information, respectively. The fusion technology

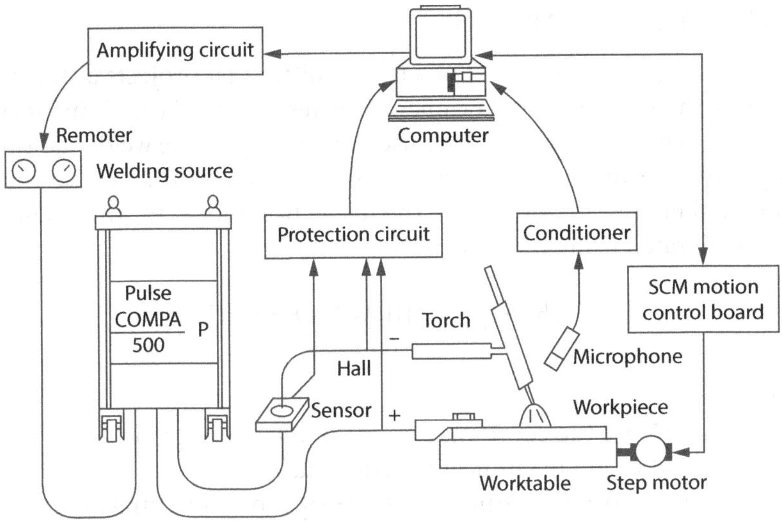


**Figure 10.4** Multisensor welding technology [10].

provides a better way to collect the information in a combined manner and the information is weighted using the weighted mean method. Therefore, the multisensor technique gives noteworthy information about the weld status because of the use of various sensors. Due to several uncertainties in the welding process, the process also has some disadvantages such as the magnetization affects the visual sensor from the weld pool being recorded properly. The noise circumstances can affect the quality of sound recorded by the sound sensors [10]. Figure 10.4 depicts the experimental setup of the multisensor technology.

## 10.4 Online Vision Inspection

This online vision inspection system deals with the usage of an illumination-based image visualization system to acquire the weld pool parameters and weld imperfection from a multilayered welding process. The laser profiles are analyzed using appropriate algorithms. Nowadays, automated robot welding processes are often used to increase the productivity rate and weld bead quality. Furthermore, the usage of seam tracking enhances the monitoring of the welding process more efficiently than the traditionally used advanced sensor-based automated welding process which simply works on preprogrammed information.

### 10.4.1 Inspection Sensors

The inspection sensors consist of a laser emitter that projects a structured light over the weld bed and a camera that records the laser stripe formed by the laser emitter. These data are used to investigate the weld quality. The filling profile of the weld groove is captured using a camera. The camera is mounted behind the weld gun. The inspection model for this technique can be calibrated using the equation

$$[ab1] = [x1x21][Ac/CcBc/Cc1]$$

where

(i) a and b are the coordinate points of the image plane,
(ii) x1 and x2 are magnificent coefficients, and
(iii) $\alpha$ and $\beta$ are the camera principle point coordinates.

### 10.4.2 Vision Inspection Contents

The vision sensors acquire the dimensions of the variables such as bead thickness, penetration distance downward, and re-enforcement elevation. From these dimensions, the quality of the weld can be determined and its demonstrations are presented in Figure 10.5. Furthermore, the data give information about the displacement of the weldment, undercut, and misalignment caused, if any.

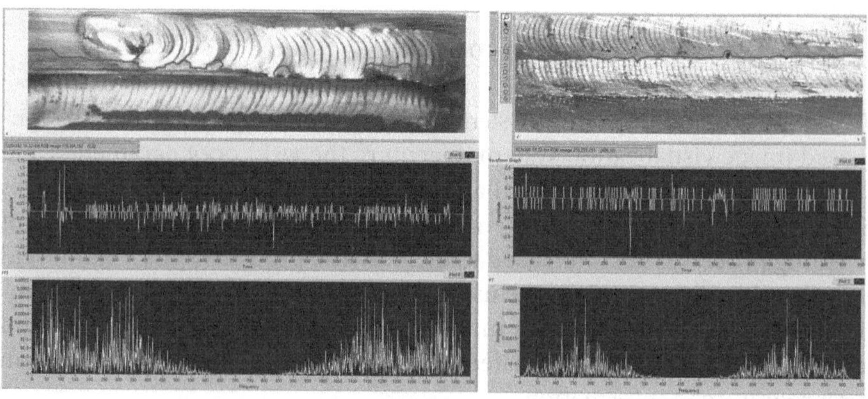

**Figure 10.5** The monitoring of the weld bead using a vision system [11].

### 10.4.3 Camera Calibration Techniques

As visual inspection techniques are used in automated welding processes to enhance the quality of the weld, it is crucial to calibrate the vision inspection system in order to avoid errors. The camera to be calibrated is made to observe a plain planar surface at different orientations and printed using a laser printer. The tests are conducted both manually and using a computer simulation. This particular calibration technique is very easy for a person to calibrate on his/her own. These calibrations are done using geometrical interpretation and closed-form solution. Figure 10.6 below shows the pictorial representation of the camera calibration using different orientations of the planar surface [12].

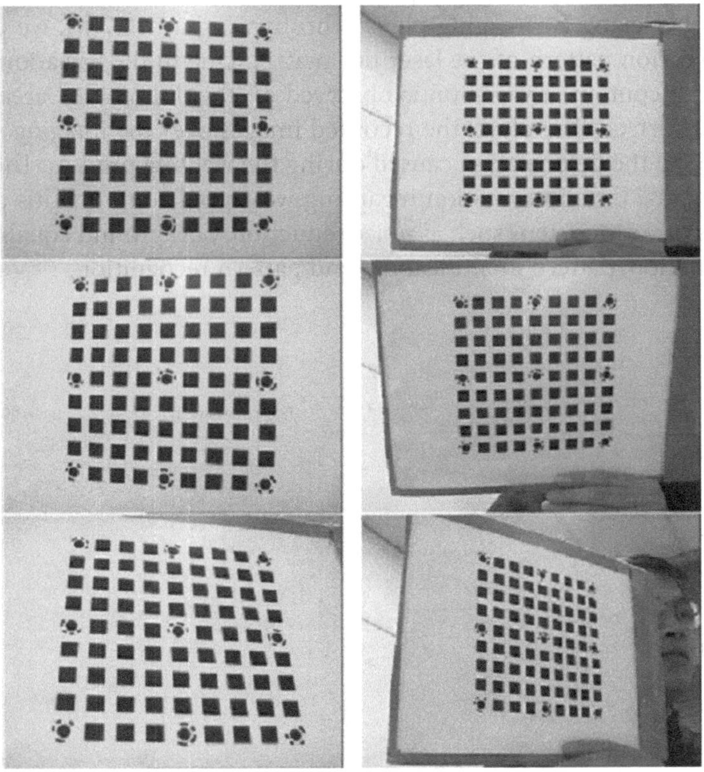

**Figure 10.6** Camera calibration using different orientations [12].

## 10.5 3 Dimensional Image Processing of Weld Pool Geometry Using Dot Matrix

In the trending automated GTAW process, three-dimensional sensing plays a vital role in observing the actual time weld pool geometry during the course. There are many methods proposed for the monitoring of the weld pool geometry using various sensors which were discussed earlier in the above topics. Furthermore, techniques are employed in the monitoring in which, image processing through laser dot matrix is effective in observing good results in processing the images of a three-dimensional contour of the weld pool. At first, a dot matrix is illuminated directly on the area of the weld pool. The reflected image is then made to an incident on a mirror. The camera captures the reflected image and the image is sent for preprocessing. The reflected picture is then distinguished from the noise disturbances.

Figure 10.7(a) shows the test setup for the GTAW method using a laser dot matrix to record the weld pool profile in actual time. Later on, it is intercepted using an imaging plane. Through Figure 10.7(b), we can see the projection pattern of the laser dot matrix. Due to the variation in the weld pool contour, a variation is observed on the illuminated area of the concave surface. Therefore, the recorded images over the imaging surface vary due to the deformation caused during the welding process. This gives us the actual time information regarding weld pool contour. This process consists of various steps such as noise reduction, background equalization, segmentation, pattern identification, and pattern recognition.

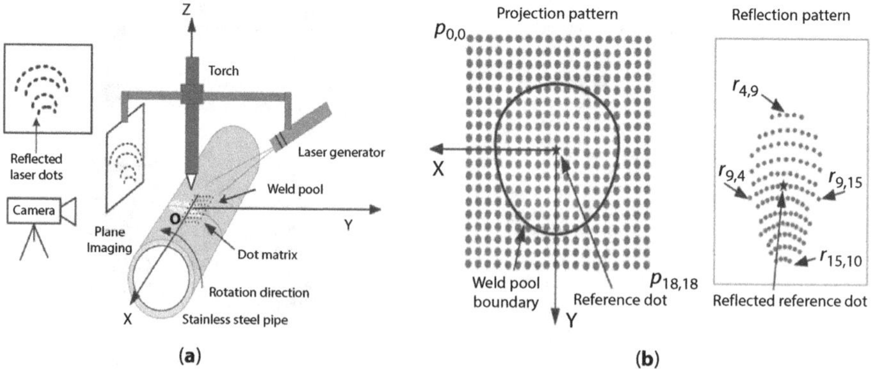


**Figure 10.7** (a) Test setup for the GTAW. (b) Projection pattern of the laser dot matrix [12].

### 10.5.1 Image Processing

Image processing consists of three different operations:

(i) Noise reduction(preprocessing)
(ii) Background equalization
(iii) Segmentation

**Noise Reduction**

Noise reduction involves the removal of unwanted noise caused during the welding process. Noise during the welding process creates fake dots which creates confusion while processing the image. In order to get rid of these fake dots, the wavelet soft thresholding method is employed.

**Background Equalization**

Background equalization is a procedure that is done to optimize the imaging patterns for effective segmentation. Therefore, a traditional histogram equalization operation is adopted to enhance the imaging pattern for more visibility.

**Segmentation**

Simple binary segmentation is an effective procedure and it is the final operation of image processing. The reflected dots have to be recovered effectively and have to be distinguished from the fake dots. So, a type of algorithm called "robust algorithm" is used to identify the fake dots from the original dots reflected from the weldment. The images are then processed and evaluated based on the loss of reflected dots vs. the gain of fake dots [13].

## 10.6 Real-Time Three-Dimensional Measurement of Topside and Backside Width of Weldment

As discussed earlier, weldment quality depends on the penetration of the weld. Therefore, it is crucial to monitor the penetration width during the process of welding. However, it is nearly impossible to track the backside width of the weld pool in real time. Therefore, the width of the backside weld pool is calculated in accordance with the data acquired from the topside weld pool width. A skilled laborer easily calculates the geometry of the backside weld pool by just looking at the topside geometry. The experimental setup for this process consists of an SCM (single-chip microcontroller) for controlling the speed of the welding and it has a camera for recording the geometry of the solidified weld pool in real time. The Single chip microcontroller setup and its flow of work nature shown in Figure 10.8. Filler metals used in the GTAW create two types of weld shapes: concave and convex shapes. The double-sided weld pool control is a difficult process to achieve and hence a 2D figure processing algorithm is adopted in this process for getting the plane profile variables of the solidified weld pool. A special algorithm called SFS (shape from shading) is adopted to interpret the surface height of the weld pool. The schematic representation shows the experimental setup of the process. The experimental setup shown in Figure 10.8 for this process consists of an SCM.

The visual analysis of the welding is done with and without filler metal and the shape of the weldment is analyzed. Weldment acquires a convex shape while using filler wire and a concave shape without filler wire which is represented in Figures 10.9 (a) & (b).

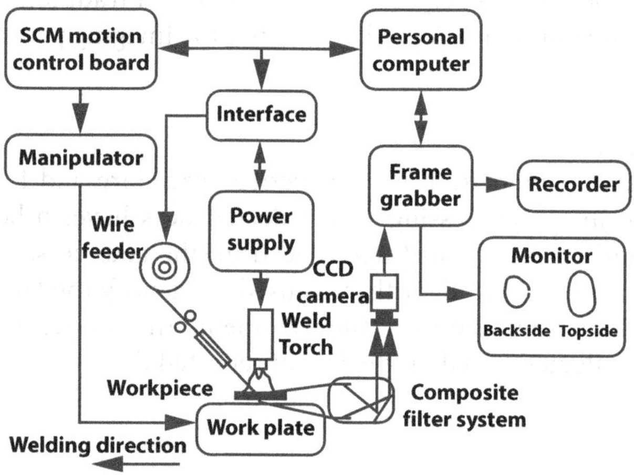


**Figure 10.8** Single chip microcontroller [14].

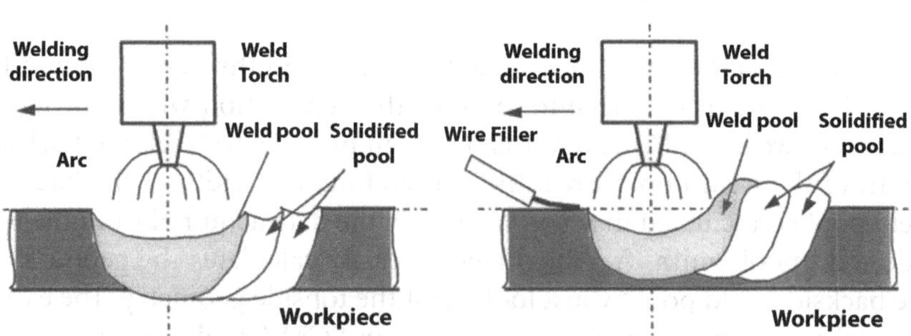


**Figure 10.9** Weldment: (a) Concave and (b) convex [14].

The weld pool images were taken for different imaging currents and were photographed. The results were compared with different imaging currents (30, 40, 50, and 60 A) in which the edge extraction was difficult for the imaging current at 60 A. This means that the edge extraction of weldment is inversely proportional to the imaging current's magnitude. That is, lowering the imaging current results in higher-quality edge extraction. The solidified metal, structure of the weld pool, and upturned images of the arc center are easily distinguished for the imaging current of 30A.

Figure 10.10 depicts the photographed images of the weld pool under various imaging currents. Here, Figure 10.10(e) is formed under the imaging current of 60 A and the image clarity is not good compared with Figure 10.10(d) which is taken at 30 A. The image processing has various steps involved such as converting the 2D figure processing into 3D figures. The lengths (L), width (W), and half-span ratio (H) are output parameters measured in the experiment. The vision techniques are classified into three different types contingent on the elements (CCD, light supply, and object). A neural network model links the welding parameters with 2D form variables, topside, and backside width. The results are used to construct a double-sided shape control method. The whole control procedure only takes a few moments. It is difficult to analyze the width of the backside of the weld pool and topside width extraction is a time-consuming process. Therefore, various methods are used to convert the two-dimensional images to three-dimensional images such as a generalized reflectance

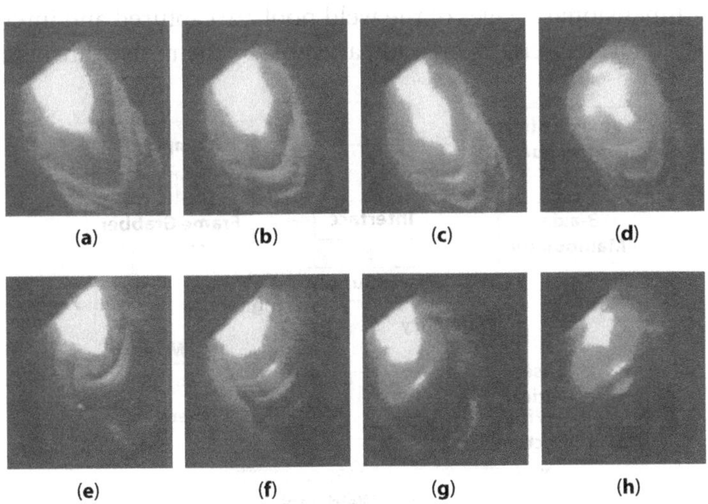


**Figure 10.10** Photographs of weld pools for different currents 30 A, 40 A, 50 A & 60 A (a-d) with edge extrusion; (e-h) without edge extrusion.

map model, shape from shading method, and surface reconstruction with the help of a synthetic figure. Surface heights of the weldment are reconstructed and converted into three-dimensional surfaces [14].

## 10.7 Dynamic Estimation of Weld Pool Geometry

As discussed in the earlier topics, the welding parameters will significantly affect the congruous appearance of the weld face. It not only affects its contour but also its microstructural properties. The penetration of the weld determines the strength of the weldment. Weld infiltration is of two types, fractional and full penetration. In this topic, an online nonlinear vibrant assessment is used for observing the weld infiltration. To achieve accuracy of the interpretation, the geometry of the neighboring weld face is also taken. Here, the method uses a pulse laser illuminator and an ultra-high shutter speed imaging system. The topside surface of the molten weld is photographed in 80 ms. Welding current is sustained between 10 and 200 A. The laser pulse is harmonized with the camera to enhance power during irradiation. The laser is irradiated in 3 ns. The welding torch and camera are mounted on a multiaxial manipulator. The image quality produced by the three-axial manipulator is higher than the previously used coaxial system. The camera is fixed at a 45° angle. The video signals are digitized by a device called frame grabber which converts the signals into 8-bit digital image matrices. Figure 10.11 shows the experimental setup of the dynamic estimation method.

Two-dimensional images of the weld pool are captured and interpreted. The topside surface, length, breadth, and area by themselves are insufficient

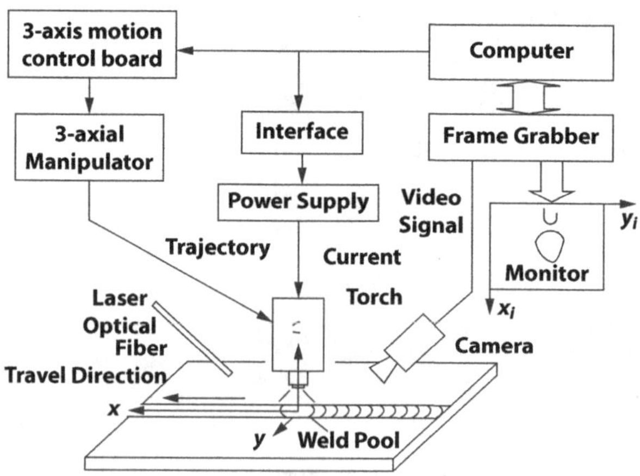


**Figure 10.11** Three-axial dynamic estimation of weld pool geometry [15].

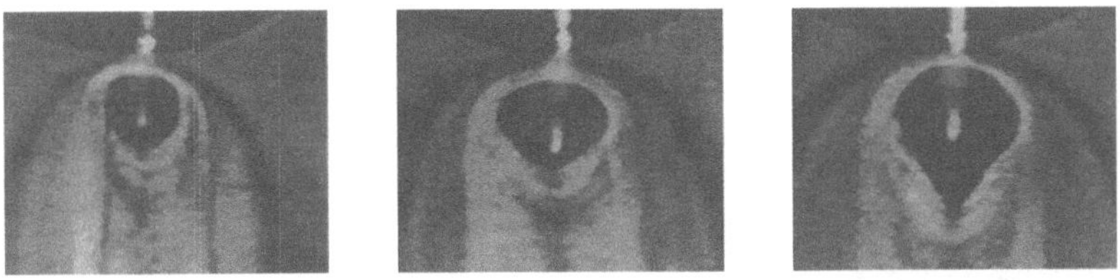


**Figure 10.12** Weld pool images [15].

to characterize the weld pool's features. Therefore, there is a need for the characteristic parameters which is used for the analysis of the geometry of the weld pool. Even though characteristic parameters play a crucial role in the analysis of the weld pool geometry, a moderate amount of parameters is sufficient rather than choosing too many parameters which will complicate the output control. Weld pool pictures are interpreted and illustrated in Figure 10.12 [15].

## 10.8 Visual Sensing Based on Supervised Machine Learning Technique

It is decisive to measure the top width as well as the back width of weld pool geometry for measuring its penetration depth. This method uses a supervised machine learning technique to capture the images of the top and bottom side weld pool images. CCD cameras and mirrors are fixed in order to capture the images of the topside and bottom side images of the weld pool. A mirror M1 is placed at 45° of tilt angle. Another mirror M2 is kept directly beneath the workpiece to capture the image of the bottom side weld pool. As the output parameters of the welding are nonlinear, the measurement of the penetration is not always 100% accurate. This method employs a nonlinear mapping to infinitely obtain the approximate objective mapping. Figure 10.13 shows the experimental setup of the pulsed GTAW process. The visual information of the penetration is extracted and illustrated in figure 3.8.2. It is well understood that penetration depth varies according to the length, width, and area of the topside weld pool. It shows three different visual extractions: partial, full penetration, and lack of fusion. Figure 10.14 shows different topside weld pool geometries taken at different time intervals such as 3, 6, 12, 15, 25, 31, 42, and 48s. The image also shows a lack of penetration, partial penetration, and full penetration from 0 to 65 s [16].

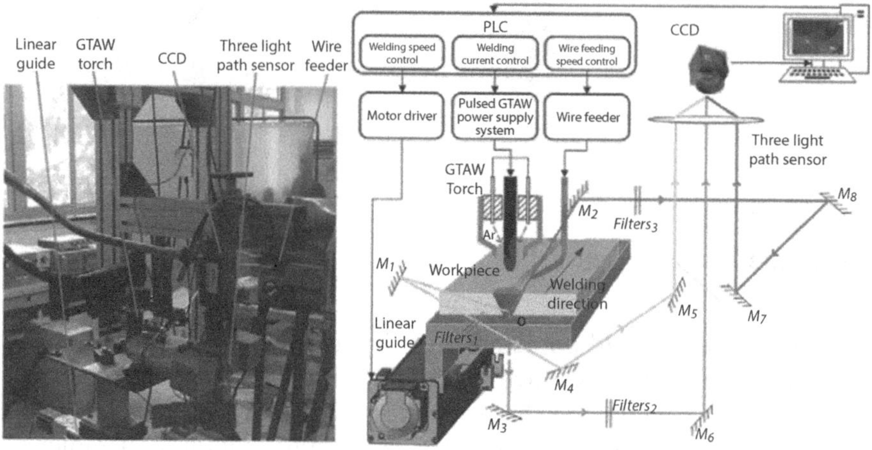


**Figure 10.13** Visual monitoring and the pulsed GTAW experimental setup [16].

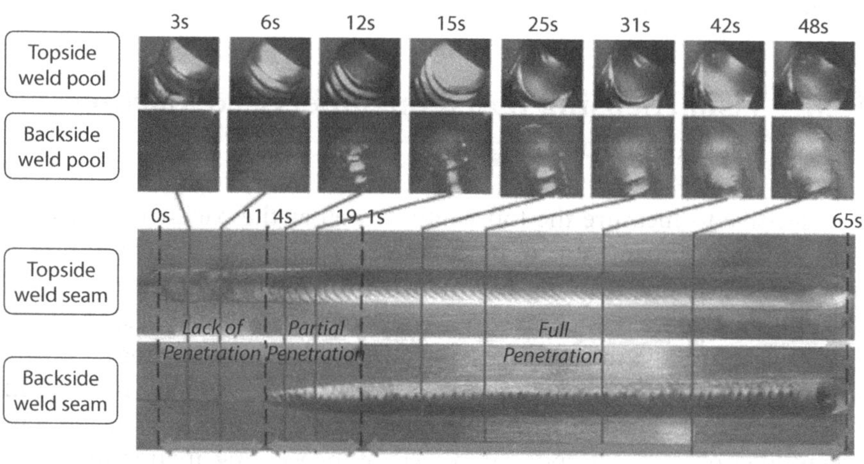


**Figure 10.14** Visual extraction of weld penetration of top and bottom weld pool [16].

## 10.9 Hybrid Network Model Using Convolutional Neural Network and Long Short-Term Memory

In the previous research studies, only single or not more than two sensors were used, whereas the multisensor data from the sound, light, electricity, and other sources may be simultaneously acquired and fused using MDF technology. It may execute more thorough and trustworthy welding eminence monitoring by utilizing the complementarity of various

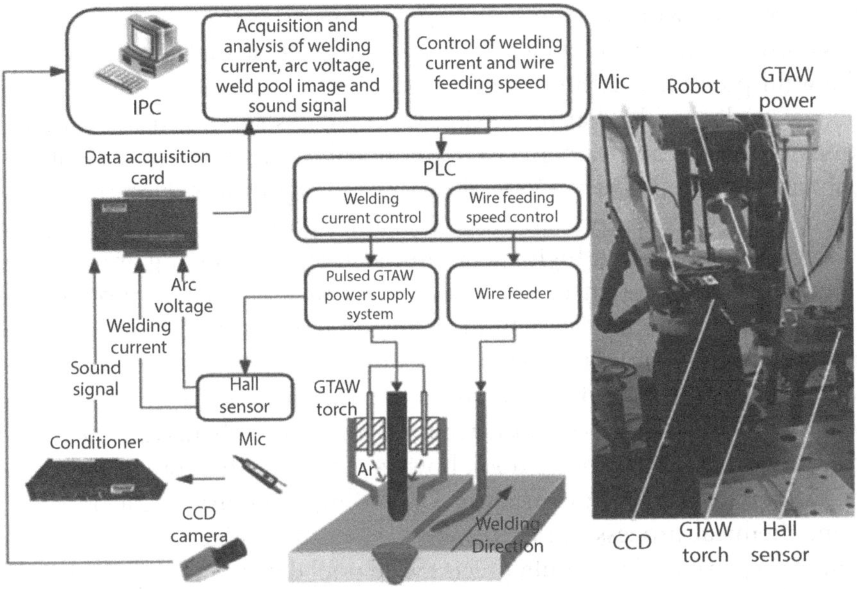


**Figure 10.15** Experimental setup of multisensory GTAW with visual monitoring [17].

sensor information. MDF technology is used to achieve instantaneous welding quality monitoring. The GTAW process, moreover, has a very hysterical tendency. Since it is too late to take action once welding problems have been discovered, the hysterical character of the population is the true issue that needs to be addressed for welding quality monitoring. Figure 10.15 shows the experimental multisensory GTAW with visual monitoring. Convolutional neural network and long-short term memory is adopted in this method. The prediction of the welding quality of the aluminum alloy using the pulsed GTAW process can be fused with the multisensor information of the dynamic welding process. The different welding states such as penetration, lack of fusion, and sag depression can be predicted using this hybrid technique [17].

## 10.10 Conclusion

Various kinds of three-dimensional sensing technologies are discussed in the above topics for the interpretation of the geometry of solidified weld in actual time. It is crucial that the actual time sensing of the welding has to be done to examine and maintain the eminence of the weldment in accordance with its penetration properties. Therefore, different kinds of vision sensing technologies were adopted to get actual time pictures of the weld pool for examining and controlling the geometry, and penetration of molten weld. Various vision technologies such as using different kinds of sensors and cameras were discussed in this chapter. Outputs are enhanced using different kinds of processing algorithms. Desirable sensing outcomes depend on efficient processing algorithms. Weld pool reconstruction methods and high-speed precise image processing are required to give actual time control over the welding phenomena. To monitor the welding technique in a variety of demanding application situations that are usual for welding, it is necessary to increase/assure the sensors' resilience in addition to their speed and accuracy. The aforementioned efforts will provide real-time, low-cost, accurate, and comprehensive penetration control. The GTAW process does, however, come with a lot of unknowns because of a variety of disturbances, including metallurgy, heat transmission, chemical processes, plasma physics, electromagnetic, deformation, and seam gap size. As a result, precise weld pool detection can occasionally be challenging. Additionally, the re-establishment of the parameter model is necessary for relevant tests and investigation when the welding substrate and associated circumstances change. Each of them creates a barrier to the creation of commercial products. Therefore, additional research about weld pool operation, advancements in sensing technologies, and the development of robust models are needed.

## References

1. Wang, X., Three-dimensional vision-based sensing of GTAW: A review. *Int. J. Adv. Manuf. Technol.*, 72, 333–345, 2014.
2. Rathod, D.W., Comprehensive analysis of gas tungsten arc welding technique for Ni-base weld overlay, in: *Advanced Welding and Deforming*, p. 105, 2021.
3. Stewart, M., *Surface Production Operations: Volume 5: Pressure Vessels, Heat Exchangers, and Aboveground Storage Tanks: Design, Construction, Inspection, and Testing*, Gulf Professional Publishing, USA, 2021.
4. Gunn, R.N., Wood head Publishing Series in Metals and Surface Engineering, in: *Duplex Stainless Steels*, pp. 133–145, Woodhead Publishing, United Kingdom, 1997.
5. Zhang, Y. (Ed.), *Real-time weld process monitoring*, CRC Press, Elsevier, USA, 2008.
6. Abebe, D.G. and Bogale, T.M., Optimization of process parameters of tig welding on 304 stainless steel using taguchi method and fuzzy logic approach. *Eng. Res. Express*, 5, 045045, 2023.

7. Xu, Y., Lv, N., Fang, G., Lin, T., Chen, H., Chen, S., Han, Y., Sensing technology for intelligentized robotic welding in arc welding processes, in: *International Conference on Robotic Welding, Intelligence and Automation*, Springer, Cham, pp. 411–423, 2014, October.
8. Banerjee, P., Govardhan, S., Wikle, H.C., Liu, J.Y., and Chin, B.A., Infrared sensing for on-line weld geometry monitoring and control. *ASME. J. Eng. Ind*, 117, 3, 323–330, 1995.
9. Fang, C.K., Kannatey-Asibu, E., Barber, J., Acoustic emission investigation of cold cracking in Gas Metal Arc Welding of AISI 4340 steel. *Weld. J. -Including Weld. Res. Suppl.*, 74, 6, 177–184, 1995.
10. Chen B., Wang J. and Chen S. A study on application of multi-sensor information fusion in pulsed GTAW. *Industrial Robot*, 37, 2, 168–176, 2010
11. Li, Y., Li, Y.F., Wang, Q.L., Xu, D., Tan, M., Measurement and defect detection of the weld bead based on online vision inspection. *IEEE Trans. Instrum. Meas.*, 59, 7, 1841–1849, 2009.
12. Zhang, Z., A flexible new technique for camera calibration. *IEEE Trans. Pattern Anal. Mach. Intell.*, 22, 11, 1330–1334, 2000.
13. Zhang, W.J., Zhang, X., Zhang, Y.M., Robust pattern recognition for measurement of three dimensional weld pool surface in GTAW. *J. Intell. Manuf.*, 26, 4, 659–676, 2015.
14. Zhao, D.B., Yi, J.Q., Chen, S.B., Wu, L., Chen, Q., Extraction of three-dimensional parameters for weld pool surface in pulsed GTAW with wire filler. *J. Manuf. Sci. Eng.*, 125, 3, 493–503, 2003.
15. Zhang, Y.M., Li, L., Kovacevic, R., Dynamic estimation of full penetration using geometry of adjacent weld pools. ASME. *J. Manuf. Sci. Eng.*, 119 4A, 631–643, 1997.
16. Zhang, Y. M., Li, L., and Kovacevic, R. Dynamic estimation of full penetration using geometry of adjacent weld pools. *ASME. J. Manuf. Sci. Eng.* 119(4A), 631–643, 1997.
17. Chen, C., Xiao, R., Chen, H., Lv, N., Chen, S., Prediction of welding quality characteristics during pulsed GTAW process of aluminum alloy by multisensory fusion and hybrid network model. *J. Manuf. Processes*, 68, 209–224, 2021.

# 11

# Simulation of Sulfur Recovery Unit Using Aspen Plus

**Krati Hardya[1,2*], V. K. Chandaliya[3], Pratik Swarup Dash[3], Barun Kumar Nandi[1], Deepak Kumar[4], Bishub Choudhury[2] and Vishnu Kuntal[2]**

[1]*Department of Fuel and Mineral Engineering, Indian Institute of Technology (ISM), Dhanbad, Jharkhand, India*
[2]*Department of Mechanical Engineering, Dronacharya Group of Institutions, Greater Noida, Uttar Pradesh, India*
[3]*Coal and Coke Making Research Group, Research & Development and Scientific Services, Tata Steel Ltd., Jamshedpur, Jharkhand, India*
[4]*Coke By-Product Plant, Tata Steel Ltd., Jamshedpur, Jharkhand, India*

## *Abstract*

Strict environmental legislation has led to close monitoring and control of industrial emissions into the environment. Sulfur being among the major environmental pollutants (mainly in the form of $SO_X$) needs restriction over its emissions from industries as per environmental norms. In the steel industry, the carbonization process produces coke and coke oven gas (COG). COG has a substantial quantity of sulfur as hydrogen sulfide ($H_2S$) besides other impurities. These impurities are valuable when recovered from the gas and also impose less threat to the environment. The Claus process recovers sulfur from gas in the form of elemental sulfur.

This work aimed to get a simulation model that can suitably characterize sulfur recovery unit (SRU). A simulation model was developed using Aspen Plus V8.8. Two case studies were taken up for validation of the model, while another was taken up to determine the impact of operating parameters on the recovery of sulfur and tail gas ratio. Feedforward control and feed-backward control were incorporated in the simulation to control primary air requirement and the proportion of the $H_2S/SO_2$ ratio in the tail gas stream.

**Corresponding author*: kratihardya@gmail.com

R. Thanigaivelan, Pradeep Kumar Krishnan, Kamalakanta Muduli and Santosh Kumar Tamang (eds.) New Materials, Processing and Manufacturability: Fabrication and Processing of Advanced Materials, (207–226) 

Simulation results were in accordance with the case studies considered and analysis by the third case study showed the requisite conversion of $H_2S$ to sulfur while maintaining the proportion of $H_2S/SO_2$ as 4 to 5 in the tail gas stream, as per plant requirement. The effect of feed stream conditions, i.e., acid gas to air molar flow ratio and concentration of $H_2S$ in an acid gas stream on sulfur recovery and tail gas ratio were also analyzed. The optimum sulfur recovery was obtained for acid gas to air molar flow ratio in the range of 0.715 to 0.720 and an $H_2S$ concentration of 16.18 kmol/hour in the acid gas stream for designed conditions of SRU at Tata Steel.

***Keywords*:** Claus process, claus kiln, claus reactor, carbonization, sulfur recovery unit, tail gas ratio

## 11.1 Introduction

Carbonization is a multistep process that produces a variety of solids, liquids, and gaseous products. Coke oven gas is an important gaseous byproduct of the carbonization process because it contains many valuable products and impurities. It contains corrosive compounds of sulfur besides other impurities [1]. The separation of these impurities is strongly required for environmental protection as well as to protect the equipment against corrosion. Due to the regulations, venting or flaring hydrogen sulfide to the surroundings is now unacceptable; therefore, it is necessary to convert it into elemental sulfur. Therefore, a sulfur recovery unit (SRU) is necessarily required for removing sulfur from COG [2]. The Claus process is an appropriate technique for extracting elemental sulfur from hydrogen sulfide.

The Claus process consists of two primary phases: a thermal phase and a catalytic phase. In the thermal phase, hydrogen sulfide is partially oxidized at elevated temperatures, resulting in the production of sulfur dioxide, water, and elemental sulfur. During the second stage, sulfur production is further augmented as unreacted hydrogen sulfide reacts with the sulfur dioxide produced, all catalyzed by materials like titanium dioxide or alumina at lower temperatures. Hydrogen sulfide being identified as one of the major environmental pollutants drew the attention of researchers toward process optimization to maximize sulfur recovery. Scholars have worked in the area of modeling, developing kinetic and mathematical models for a better understanding of the Claus process. However, most of the research work is focused on the thermal stage. Manenti *et al.* studied the waste heat boiler (WHB) and Claus thermal reaction furnace using a multi-scale modeling approach. Detailed reaction kinetics considering 2400 reactions and 140 species, including the recombination effect in WHB, were used for modeling. The process parameters were optimized to achieve maximum acid gas conversion, elemental sulfur recovery, and steam generation. The process parameters were adjusted to maximize the recovery of elemental sulfur, conversion of acid gas, and generation of steam [3, 4]. Zarei *et al.* used equilibrium and kinetic models to model the thermal section. The results were in accordance with the industrial data with a mean absolute error of 6.5% [5]. Monnery *et al.* also adopted a modeling approach to achieve maximum sulfur recovery [6]. Ghahraloud *et al.* developed kinetics for hydrogen sulfide conversion at elevated temperatures which very well correlated with experimental data [7]. Monnery *et al.* also did a comparative study of the predicted concentration of the main constituents at the reaction furnace outlet based on equilibrium and empirical models. The equilibrium model failed to provide a better estimate of sulfur recovery compared to the empirical model because the properties of the outlet stream from the reaction furnace are mostly governed by kinetic laws and not equilibrium conditions [8]. Nabgan *et al.*, on the other hand, simulated the process using Aspen Hysys software for the characterization of SRU. They studied the impact of air feed composition and molar flow on hydrogen sulfide conversion [9]. Pahlavan and Fanaei conducted a simulation of the Claus reaction furnace using a kinetic model. The results of the model accurately agreed with the results predicted by the PROMAX simulator [10]. Rezazadeh and Rezvantalab investigated the effect of the inlet gas stream effects on the Claus reaction furnace using MATLAB. An increase in the concentration of $H_2S$ in the incoming gas would lead to a higher $H_2S$ conversion rate, thus improving the overall efficiency of the furnace [11]. Asadi *et al.* also studied the effect of $H_2S$ concentration, $H_2S/CO_2$ ratio and fraction AG splitter flow in three different concentrations of oxygen on sulfur recovery using the TSWEET simulator. Maximum sulfur recovery was obtained when $H_2S$ concentration was in the range of 25 to 30 mol % with sulfur recovery above 98% in all three concentrations of oxygen [12].

Research has also been carried out to increase the sulfur recovery percentage by decreasing COS and $CS_2$ formation in the thermal stage. Karan *et al.* have carried out high-temperature studies, viz., experimental and modeling on the key reactions leading to COS formation. They found that the key reactants leading to COS formation are CO, $H_2S$, and $S_2$. They developed reaction kinetics for both COS formation and decomposition with the help of experiments done at high temperatures [13, 14]. Karan *et al.* also developed the reaction kinetics for the formation of $CS_2$ in the reaction furnace. A kinetic model was formulated to understand the formation of $CS_2$ in the $CH_4$-$S_2$ system and the $CH_4$-$H_2S$ system [15].

Sulfur recovery efficiency is also affected by $H_2S$ and COS formation in the WHB [16]. Reaction forming COS and $H_2S$ occurs at appreciable rates in the range of 900°C to 1300°C [17]. Nasato *et al.* analyzed that by rapid quenching of reaction gases, reassociation of sulfur with hydrogen, and formation of undesirable carbonyl sulfide in the boiler tubes can be prevented [18]. Karan *et al.* further improved the model proposed by Nasato *et al.* by including the radiative heat transfer effect in modeling WHB, which led to faster quenching of reactions and hence suppressing COS and $H_2S$ formation in the boiler tubes [19]. Abedini *et al.* worked on the catalytic stage of SRU. They modeled and simulated the catalytic stage of SRU using MATLAB. Modeling was done using energy balance and molar flow relations achieving sulfur recovery of 1200 kg/day. The sulfur amount collected in the catalytic bed was also measured with the help of the proposed model [20].

The present study was conducted to obtain a simulation model using Aspen Plus V8.8 that can suitably characterize Claus SRU. The simulation model was formulated on the basis of actual design data and operating process parameters collected from SRU of Tata Steel Ltd. Jamshedpur, India. The study was taken up with the motive to enhance sulfur recovery from this SRU. Similar studies were also carried out in the past at Tata Steel Ltd. to increase the recovery of sulfur. However, the studies were focused on the analysis of Claus catalyst and SRU operating conditions. Simulation approaches were also tried but a simulation model that can suitably characterize Claus SRU at Tata Steel Ltd. was not available so far. An effective simulation model was developed in this work which had been utilized to analyze the effect of feed stream parameters on sulfur recovery and tail gas ratio.

## 11.2 Process Description

The Claus process is an effective technique employed for the retrieval of elemental sulfur. It produces high-quality sulfur with appreciable conversion of hydrogen sulfide. The Claus process considered here is a straight-through type consisting of kiln and catalytic reactors. The feed stream is led to the burner located at the upper part of the kiln, where the conversion of hydrogen sulfide to elemental sulfur and sulfur dioxide takes place according to reactions 11.1 and 11.2 given below:

$$H_2S + {}^3/_2 O_2 \rightarrow SO_2 + H_2O \ldots\ldots\ldots \Delta H = -518 \text{ kJ mol}^{-1} \quad (11.1)$$

$$2H_2S + SO_2 \rightarrow {}^3/_2 S_2 + 2H_2O \ldots\ldots\ldots \Delta H = +47 \text{ kJ mol}^{-1} \quad (11.2)$$

The WHB is used to cool the product gasses for recovering heat, and then the sulfur is condensed to prevent condensation in pipelines before entering the catalytic stage. In the second stage, sulfur production is further enhanced by the reaction between unreacted $H_2S$ and produced $SO_2$. The reaction occurs over Al or $TiO_2$ based catalyst at lower temperatures (250°C to 300°C) over two series catalytic reactors, according to reaction 11.3. The produced sulfur vapors in the catalytic stage are then condensed out to obtain liquid sulfur.

$$2H_2S + SO_2 \leftrightarrow {}^3/_8 S_8 + 2H_2O \ldots\ldots\ldots \Delta H = -108 \text{ kJ mol}^{-1} \quad (11.3)$$

During the catalytic stage, primarily $S_8$ is generated through an exothermic reaction, while in the thermal stage, the main product is $S_2$, and the reaction is endothermic [21]. The complete process can be summarized by the following overall reaction:

$$3H_2S + 1.5O_2 \rightarrow {}^3/_n S_n + 3H_2O \ldots\ldots\ldots \Delta H = -626 \text{ kJ mol}^{-1} \quad (11.4)$$

## 11.3 Methodologies

### 11.3.1 Description of Simulation Flow Sheet

The simulation was performed using Aspen Plus V8.8 for modeling the SRU of Tata Steel Limited, Jamshedpur. SOLIDS property package was used in the simulation to incorporate various allotropic forms of sulfur. The process flow diagram of SRU in Aspen Plus is shown in Figure 11.1. Acid gas stream, fuel gas (COG) stream, main air stream for COG combustion, and air stream for partial acid gas combustion were provided at the inlet of SRU. Partial oxidation of $H_2S$ and ammonia decomposition occurred in the furnace (KILN), modeled as a Gibbs reactor. The high-temperature furnace outlet stream went into the WHB (WHB), where the stream was cooled down by the conversion of water to steam, before being fed into the first stage Claus reactor (CLAUS1), modeled as a plug flow reactor. The product stream from the reactor was cooled down in the condenser and went to the separator block (SEPRTR1), where the sulfur vapors formed

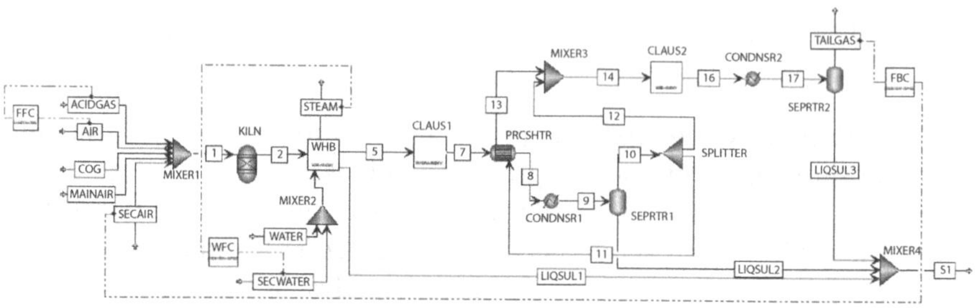


**Figure 11.1** Process flow diagram of SRU in Aspen Plus.

in the reactor were condensed out. The vapor stream from SEPRTR1 was split into two fractions, of which, one was indirectly heated in the process gas heater and the other was sent without reheating. Both the split streams were mixed and entered into the second stage Claus reactor (CLAUS2), also modeled as a plug flow reactor. Claus reaction kinetics was used to model both the plug flow reactors. The outlet stream of the reactor was cooled down and sulfur vapors were again condensed out in the separator block (SEPRTR2). Vapor stream (TAILGAS) from SEPRTR2 was recycled back to coke oven gas suction mains. In order to minimize the early corrosion of the pipeline carrying the tail gas, the $H_2S/SO_2$ ratio in the tail

**Table 11.1** Operating conditions of subunits of SRU.

| SRU subunit | Block name | Block type | Temperature (°C) | Pressure (kPa) |
|---|---|---|---|---|
| **Claus kiln** | KILN | RGibbs | 1100 | 135.0 |
| **Waste heat boiler** | WHB | HeatX | 250 ($T_{out}$) | - |
| **Claus reactor 1** | CLAUS1 | RPlug | - | - |
| **Claus reactor 2** | CLAUS2 | RPlug | - | - |
| **Process gas heater** | PRCSHTR | HeatX | 194 ($T_{out}$) | - |
| **Sulfur condenser** | CONDNSR1 | Heater | 135 | 108.4 |
| **Sulfur condenser** | CONDNSR2 | Heater | 135 | 103.0 |
| **Sulfur separator 1** | SEPRTR1 | Flash2 | 130 | 107.4 |
| **Sulfur separator 2** | SEPRTR2 | Flash2 | 130 | 102.4 |

gas stream was kept close, but not below 4:1. Sulfur was recovered in the process in the form of liquid sulfur and sent to storage tank.

Sulfur allotropes formed during the process were converted to mono-atomic sulfur for its easy condensation. Feedforward control and feed-backward control were incorporated in the simulation to control the amount of air required for acid gas combustion and to simultaneously maintain the tail gas ratio within the plant's desired limits of 4:1 in the

**Table 11.2** Inlet stream flow composition (mole fraction) and properties.

| Components | Acid gas | Air | COG | Main air |
|---|---|---|---|---|
| $H_2S$ | 0.10528 | 0.00000 | 0.00000 | 0.00000 |
| $NH_3$ | 0.33263 | 0.00000 | 0.00000 | 0.00000 |
| HCN | 0.01848 | 0.00000 | 0.00000 | 0.00000 |
| $CO_2$ | 0.18397 | 0.00000 | 0.04039 | 0.00000 |
| $H_2O$ | 0.35676 | 0.00000 | 0.00000 | 0.00000 |
| H2 | 0.00000 | 0.00000 | 0.56541 | 0.00000 |
| CO | 0.00000 | 0.00000 | 0.09087 | 0.00000 |
| $O_2$ | 0.00000 | 0.21000 | 0.00606 | 0.21000 |
| $N_2$ | 0.00000 | 0.79000 | 0.01817 | 0.79000 |
| $CH_4$ | 0.00000 | 0.00000 | 0.26251 | 0.00000 |
| $C_2H_6$ | 0.00000 | 0.00000 | 0.01660 | 0.00000 |
| Benzene | 0.00109 | 0.00000 | 0.00000 | 0.00000 |
| Toluene | 0.00093 | 0.00000 | 0.00000 | 0.00000 |
| Xylene | 0.00087 | 0.00000 | 0.00000 | 0.00000 |
| Total flow (kmol/hour) | 152.168 | 209.969 | 13.591 | 56.062 |
| Temperature (°C) | 100 | 130 | 26 | 70 |
| Pressure (kPa) | 141.325 | 138.000 | 148.264 | 140.525 |
| Vapor fraction | 1 | 1 | 1 | 1 |
| Liquid fraction | 0 | 0 | 0 | 0 |

outlet stream. A feed water controller loop was also incorporated to maintain the temperature of the steam generated in WHB in the specific range.

Operating conditions of major subunits of SRU are defined in Table 11.1 and inlet stream flow composition and properties are presented in Table 11.2.

### 11.3.2 Validation of Simulation Model

The simulation model was initially verified using operational data obtained from an SRU located in the South Pars region of Iran. Validation of kinetics used in the reactor was carried out by comparing the simulation outcomes with the actual industrial data at the outlet of the first Claus reactor. Since the Claus reaction occurs at a low-temperature range of 200°C to 320°C, the majority of the sulfur formed via the Claus reaction is in the form of $S_8$ and $S_6$ with a small amount of $S_2$. Therefore, kinetics for the reactions forming $S_6$ and $S_8$ have been required for modeling the Claus reactor. Reactions used for modeling the reactor are as presented by reactions 11.5 and 11.6.

$$2H_2S + SO_2 \leftrightarrow 1/_2 S_6 + 2H_2O \tag{11.5}$$

$$2H_2S + SO_2 \leftrightarrow 3/_8 S_8 + 2H_2O \tag{11.6}$$

The model established by Abaskuliev *et al.* for reaction kinetics was obtained with slight variations done by Zughbi and Razzak [23]. The kinetic model for Claus reaction by Abaskuliev *et al.*, is represented by Equation 11.7 [22]. Kinetics for the formation of $S_6$ and $S_8$ sulfur species, as proposed by Zughbi and Razzak, is represented by Equations 11.8 and 11.9 [23].

$$R(C,T) = 7.3919 * 10^4 e^{-\left(\frac{30594}{8.314T}\right)} \frac{T^{1.5} C_{H_2S} C_{SO_2}^{0.5}}{\left(1 + 46.56TC_{H_2O}\right)^2} \tag{11.7}$$

$$R(C,T) = 7.3919 * 10^4 e^{-\left(\frac{30594}{8.314T}\right)} \frac{T^{1.5} C_{H_2S} C_{SO_2}^{0.7}}{(1 + 46.56TC_{H_2O})^2} \tag{11.8}$$

$$R(C,T) = 7.3919 * 10^4 e^{-\left(\frac{30594}{8.314T}\right)} \frac{T^{1.6} C_{H_2S} C_{SO_2}^{0.7}}{(1 + 46.56 T C_{H_2O})^2} \tag{11.9}$$

The simulation model was further validated with the industrial data available for SRU at Tata Steel Ltd. The simulation model was used to compare the flow rates of individual components, the quantity of sulfur recovered, the tail gas ratio, and the process gas temperatures at the inlet and outlet of the SRU subunits with the corresponding values from the industrial data.

## 11.4 Results and Discussion

The accuracy of the model was validated at first before proceeding for further analysis. Reaction kinetics used in modeling the Claus reactor was based on empirical correlations, while the kiln was modeled using Gibbs free energy of minimization. Therefore, the model was first validated by comparing the simulation results at the kiln and the reactor outlet with the industrial data available from SRU in the South Pars region of Iran [24]. Table 11.3 shows the comparison between simulation data and the industrial data at the kiln and the reactor outlet [24]. However, the stream flow composition and the properties at the outlet of various SRU subunits as per the simulation model, for SRU at Tata Steel Ltd., are tabulated in Table 11.4.

The simulation results were in accordance with the industrial data, as can be inferred from Figures 11.2 and 11.3. The amount of $H_2O$ and $CO_2$ was higher than the other components in the feed stream as well as at the kiln outlet and the reactor outlet. The total sulfur molar flow differed by about 7% at the reactor outlet, and the value of the coefficient of determination ($R^2$) was more than 0.99 for both cases, representing a better fit of the simulation data. Therefore, it can be concluded that the Gibbs reactor used for modeling the reaction furnace and the Claus reaction kinetics used for modeling the Claus reactor were quite accurate.

The accuracy of the model was further validated by comparing the simulation data with the data available for SRU at Tata Steel Ltd. Comparative analysis between the plant and the simulation model with respect to the designed conditions has been tabulated in Table 11.5. The flow rate of components, amount of sulfur recovered, tail gas ratio, and temperature of the process gas at the inlet and outlet of SRU subunits obtained were compared. Good compatibility between the plant and the simulation model

**Table 11.3** Comparison between simulation and industrial data (mole fraction).

| Components | Simulation data | | Industrial data | |
|---|---|---|---|---|
| | KILN outlet | Reactor outlet | KILN outlet | Reactor outlet |
| $H_2S$ | 0.041828 | 0.004699 | 0.034518 | 0.016135 |
| $CO_2$ | 0.245534 | 0.258099 | 0.232488 | 0.266593 |
| $H_2O$ | 0.231609 | 0.282103 | 0.241455 | 0.276673 |
| $SO_2$ | 0.024048 | 0.005638 | 0.027350 | 0.008052 |
| COS | 0.003274 | 0.003441 | 0.018299 | 0.001014 |
| $CS_2$ | 5.22E-06 | 5.47E-06 | 0.001602 | 0.000179 |
| $H_2$ | 0.01119 | 0.011763 | 0.009838 | 0.010438 |
| CO | 0.019419 | 0.020414 | 0.016950 | 0.017984 |
| $O_2$ | 8.21E-13 | 8.63E-13 | 0.000000 | 0.000000 |
| $N_2$ | 0.371451 | 0.390478 | 0.372751 | 0.395491 |
| $CH_4$ | 2.90E-11 | 3.04E-11 | 0.000000 | 0.000000 |
| $C_3H_8$ | 9.11E-34 | 0.000000 | 0.000000 | 0.000000 |
| $C_2H_6$ | 9.19E-23 | 0.000000 | 0.000000 | 0.000000 |
| $S_2$ | 5.16E-02 | 0.000000 | 0.044749 | 0.000224 |
| $S_6$ | 3.47E-09 | 0.002166 | | |
| $S_8$ | 1.77E-12 | 0.005738 | 0.000000 | 0.007217 |
| | | | | |
| Total flow (kmol/hour) | 1285.98 | 1223.28 | 1280.74 | 1207.10 |
| Temperature (°C) | 1000 | 304.1 | 1000 | 307.6 |
| Pressure (kPa) | 150 | 130 | 150 | 134 |

**Table 11.4** Streams flow composition (mole fraction) and properties.

| Components | 2 | 5 | 7 | 14 | 16 | S1 | Tail gas |
|---|---|---|---|---|---|---|---|
| $H_2S$ | 0.01689 | 0.01679 | 0.00624 | 0.00644 | 0.00485 | 4.79E-05 | 0.00486 |
| $NH_3$ | 4.17E-07 | 4.15E-07 | 4.11E-07 | 4.23E-07 | 4.23E-07 | 4.18E-09 | 4.24E-07 |
| HCN | 3.12E-10 | 3.10E-10 | 3.07E-10 | 3.16E-10 | 3.16E-10 | 1.83E-11 | 3.17E-10 |
| $CO_2$ | 0.07491 | 0.07450 | 0.07372 | 0.07603 | 0.07591 | 0.00021 | 0.07612 |
| $H_2O$ | 0.32657 | 0.32477 | 0.33178 | 0.33797 | 0.33901 | 0.13413 | 0.33958 |
| $SO_2$ | 0.00723 | 0.00719 | 0.00193 | 0.00199 | 0.00120 | 4.10E-05 | 0.00120 |
| COS | 0.00035 | 0.00035 | 0.00035 | 0.00036 | 0.00036 | 4.12E-06 | 0.00036 |
| $CS_2$ | 2.07E-07 | 2.06E-07 | 2.04E-07 | 2.10E-07 | 2.09E-07 | 2.58E-08 | 2.10E-07 |
| H2 | 0.02713 | 0.02698 | 0.02670 | 0.02754 | 0.02750 | 2.04E-06 | 0.02757 |
| CO | 0.01267 | 0.01260 | 0.01247 | 0.01287 | 0.01285 | 7.59E-06 | 0.01288 |
| $O_2$ | 9.56E-12 | 9.51E-12 | 9.41E-12 | 9.71E-12 | 9.69E-12 | 6.09E-15 | 9.72E-12 |
| $N_2$ | 0.52872 | 0.52581 | 0.52036 | 0.53668 | 0.53583 | 0.00031 | 0.53729 |
| $CH_4$ | 3.26E-11 | 3.24E-11 | 3.21E-11 | 3.31E-11 | 3.30E-11 | 3.69E-14 | 3.31E-11 |
| $C_2H_6$ | 8.04E-23 | 0 | 0 | 0 | 0 | 0 | 0 |
| S | 5.14E-07 | 0.01100 | 0.02645 | 0.00013 | 0.00250 | 0.86526 | 0.00014 |
| $S_2$ | 0.00553 | 0 | 0 | 0 | 0 | 0 | 0 |
| $S_6$ | 5.63E-13 | 0 | 0 | 0 | 0 | 0 | 0 |
| $S_8$ | 1.23E-17 | 0 | 0 | 0 | 0 | 0 | 0 |
| Benzene | 0 | 0 | 0 | 0 | 0 | 0 | 0 |
| Toluene | 0 | 0 | 0 | 0 | 0 | 0 | 0 |
| Xylene | 0 | 0 | 0 | 0 | 0 | 0 | 0 |
| | | | | | | | |
| Total flow (kmol/hour) | 450.9024 | 453.3954 | 458.1488 | 444.2069 | 444.9091 | 15.14802 | 443.7029 |
| Temperature(°C) | 1100 | 250 | 266.0859 | 212.2547 | 214.723 | 130.0249 | 130 |
| Pressure (kPa) | 135 | 113.4 | 111.4 | 105.4 | 104 | 102.4 | 102.4 |
| Vapor fraction | 1 | 1 | 1 | 1 | 1 | 0.001 | 1 |
| Liquid fraction | 0 | 0 | 0 | 0 | 0 | 0.999 | 0 |

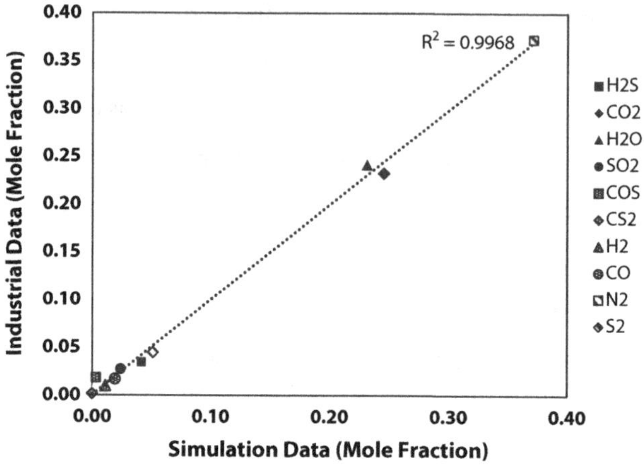


**Figure 11.2** Validation of simulation vs. industrial data at kiln outlet.

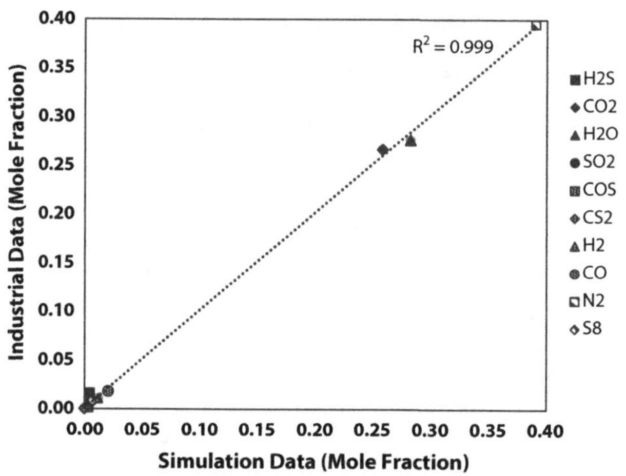


**Figure 11.3** Validation of simulation vs. industrial data at reactor 1 outlet.

was observed with a deviation of 2% in the amount of sulfur recovered. Therefore, it was concluded that the model precisely replicated the Claus process.

The simulation model developed here could suitably characterize the Claus SRU. Feed stream parameters affecting the efficiency of SRU were then analyzed to optimize the sulfur recovery. Therefore, sensitivity analysis was carried out on feed stream conditions to optimize the value of flow

**Table 11.5** Comparison between the simulation and the plant data (designed conditions).

| Parameters | Plant data | Simulation data |
|---|---|---|
| Acid gas (kg/hour) | 3734 | 3734 |
| Air (Nm$^3$/hour) | 4819 | 4767.67 |
| COG (Nm$^3$/hour) | 300 | 300 |
| SEC air (Nm$^3$/hour) | 0 | 85 |
| Kiln temp (°C) | 1100 | 1100 |
| Kiln press bar (abs) | 1.35 | 1.35 |
| Claus 1 inlet temp (°C) | 250 | 250 |
| Claus 1 outlet temp (°C) | 270 | 266.08 |
| Claus 2 inlet temp (°C) | 210 | 212.2 |
| Claus 2 outlet temp (°C) | 220 | 214.7 |
| Sulfur recovery (kg/hour) | 412 | 420 |
| Tail gas ratio | 3.75 | 4.11 |

parameters for maximum sulfur recovery, simultaneously maintaining the tail gas ratio. The ones considered for this study were the following:

- Acid gas/air molar flow ratio (with a constant acid gas flow rate)
- Acid gas/air molar flow ratio (with a constant airflow rate)
- $H_2S$ concentration in acid gas stream

### 11.4.1 Effect of Acid Gas/Air Molar Flow Ratio on Sulfur Recovery and Tail Gas Ratio

The ratio of acid gas to air is a significant parameter that influences the conversion of hydrogen sulfide into elemental sulfur. Initially, the analysis was conducted by maintaining a constant acid gas flow rate while adjusting the airflow rate. This was done to maintain the constant amount of sulfur entering the system. In the second case, the acid gas flow rate was varied keeping the constant airflow rate.

It was found from the analysis that the amount of sulfur recovered increased up to a maximum value and then decreased with a further increase in the acid gas/air molar flow ratio, while the tail gas ratio increased continuously for both cases. The variation in trend was so obtained because, for a constant rate of acid gas, an increase in airflow resulted in more conversion of $H_2S$ to $SO_2$ in the kiln and subsequently conversion to elemental sulfur in the catalytic reactor. However, sulfur recovery percentage decreased with elevated airflow rate (lower acid gas/air ratio) because the enhanced supply of air converted an ample amount of $H_2S$ to $SO_2$, leaving behind less amount of unreacted hydrogen sulfide to carry out Claus reaction in the reactor. Hence, decrement in the recovery of sulfur and tail gas ratio (lower amount of $H_2S$ and higher amount of $SO_2$ in tail gas) was obtained with a further increase in the amount of air supplied [Figure 11.4(a)]. Results showed that the maximum percentage of the sulfur recovery obtained was 83.18% for an acid gas/air molar flow ratio of 0.69 when acid gas flow was kept constant.

Comparable findings were acquired when keeping the airflow rate and acid gas flow rate constant. Sulfur recovery consistently rose until reaching a peak value, after which it declined with an increase in the acid gas flow rate. The variation observed was due to the presence of a sufficient amount of air initially in the flow stream, which led to the conversion of hydrogen sulfide to sulfur dioxide and elemental sulfur increased. Contrary, on further increasing the acid gas flow rate, the air supply became deficient for

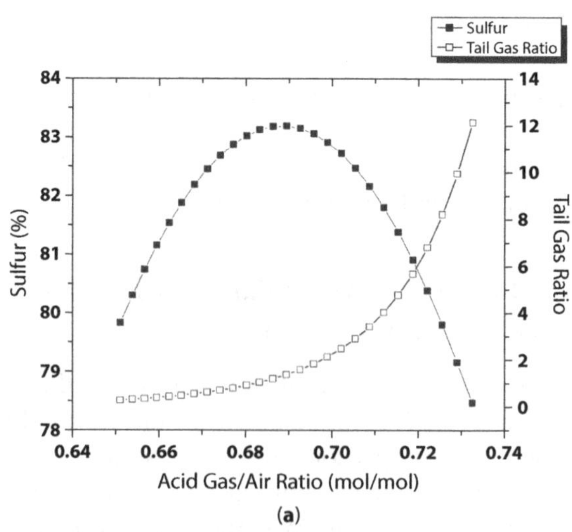


**Figure 11.4** (a): Effect of acid gas/air molar flow ratio on sulfur recovery and tail gas ratio with a constant acid gas flow rate.

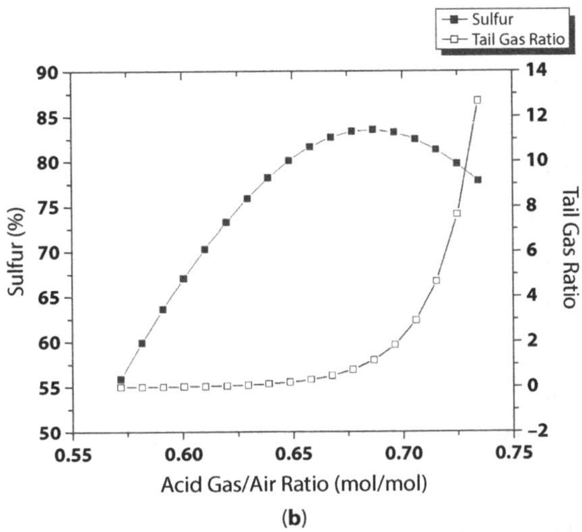


**Figure 11.4** (b): Effect of acid gas/air molar flow ratio on sulfur recovery and tail gas ratio with constant airflow rate.

the effective conversion of hydrogen sulfide to sulfur dioxide and subsequently to elemental sulfur. Consequently, an elevation in the tail gas ratio (indicating higher $H_2S$ and lower $SO_2$) was noted as the quantity of acid gas provided increased, while the amount of sulfur recovered decreased [Figure 11.4(b)].

Results showed that the maximum percentage of the sulfur recovery obtained with a constant acid gas flow rate and constant airflow rate was 83.18 and 83.6%, respectively, for an acid gas/air molar flow ratio of 0.686. However, the tail gas ratio obtained in both cases was 1.2, which was beyond the plant's desired range of 4 to 5. Therefore, in order to maintain the tail gas ratio in the desired range, the acid gas to air ratio should be kept in a close range of 0.715 to 0.72. This resulted in a very slight decrease in sulfur recovery from 82 to 83% (maximum) to 81 to 82% as shown in Figures 11.4(a and b).

### 11.4.2 Effect of $H_2S$ Concentration in Acid Gas Stream on Sulfur Recovery and Tail Gas Ratio

$H_2S$ concentration in the acid gas stream was varied to analyze the maximum amount of sulfur that can be recovered for a given amount of airflow (213.71 kmol/hour), as per plant-designed conditions. Simultaneously, its effect on the tail gas ratio was also analyzed.

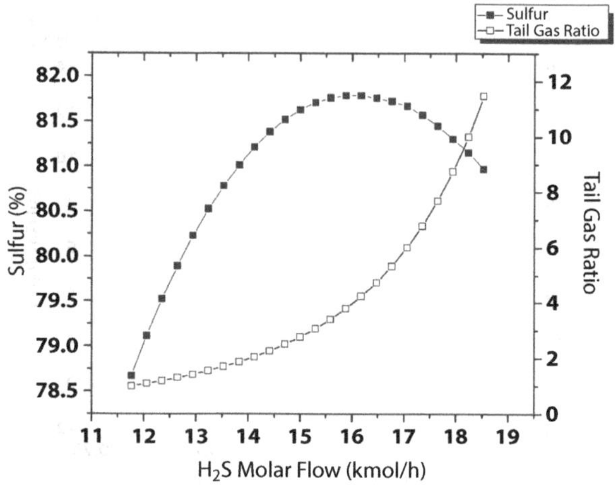


**Figure 11.5** Effect of $H_2S$ concentration in an acid gas stream on sulfur recovery and tail gas ratio.

The sulfur recovery amount was found to rise with $H_2S$ concentration in the feed stream to a maximum value and then fell with a further increase in $H_2S$ concentration. The increase in hydrogen sulfide molar flow rate (acid gas flow rate) led to an increase in the amount of air required for its effective conversion to sulfur dioxide. However, the airflow rate became deficient after a certain value for hydrogen sulfide concentration in acid gas flow. The optimum sulfur recovery was obtained for an $H_2S$ molar flow of 16.18 kmol/hour. The optimum value of sulfur recovered was 81.78% with a tail gas ratio of 4.23 as shown in Figure 11.5.

## 11.5 Conclusion

In the present work, Aspen Plus simulation software was used to simulate the Claus SRU. By comparing the simulation results to industry data from South Pars Refinery, Iran, and Tata Steel Ltd., India, the accuracy of the model was validated. Sensitivity analysis was conducted to investigate the effect of feed stream conditions on the efficiency of sulfur recovered and the tail gas ratio. The effect of acid gas to air molar flow ratio and the concentration of $H_2S$ in the acid gas stream were analyzed. The sensitivity analysis results depicted that the maximum amount of sulfur was recovered for acid gas to air molar flow ratio in the range of 0.67 to 0.69, but the tail gas ratio was below the desired range of 4 to 5. Therefore, it was concluded that, for the optimum sulfur recovery, the acid gas to air ratio should be kept in the range of 0.715 to 0.720. This resulted in a slight decrease in sulfur recovery efficiency from 83 to 81%. It was also observed that with the increase in $H_2S$ concentration, the sulfur recovery rate increased to a maximum value and then gradually decreased. The optimum sulfur recovery of 81.78% was obtained for an $H_2S$ concentration of 16.18 kmol/hour in the acid gas stream.

## Acknowledgments

We would like to express our gratitude toward the authorities of Tata Steel Ltd. and IIT (ISM) Dhanbad for providing us with the opportunity and the requisite facilities to complete the research work.

## References

1. Miura, K., Mae, K., Inoue, T., Yoshimi, T., Nakagawa, H., Hashimoto, K., Simultaneous Removal of COS and H2S from Coke Oven Gas at Low Temperature by Use of an Iron Oxide. *Ind. Eng. Chem. Res.*, 31, 415–419, 1992.
2. Kazempour, H., Pourfayaz, F., Mehrpouya, M., Modeling and Multi-Optimization of Thermal Section of Claus Process based on Kinetic Model. *J. Nat. Gas Sci. Eng.*, 38, 235–244, 2017.
3. Manenti, F., Papasidero, D., Bozzano, G., Ranzi, E., Model-based Optimization of Sulfur Recovery Units. *Comput. Chem. Eng.*, 66, 1–8, 2014.
4. Manenti, F., Papasidero, D., Frassoldati, A., Bozzano, G., Pierucci, S., Ranzi, E., Multi-scale Modeling of Claus Thermal Furnace and Waste Heat Boiler using detailed Kinetics. *Comput. Chem. Eng.*, 59, 219–225, 2013.
5. Zarei, S., Ganji, H., Sadi, M., Rashidzadeh, M., Thermo-Kinetic Modeling and Optimization of the Sulfur Recovery Unit Thermal Stage. *Appl. Therm. Eng.*, 103, 1095–1104, 2016.
6. Monnery, W.D., Hawboldt, K.A., Pollock, A., Svrcek, W.Y., New experimental data and kinetic rate expression for the Claus reaction. *Chem. Eng. Sci.*, 55, 5141–5148, 2000.
7. Ghahraloud, H. *et al.*, Modeling and Optimization of an Industrial Claus process: Thermal and Catalytic Section. *J. Taiwan Inst. Chem. Eng.*, 76, 1–9, 2017.
8. Monnery, D., Svrcek, Y., Behie, A., Modelling the Modified Claus Process Reaction Furnace and the Implications on Plant Design and Recovery. *Can. J. Chem. Eng.*, 71, 711–724, 1993.

9. Nabgan, W., Abdullah, T.A.T., Nabgan, B., Ripin, A., Kidam, K.B., Saeh, I., Moghadamian, K., A Simulation of Claus Process Via Aspen Hysys for Sulfur Recovery. *Chem. Prod. Process Model.*, 11, 1–6, 2016.
10. Pahlavan, M. and Fanaei, M.A., Modeling and Simulation of Claus Unit Reaction Furnace. *Iran. J. Oil Gas Sci. Technol.*, 5, 42–52, 2015.
11. Rezazadeh, R. and Rezvantalab, S., Investigation of Inlet Gas Streams Effect on the Modified Claus Reaction Furnace. *Adv. Chem. Eng.*, 3, 6–14, 2013.
12. Asadi, S., Pakizeh, M., Chenar, M., Shanbedi, M., Amiri, A., Effect of H2S Concentration on the Reaction Furnace Temperature and Sulphur Recovery. *Int. J. Appl. Eng. Res.*, 1, 4, 961–972, 2011.
13. Karan, K., Mehrotra, A.K., Behie, L.A., COS-Forming Reaction between CO and Sulfur: A High-Temperature Intrinsic Kinetics Study. *Ind. Eng. Chem. Res.*, 37, 4609–4616, 1998.
14. Karan, K., Mehrotra, A.K., Behie, L.A., A High-Temperature Experimental and Modeling Study of Homogeneous Gas-Phase COS Reactions Applied to Claus Plants. *Chem. Eng. Sci.*, 54, 2999–3006, 1999.
15. Karan, K. and Behie, L.A., CS2 Formation in the Claus Reaction Furnace: A Kinetic Study of Methane-Sulfur and Methane-Hydrogen Sulfide Reactions. *Ind. Eng. Chem. Res.*, 43, 3304–3313, 2004.
16. Sames, J.A., Paskall, H.G., Brown, D.M., Chen, M.S.K., Sulkowski, D., Field Measurements of Hydrogen Production in an Oxygen-Enriched Claus Furnace. Proceedings Sulfur 1990. *International Conference, Cancun, Mexico; British Sulphur Corp. Ltd*, pp. 89–105, 1990.
17. Nasato, L.V., *Modelling of Quench Times in the Waste Heat Boiler of a Modified Claus Plant. M. Eng. Thesis*, University of Calgary, Calgary, Alberta, Canada, 1993.
18. Nasato, L.V., Karan, K., Mehrotra, A.K., Behie, L.A., Modeling Reaction Quench Times in the Waste Heat Boiler of a Claus Plant. *Ind. Eng. Chem. Res.*, 33, 7–13, 1994.
19. Karan, K., Mehrotra, A.K., Behie, L.A., Including Radiative Heat Transfer and Reaction Quenching in modeling a Claus Plant Waste Heat Boiler. *Ind. Eng. Chem. Res.*, 33, 2651–2655, 1994.
20. Abedini, R., Salooki, M.K., Ghasemian, S., Modeling and Simulation of Condensed Sulfur in Catalytic Beds of Claus Process: Rapid Estimation. *Chem. Eng. Res. Bull.*, 14, 110–114, 2010.
21. Sassi, M. and Gupta, A.K., Sulfur Recovery from Acid Gas Using the Claus Process and High Temperature Air Combustion (HiTAC) Technology. *Am. J. Environ. Sci.*, 4, 502–511, 2008.
22. Abaskuliev, D.A., Guseinov, N.M., Mekhraliev, A., *Ch. Mathematical Model of the Catalytic Convertor of a Claus Unit, Gazov. Promst.*, vol. 9, 60, 1990.

23. Zughbi, H.D. and Razzak, S.A., Simulation of Flow and Chemical Reactions in a Claus Sulfur Converter. *Ind. Eng. Chem. Res.*, 44, 9828–9839, 2005.
24. Nabikandi, N.J. and Fatemi, S., Kinetic Modelling of a Commercial Sulfur Recovery Unit based on Claus Straight Through Process: Comparison with Equilibrium Model. *J. Ind. Eng. Chem.*, 30, 50–63, 2015.

# 12

# Fabrication of Magnesium Metal Matrix Nanocomposites Using Ultrasonic-Assisted Stir-Casting Method

**Parthiban K.*, Balaji E., Durairaj S. and Ashok Kumar R.**

*Department of Mechanical Engineering, A K T Memorial College of Engineering and Technology, Kallakurichi, Tamil Nadu, India*

***Abstract***
This chapter discusses the most up-to-date findings from studies conducted on the ultrasonic-assisted stir-casting method for creating metal matrix nanocomposites (MMNC) using a magnesium material. The limited mechanical properties of pure metals and alloys rule them out of use in many engineering contexts. Incorporating suitable reinforcements into a metallic matrix has been found to improve ultimate tensile strength, specific strength, wear properties, and porosity when compared to conventional and monolithic engineered materials employed in aerospace and automotive applications. Furthermore, in the future, Mg-MMNCs will be used to create components with functional features. The high surface-to-volume ratio of nanoparticles in a magnesium substrate has a significant impact on its mechanical characteristics, even at minimal concentrations. Ceramic nanoparticles have been widely and readily available for a period of years, allowing for the modification of their mechanical characteristics. To influence the resultant properties, the particulate chemical composition, concentration, process parameters, and particle size must be optimized for the specific alloy. The distribution of extremely minute particles in the melt is a significant challenge. The particles are held together in clusters by strong van der Waals forces because of their enormous surface area. Furthermore, the particles have low wettability when exposed to magnesium melt. Nanoparticles can be separated from each other and spread out uniformly with the help of ultrasonic-assisted stirring.

**Corresponding author*: parthisha6@gmail.com

R. Thanigaivelan, Pradeep Kumar Krishnan, Kamalakanta Muduli and Santosh Kumar Tamang (eds.) New Materials, Processing and Manufacturability: Fabrication and Processing of Advanced Materials, (227–246) © 2024 Scrivener Publishing LLC

***Keywords*:** Magnesium alloy, ultrasonication, ceramic reinforcing nanoparticles, metal matrix nanocomposites

## 12.1 Introduction

Metal matrix composites initially appeared as cutting-edge technology for technologically advanced military systems, and recent developments in the fields of production, engineering, and services have made them more affordable and readily available than ever before, especially for defense applications [1]. Owing to their low density, superior castability, higher specific strength, anticorrosion characteristics, superior electrical and thermal conductivity, etc., magnesium alloys offer great potential for use in aviation and motor vehicle applications [2]. Despite all of these desirable qualities, they are not ideal for several crucial technical applications because of their lack of strength, porosity, low ductility, etc. Therefore, magnesium-based metal matrix nanocomposites (MMMCs) made with a suitable process can solve most of the problems with Mg alloys [3]. When it comes to mass manufacturing of magnesium composite components in industries, cast magnesium alloys have an economic edge over wrought magnesium alloys because of their quicker processing cycles and lower assembling costs [4]. Casting is the predominant manufacturing technique for nearly 98% of structural magnesium applications [5]. Rapid oxidation, caused by oxygen's strong affinity for magnesium, presents difficulties for scientists and researchers involved in the manufacturing of MMMCs. Magnesium composite castings may be made using a number of different methods, although the gravity sand and permanent mold process are particularly popular for advanced military applications [6]. The performance of composites suffers because nanoparticles are more difficult to distribute and form agglomeration during processing than microparticles. Therefore, the most suitable method for producing composites in a highly effective manner for mass production is vacuum casting supported by ultrasonic vibrations, which allows for the elimination of these issues [7]. In addition, it is useful for making MMMCs that are eco-friendly, have no pores, and have a homogeneous distribution of reinforcing particles.

It is possible to modify the microstructure and hence the characteristics of all the alloys by adding ceramic nanoparticles to the melt either during or prior to casting. There are a number of methods that may be used to facilitate the even distribution of reinforcements in the melt. When ceramic particles or fibers are integrated into metallic matrices, Orowan strengthening or Hall-Petch boundary grain strengthening is applied.

Ultrasonic treatment (UST) is an alternative to melt shearing for dispersing the particles. Molten metals cause ceramic nanoparticles to cluster together due to their high surface energy. Also, it can be challenging to get a magnesium melt to stick to the ceramic surface of the particles. The UST can break up the agglomerates and distribute the particles more uniformly throughout the melt. The melt of a magnesium alloy can be treated with ultrasonic waves to bring about a number of effects.

### 12.1.1 Ultrasound Equipment

Molten metals can be ultrasonically treated with a wide variety of transducers, generators, and acoustic horn materials and designs. The transducers are typically magnetostrictive or piezoceramic. The former has resulted in having to be kept cool enough to prevent the melt, which can be accomplished by a longhorn. The amplitude associated with the ultrasound that it generates may be kept consistent thanks to these technologies. For instance, making use of a horn in molten metal might alter the resonance frequency, which can cause issues. The reason for this is that Young's modulus changes with temperature and as the material stretches. Water-cooled magnetostrictive plates are the basis of magnetostrictive transducers. This means that there is no chance that these devices will get too hot. The frequency of the horn's vibrations can alter depending on the temperature, although their designs account for this. Magnetostrictive ultrasonic transducers may have their frequency adjusted by the controller to accommodate changes in the resonance of the system. These technologies are therefore more resilient than piezoceramic ones.

In most cases, the acoustic transducer will also require an extension and a sonotrode. Between the two of them, a booster can be placed if necessary. As was previously noted, the acoustic horn's resonance frequency is defined by the material's Young's modulus. The resonance frequency f is a function of the wavelength l and the speed of sound c, as shown in Equation 12.1. The Young's modulus (E) and the density (r) are factored in as material properties in the following equation:

$$\lambda = \frac{C}{f} = \frac{1}{f}\sqrt{E/\rho} \tag{12.1}$$

It is usually vital to achieve a balance between competing needs when deciding on a sonotrode material for magnesium melts. The substance must be insoluble in magnesium melt and have a high melting point.

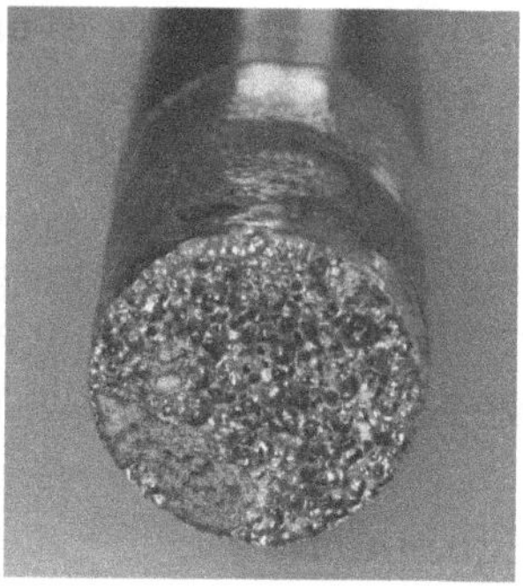

**Figure 12.1** Sonotrode, diameter: 35 mm.

Having a Young's modulus that is not too affected by temperature is also beneficial. In addition to high strength, industrial applications need low sensitivity to fluctuating temperatures alongside excellent fatigue strength [8]. Magnesium alloys frequently include niobium and titanium. Niobium is a desirable material owing to its high melting point and low change in Young's modulus with temperature. On the other hand, titanium is practically insoluble in a magnesium melt and remains remarkably stable in it. As can be seen in Figure 12.1, a sonotrode made of titanium alloy is used.

### 12.1.2 Ultrasonication-Assisted Stir-Casting Technique (UASCT)

The accumulation trends of the nanoparticles have a meaningful impact on the homogeneous distribution of hard nano granules in a molten Mg matrix. Owing to the larger ratio between surface area and volume ratio as well as the restricted matrix wettability of the nanoparticles, distributing nano-reinforcement particles using liquid metallurgical processes such as compo casting and stir casting is problematic. Researchers have adapted a variety of metallurgical processes in liquid state as well as devised the stir-casting process with ultrasonic assistance. Vinod *et al.* [9] suggested the quality of MMNCs produced is determined by the dispersion of nanocomponents in the matrix. Gnanavelbabu *et al.* [10] demonstrated the casting method aided with an ultrasonic methodology that showed better delivery as well as nanoparticle dispersion over the matrix and it will have superior properties when contrasted with the traditional stirring procedures. Composites made with nanoparticles having notable and homogeneous structured grains also have better mechanical and tribological characteristics. Li *et al.* [11] listed several advanced material processing methodologies that have recently evolved to modify the microstructure of Mg-based composites. Prasad Reddy *et al.* [12] suggested that the ultrasonic-assisted casting approach is gaining popularity due to its capacity to optimize the grain structure and improve particle dispersion. The period for sonication, the pressure of the ultrasonic wave, the weight percentage of the nanoparticle, the heat energy required for processing, and the intensity of the ultrasonic sound wave are important for ceramic nanoparticles scattering in the molten liquid.

A conceptual line layout for a sonication-aided stir-casting system is indicated in Figure 12.2 (Zhou *et al.* [13]). The schematic has a sonotrode, transducer, generator, and argon gas supply unit. Figure 12.3 depicts the ultrasonication process and nucleation growth (Lan *et al.* [14]). High-intensity acoustic streaming and transient cavitation are produced during sonication in molten material. Dispersing nanoparticles that are being grouped, eliminating porosity, and forming a homogeneous arrangement are all aided by transient cavitation and sonic streaming. Ultrasonic cavitation within the molten melt creates a hot spot of transient micro size with extraordinarily enhanced pressure and temperature. These micro hot spots effectively break apart the clustered nanoparticles. In the molten melt, it also improves the wettability of reinforced particles. Harichandran and Selvakumar [15] predicted, at the same time, acoustic streaming boosts nanoparticle mobility and uniform dispersion in the molten melt.

The ultrasonic intensity affects the dispersion of reinforcing particles of nano-size at the matrix of metal and refinement of grain. Srivastava and Chaudhari [16] established a relationship between the intensity of

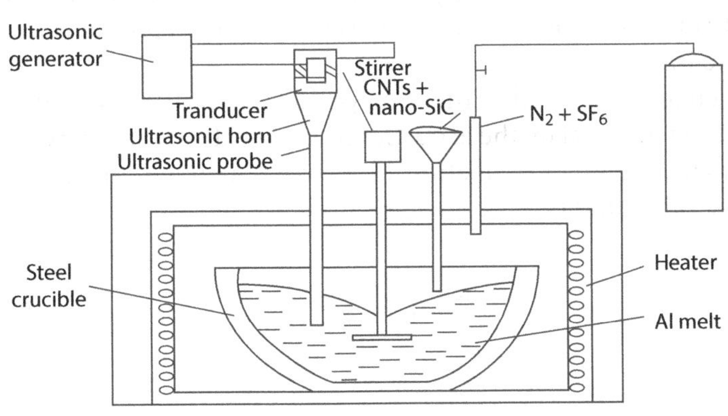


**Figure 12.2** Schematic line diagram of ultrasonically assisted casting setup [13].

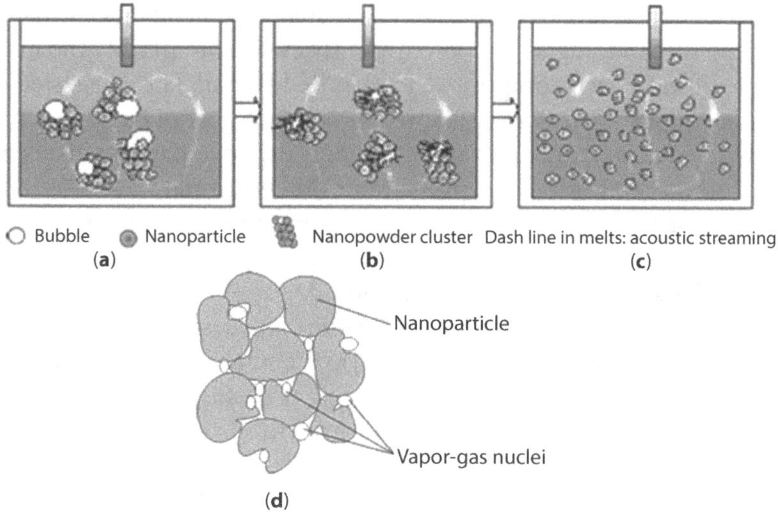


**Figure 12.3** (a) Initial ultrasonication process, (b) nucleation starts, (c) acoustic streaming, and (d) schematic view of vapor gas nuclei [14].

ultrasonic waves and process variables. The relationship is shown in Equation 12.2 as given below.

$$I = \frac{\rho c(2\pi fA)^2}{2} \tag{12.2}$$

Here,

ρ is the molten metal density (g/cc), c is the travel velocity of sound in the melt (m/s), f is the frequency (kHz), A is the amplitude, and I is the ultrasonic intensity ($kW/cm^2$).

The wetted nano-sized reinforcing particles are carried and distributed into the molten metal by the acoustic streaming effect caused by cavitation in the molten metal. The equation 12.3 gives the molten metal pressure of the ultrasonic sound ($P_e$).

$$P_e = \left( \frac{2P\rho_m c_m}{A_{probe}} \right)^2 \tag{12.3}$$

where

$\rho_m$ is the molten metal density (g/cc), P is the ultrasonic generator output power (kW), $C_m$ is the ultrasound speed in molten metal (m/s), and

$A_{probe}$ is the probe area ($cm^2$). $P_e$ value should be greater than that of the value of the ultrasonic streaming threshold.

### 12.1.3 Nanocomposites Processing

The method of the unique technique employed in the experiments is depicted in Figure 12.4. The melt was agitated at 500 rpm for 2 min to form a homogeneous alloy mixture. The molten material was progressively mixed with preheated SiCP and hBNP nanoparticles (400°C for 30 min). Mechanical stirring continued for 5 min to predominantly combine the nanoparticles. It cuts down on the amount of time it takes to sonicate. The stirrer was removed from the melt after spreading. The probe was placed in melt for 30 mm before the ultrasonic wave generator was activated. Each test was subjected to 10 min of UST. The ultrasonic cavitation effect was generated using a titanium horn attached to an ultrasonic generator operating at 20 kHz. If the ultrasonic vibrations are strong enough, the Mg binding energy of the liquid melt will be overcome, and cavitation bubbles will emerge inside the liquid. When the cavitation bubbles burst due to a change in the pressure sequence, the particles in the vicinity scattered at random according to all the rules. The aggregation behavior of nanoparticles was drastically reduced after ultrasonic processing was applied.

Nonlinear sonication phenomena such as cavitation and acoustical streaming are triggered by a high-intensity cyclic ultrasonic pulse in Mg melt. This nonlinear mechanism was utilized to refine the matrix grain size, degassing the liquids, and homogeneously dispersing the nanoparticles. Transient cavitation is the phenomenon of creation, expansion, pulsating, and collapsing of microbubbles produced by cyclic high-intensity vibrations in a liquid matrix. Bubbles burst in less than 10 to 6 s on each

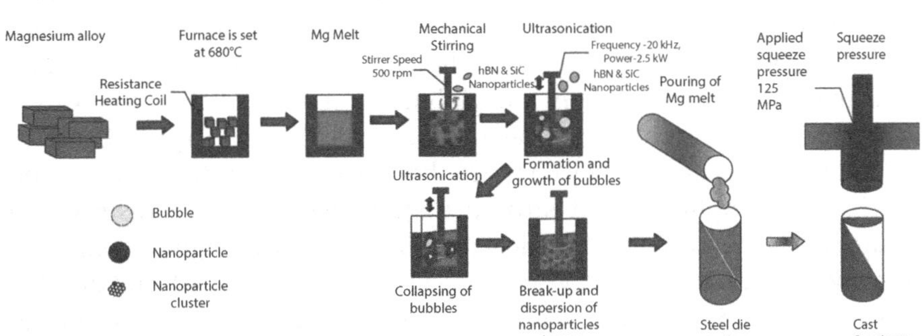


**Figure 12.4** Ultrasonication assisted stir-casting process.

cavitation cycle. It then generates transient (in microseconds) micro "hot spots" with 5000°C temperatures, 1000 atm pressure, and above 1010 K $s^{-1}$ heating and cooling rates. Transient cavitation could provide sufficient energy to disperse clustered nanoparticles more equally in liquid melts. Transient cavitation could also clear the gaseous film from nanoparticle surfaces, allowing the matrix and nanoparticles to be more wettable. Acoustic streaming could increase nanoparticle transfer in liquids, balance the temperature and melt solute concentration, and fragment the dendrites.

## 12.2 Effect of Ultrasound on Magnesium Alloys

Eskin uses UST to degas a Mg-Al-Zn alloy in his tests, which are described in [17]. It was discovered that the hydrogen concentration may have been drastically reduced to only half of what it was in the untreated alloy. Degassing was the first application of ultrasonic technology in a commercial setting for aluminum and other light metals. It relies on the fact that expanding cavitation bubbles may absorb hydrogen. Ultrasound experiments with aluminum alloys have revealed that the intensity of the ultrasound has a positive effect on the rate of degassing.

After being cast, pure magnesium, AM60, and AZ91 were examined for their grain sizes and tensile strengths with and without UST [18]. After 10 min of UST, the grains in all three materials were found to be significantly smaller. For pure magnesium, the grain size dropped from 500 to 175 μm, for AZ91 from 205 to 125 μm, and for AM60 from 200 to 90 μm. After treatment, the number of entrapped inclusions (oxides) increased while particle size decreased. The frequency, intensity, and length of treatment all contribute to the refining impact on the grain. In [19], on the microstructure and mechanical properties, the impacts of treating AZ91D to disintegrate with the US in the semi-solid state at 580°C were assessed. Ultrasonic power intensities of 3.2, 7.65, and 19.6 W/$cm^2$ (at 0.4, 0.6, and 0.8 kW, respectively) were applied to the slurry. According to experimental data, the best grain refinement and globularity happened at 0.6 kW, which is thought to be the optimal balance between lower energies that cannot refine grains and higher energies that raise the melt temperature too much. Khosro Aghayani and Niroumand [20] investigated microstructural changes, phase amount and particle distribution, and strength in a conventionally sand-cast AZ91 alloy. When UST (20 kHz; 120, 240, and 360 W) was applied for 5 min, there was a significant impact on the dimension and spheroid of Mg dendrites and intermetallic particles Mg2Si, MnFeAl (Si), and Mg17Al12. As the US influence grew, the size of individual grains and intermetallic particles decreased and became more uniform. Streaming and cavitation during UST were blamed for these microscopic alterations. The refined $Mg_{17}Al_{12}$ had a less continuous distribution, which contributed to the material's increased tensile strength. Bhingole *et al.* also utilized UST in their processing of AZ91 [21, 22]. The application of ultrasonic waves of 15-, 30-, and 45-μm amplitudes (1.34, 2.69, and 4.031 kW/cm$^2$, respectively) lasted for 1 min. As intensity increased, grain size decreased from 120 to 20 μm, while Vickers hardness increased concurrently. In the US-treated alloys, the phase $Mg_{17}Al_{12}$ seemed to be uniformly distributed along the grain boundaries.

### 12.2.1 AlN Reinforcement

Researchers Dieringa *et al.* [Eskin 1997, previously referenced] studied the effects of ultrasonic-aided processing (20 kHz and 0.3 kW) on an AM60 magnesium alloy that had been strengthened with nano-AlN. At room temperature, the material's tensile strength is much higher, the grain size is smaller, and the ductility remains intact.

By incorporating ultrasonics with a power of 0.3 kW and a frequency of 20 kHz for 5 min, 1 wt % AlN nanoparticles of the commercial magnesium alloy Elektron21 (Mg-2.8Nd-1.2Gd-0.4Zr-0.3Zn) were strengthened [23]. The nanoparticle-added Elektron21 and the unmodified Elektron21 were compared in relation to their mechanical and microstructure properties at both ambient temperature and higher temperature. This was done to see what effect the nanoparticle addition alone had. The microstructure of both materials is seen in Figure 12.5. The grain size of unreinforced

(a) (b)

**Figure 12.5** Images from optical microscopy. (a) Elektron21 and (b) Elektron21 with 1 wt % AlN nanoparticles, both were taken from [23].

Elektron21 is 80.1 μm, while the grain size of Elektron21 reinforced by 1 wt % AlN nanoparticles is 74.1 μm, suggesting a dendritic structure. In Elektron21, nanoparticles have no effect on hardness.

The same 1 wt % AIN nanoparticles were included in the die-casting alloy AM60 throughout production [24]. Because of its high ductility and outstanding room-temperature characteristics, AM60 finds widespread application in the automobile sector. AM60 is used for a variety of non-heat-sensitive automotive parts, including housings, seat shells, and steering wheels. These components are manufactured by die casting and utilized in their as-cast state. Die casting always results in a fine-grained structure with 2 to 4% porosity because of the irregular flow of melt that occurs during the filling of the mold. Because of their porosity, die-cast pieces cannot be heat-treated or welded.

The AM60's microstructure after being ultrasonically reinforced with AlN nanoparticles is similar to that of the die-cast AM60. Figure 12.6 displays the drastic reduction in particle size from 1277 to 84.9 μm. In tensile tests conducted at room temperature, Young's modulus increased by 103% to 91.2 from 44.9 MPa; in addition, the UTS increased by 115% to 235.1 from 109.3 MPa. Even though ductility is often reduced by reinforcement with microparticles or fibers, the inclusion of AlN nanoparticles in this case doubles more than the length of time to failure. When AlN nanoparticles were included, the percentage rose by 140%, from 15.4 to 6.4% [24].

The highest fraction of the calculated strengthening contributions comes from the process of grain refining. This imparts a pressure of 42.70 MPa. The addition of nanoparticles to strengthen Orowan, which stops dislocations from moving, only adds about 8.3 MPa, while other methods of strengthening add almost nothing [24]. The characteristics of AM60 and the AM60-based nanocomposite are listed in Table 12.1.

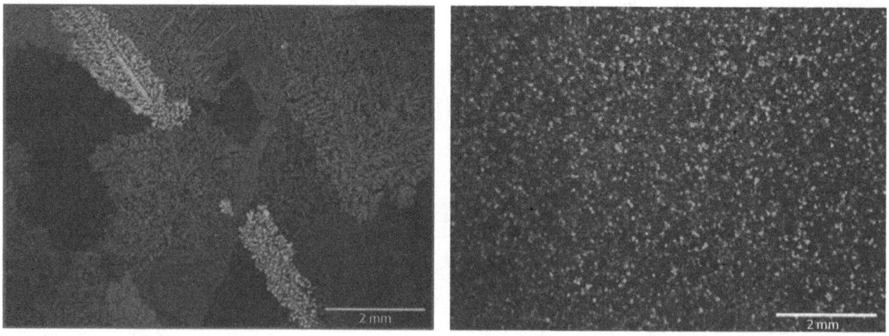


(a) (b)

**Figure 12.6** Microstructure of (a) AM60; and (b) AM60 + AlN [24].

**Table 12.1** Grain size, hardness, density, porosity proportion, and mechanical properties of the materials investigated, data from [24].

| **Property** | **AM60** | **AM60 + AlN** |
|---|---|---|
| Grain size [μm] | 1277.0 ± 301.3 | 84.9 ± 6.2 |
| Hardness [HV5] | 48.0 ± 4.0 | 46.4 ± 6.0 |
| Density [g/cm$^3$] | 1.7848 ± 0.0004 | 1.783 ± 0 |
| Porosity [%] | - | 0.919 |
| Yield strength [MPa] | 44.9 ± 6.9 | 91.2 ± 3.8 |
| UTS [MPa] | 109.3 ± 19.2 | 235.1 ± 6.4 |
| Elongation [%] | 6.4 ± 3.4 | 15.4 ± 4.2 |

### 12.2.2 $Al_2O_3$ Reinforcement

Ultrasonic-aided melt stirring was used to incorporate $Al_2O_3$ and $Al_2O_3$-AlOOH into the AZ91magnesium alloys and Elektron21 (El21), and then the alloys were permanently molded and chilled cast [25]. Both commercial alloys' chemical compositions are listed in Table 12.2. When alumina-based particles were added to the magnesium alloy melt, MgO was quickly produced by a chemical reaction. After that, aluminum started showing up in the Elektron21 alloy's eutectic zone. Submicron MgO particles are thought to increase wettability and decrease the likelihood of particle aggregation in the matrix.

Using ultrasound-assisted stirring, the creep properties of the AS41 magnesium alloy reinforced with 2 and 5 wt % $Al_2O_3$ nanoparticles were studied [26]. The AS41 melt was then infused with alumina nanoparticles with a diameter of 50 nm and agitated for 15 to 20 min. After the mixture was stirred, it was sonicated for 3 min at 20 kHz and 60-μm amplitude. Stresses of 109.2, 124.8, and 140.0 MPa were applied during indentation creep tests conducted at 175°C, 200°C, 225°C, and 250°C, respectively.

**Table 12.2** Composition of AZ91 and Elektron21 in wt %, data from [25].

| **Alloy** | **Al** | **Zn** | **Mn** | **Nd** | **Gd** | **Zr** | **Mg** |
|---|---|---|---|---|---|---|---|
| AZ91 | 8.71 | 0.66 | 0.22 | - | - | - | Bal. |
| El21 | - | 0.2–0.5 | - | 2.6–3.1 | 1.0–1.7 | Satur. | Bal. |

A higher percentage of alumina was shown to improve creep resistance. This is because sonication evenly disperses smaller Mg2Si precipitates, and the nanoparticles enhance Orowan. Based on formulae that describe how temperature and stress influence the lowest creep rates once activation energies and stress exponents have been computed, it is thought that pipe diffusion-controlled dislocation creep is the rate-controlling deformation mechanism during indentation creep.2.4.

### 12.2.3 SiC Reinforcement

The AZ91 magnesium alloy was strengthened by UST with SiC nanoparticles that are 40 nm in size [27]. When compared to unreinforced AZ91, nanocomposites with SiC contents of 0.5, 1.0, 1.5, and 2.0 wt % were studied. The optical microscope results demonstrate that the nanoparticles have a grain-refining impact. In Figure 12.7, we see a graph of the mechanical characteristics. When comparing reinforced and unreinforced AZ91, it is clear that an addition of 1.0 wt % nano-SiC yields the most significant improvements in both strength and ductility.

A $Mg_6Zn$ alloy was altered by adding 1.5 wt % of 50-nm-sized SiC nanoparticles while using UST at 17.5 kHz and 3.5 kW of power [28]. Tests of tensile strength revealed that nano-SiC additions of 40 and 34%, respectively, enhance yield and UTS while maintaining ductility. Studies using transmission electron microscopy revealed a strong connection between SiC particles and a $Mg_6Zn$ matrix. Mechanical and microstructural studies of ultrasonic-aided casting-generated nano-SiC-reinforced AS21 and AS41

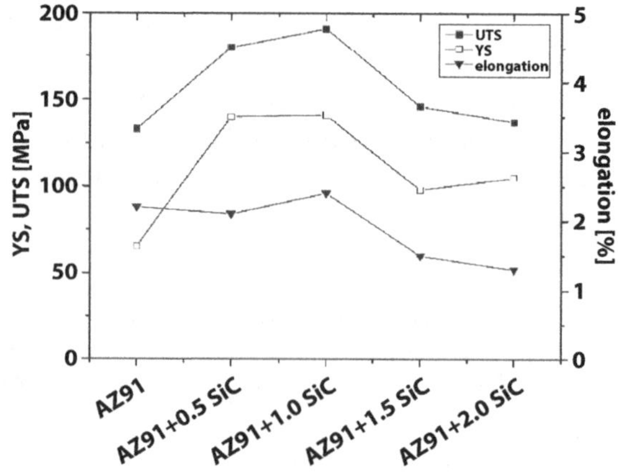


**Figure 12.7** Tensile properties of nano-SiC/AZ91 at room temperature, data from [27].

were also published by this research group [29]. The nanocomposites in both cases were compared to the unreinforced alloys after 2 wt % was added. Both the tensile strength and the refinement of the grains were greatly enhanced.

With the use of ultrasonic energy, SiC particles (50 nm in diameter) were mixed into a pure magnesium melt [30]. The temperature used was 700°C. The tensile specimens were made by pouring the molten material into a steel mold that had been heated to 350°C, followed by stirring. Table 12.3 lists the mechanical characteristics of pure magnesium reinforced with 0.5, 1, 2, and 4% nano-SiC. The 2% nano-SiC reinforcement has the highest strength and the highest ductility. Strongly reduced ductility at 4% SiC, which may possibly be attributable to a different melt treatment. Casting took place without first filtering the melt.

For reinforcement, 1-μm SiC particles were mixed into AZ91 magnesium alloy with the help of ultrasonic vibrations, while the mixture was still mostly solid [31]. Particle additions of 3 and 5 vol % were made to the materials, and their characteristics were compared to those of unreinforced AZ91 cast in the absence of stirring and ultrasonic vibration. When compared to base metal, the 3 vol % composite has much finer grain sizes. In the 3 vol % composite, the phase $Mg_{17}Al_{12}$ underwent a morphological transition, changing into fine lamellae. Most of the particles in both composites are located near the grain boundaries, while the rest are mostly concentrated in the matrix. Particle pushing during solidification is a strong possibility. Yield strength and UTS are enhanced for the 3 vol % composite, while ductility is maintained. The YS for the 5 vol % composite increased when SiC agglomerates developed in the microstructure, while the UTS and elongation decreased.

Once again, 40 nm diameter SiC was used to strengthen the magnesium alloy AZ91 at mass fractions of 0.5, 1.0, 1.5, and 2.0 wt % [32].

**Table 12.3** Mg and Mg/SiC nanocomposites' average mechanical characteristics, according to data from [30].

| Material | YS [MPa] | UTS [MPa] | E [%] |
|---|---|---|---|
| Mg | 20 | 89.6 | 14.0 |
| Mg-0.5SiC | 28.3 | 120.7 | 15.5 |
| Mg-1SiC | 30.3 | 124.1 | 14.2 |
| Mg-2SiC | 35.9 | 131.0 | 12.6 |
| Mg-4SiC | 47.6 | 106.9 | 5.5 |

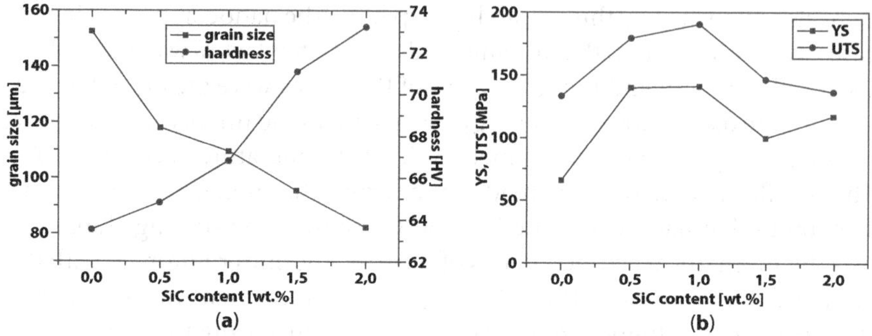


**Figure 12.8** The mechanical characteristics of AZ91 and their SiC-reinforced nanocomposites, as well as data on grain size, hardness, and both (a) and (b) [32].

Particles were added after the melt had been heated to 650°C and while it was being treated to ultrasound for 10 min at a power of 4 kW and a frequency of 20 kHz. A comparison of AZ91 and these nanocomposites in terms of particle size and hardness is displayed in Figure 12.8(a). As the percentage of SiC in the material rises, the grains become smaller and the material becomes harder Figure 12.8(b) shows that after increasing the SiC concentration from 0 to 1 wt %, the material's strength drops again. The yield strength of the composite with 1 wt % nano-SiC is more than double that of its unreinforced counterpart AZ91.

SiC nanoparticles of 50 nm in size were used to reinforce the AZ91 magnesium metal once again [33]. The frequency was 20 kHz, the power was 0.6 kW, and the therapy lasted 30 min. Material grain size was decreased from 90 m in AZ91 to 60 m in 0.5 wt % Si-reinforced AZ91. SiC nanoparticles' heterogeneous nucleation action is responsible for the dramatic grain refinement. This is because the misfit between the SiC(111) crystal face and the Mg(0001) crystal face is quite tiny at 3.96%. Table 12.4 lists the mechanical qualities. It's clear that not only has strength increased, but ductility as well, by an immense 83 percent.

While Mg-8Zn was reinforced with 3.0 wt % zinc, Mg-4Zn and Mg-6Zn were strengthened with 1.5 wt % SiC of 50 nm average diameter.

**Table 12.4** Data from [33] about the mechanical characteristics of AZ91 and its 0.5 SiC nanocomposite.

| Material | YS [MPa] | UTS [MPa] | E [%] |
|---|---|---|---|
| AZ91 | 104 | 174 | 3.6 |
| AZ91 + 0.5 SiC | 124 | 216 | 6.6 |

**Table 12.5** Mg-Zn alloys' mechanical characteristics and its SiC-reinforced nanocomposites' data are found in [34].

| Material | YS [MPa] | UTS [MPa] | E [%] |
|---|---|---|---|
| Mg-4Zn | 44 ± 2 | 112 ± 14 | 5 ± 1 |
| Mg-4Zn + 1.5 SiC | 67 ± 4 | 199 ± 6 | 10 ± 1 |
| Mg-6Zn | 51 ± 4 | 136 ± 19 | 5 ± 1 |
| Mg-6Zn + 1.5 SiC | 79 ± 5 | 194 ± 15 | 7 ± 1 |
| Mg-8Zn | 81 | 152 | 3 |
| Mg-8Zn + 3 SiC | 111 | 222 | 7 |

With assistance from the United States, the SiC nanoparticles were incorporated into the melt at 700°C [34]. For the purpose of sonication, a niobium sonotrode was utilized to generate 17.5 kHz at 4.0 kW of power. Tensile tests were performed on the materials at ambient temperatures. As can be seen in Table 12.5, there were noticeable increases in yield strength, UTS, and ductility.

By churning semi-solids and distributing ultrasonically in the liquid phase, SiC nanoparticles with an average diameter of 40–50 nm were added to an 18% zinc alloy [35]. The matrix had a very uniform particle dispersion, according to optical examination. The hardness of the nanocomposite is 183 kg/mm2, which is approximately 140% greater than the 76 kg/mm2 hardness of the alloy.

Song *et al.* [36] analyzed the efficiency of SiC nanoparticle dispersion in an AZ91 melt by varying the frequency, ultrasonic power, sonotrode depth, and processing time. The optimal conditions for dispersing nanoparticles include an ultrasonic power of 2 kW, a frequency of 20 kHz, a sonotrode depth of 20 to 30 mm, and a sonication time of 120 s. This combination of settings for a 1 vol % addition of SiC particles between 30 and 300 nm in diameter has been validated by experiments. Grain refinement and Orowan strengthening both contribute to the enhanced mechanical characteristics observed in nanocomposites.

### 12.2.4 $TiB_2$ Reinforcement

$TiB_2$ nanoparticles with a 25-nm diameter were used to strengthen the AZ91D magnesium alloy [37]. Niobium sonotrodes and ultrasonic generators producing 20-kHz amplitudes of 60 μm were employed to mix the material at 700°C. It took 15 min to complete the process.

**Table 12.6** Mechanical characteristics of the $TiB_2$ nanocomposites made from the AZ91D alloy, according to [37].

| Material | YS [MPa] | UTS [MPa] | E [%] |
|---|---|---|---|
| AZ91 | 88 ± 3 | 162 ± 5 | 2.88 ± 0.12 |
| AZ91 + 1 $TiB_2$ | 104 ± 1 | 180 ± 0 | 3.33 ± 0.05 |
| AZ91 + 2.7 $TiB_2$ | 107 ± 1 | 188 ± 7 | 4.27 ± 0.61 |

As a result of this reduction in grain size, mechanical tests conducted at room temperature reveal increased strength in addition to enhanced ductility. The grain size is cut in half with a $TiB_2$ level of between 1 and 2.7%. Table 12.6 summarizes the mechanical qualities.

## 12.3 Conclusion

Ultrasonic-assisted stir casting, which helps in the dispersion of nanoparticles in the melt, is clearly highly successful in the production of MMNC. This is evident from the fact that adding nanoparticles to UST dramatically enhances the characteristics of the resulting alloys. Unfortunately, most articles just briefly mention the criteria for ultrasonic therapy, if at all. The sonotrode material, immersion depth, UST time, melting temperature, melt volume, and sonotrode volume are all variables.

For ultrasonic horns, titanium and Ti6Al4V appear to be the most promising materials. This is because both materials have a high specific strength and a high melting point: titanium has 1668°C and Ti6Al4V has 1604°C, respectively. Sonotrodes made of niobium, which has a melting point of 2477°C, are also in use. Titanium dissolves extremely poorly in magnesium, although niobium's solubility in magnesium has not been reported. The four companies whose ultrasonic generators were used to disperse nanoparticles in magnesium melts are Hielscher (Germany), Misonix Inc. (USA), Sonics and Materials (USA), and Advanced Sonics (USA). Both of them use ultrasonic waves, although the former uses 17.5 kHz and the latter uses 20 kHz. Three minutes to an hour is typical for the sonication process. We may presume that dispersion and deagglomeration proceed rapidly because the outcomes of several of the shorter USTs are already extremely good. Longer periods of churning cause the grains to become coarser. It is impossible to provide any general notion for the duration of ultrasonic therapy since the specific quantity of melt handled is never indicated in the literature.

Multiple studies have indicated that UST's addition of nanoparticles to magnesium and magnesium alloys has a startling effect on the matrix alloy's mechanical properties, even when just a tiny number of nanoparticles are added. Substantial improvement in strength can be achieved by adding as little as 1 to 2% SiC, $Al_2O_3$, AlN, or carbon-based nanomaterials. It has been demonstrated that an increase in particle count can actually lower strength. The yield strength or tensile strength of a material is doubled when just one percent by weight of AlN nanoparticles is added. Sometimes the strength boost is the result of grain refining, and sometimes it is the result of dispersion or Orowan strengthening. In the past, adding micrometer-sized particles or fibers was the only way to boost strength. However, ductility was sacrificed in every case. However, things change drastically when nanoparticles are involved. These nanocomposites are excellent candidates for application as structural materials because of their high strength-to-weight ratios and relatively high ductility. Future study and development in this area will considerably increase the variety of possible uses for these intriguing materials.

## References

1. Miracle, D.B., Metal matrix composites – From science to technological significance. *Compos. Sci. Technol.*, 65, 2526–2540, 2005.
2. Mordike, B.L. and Ebert, T., Magnesium: Properties — applications —potential. *Mater. Sci. Eng. A*, 37–45, 302, 2001.
3. Khandelwal, A., Mani, K., Srivastava, N., Gupta, R., Chaudhari, G.P., Mechanical behavior of AZ31/Al2O3 magnesium alloy nanocomposites prepared using ultrasound assisted stir casting. *Compos. Part B Eng.*, 123, 64–73, 2017.
4. Pan, F., Yang, M., Chen, X., A Review on Casting Magnesium Alloys: Modification of Commercial Alloys and Development of New Alloys. *J. Mater. Sci. Technol.*, 32, 1211–1221, 2016.
5. Kumar, A., Kumar, S., Mukhopadhyay, N.K., Introduction to magnesium alloy processing technology and development of low-cost stir casting process for magnesium alloy and its composites. *J. Magnes. Alloy.*, 6, 245–254, 2018.
6. Synthesis Techniques for Magnesium-Based Materials, in: *Magnesium, Magnesium Alloys, and Magnesium Composites*, pp. 13–38, 2011.
7. Dieringa, H. and Kainer, K.U., Magnesium Matrix Composites: State-of the-Art and what's the Future. *Adv. Mater. Res.*, 410, 275–278, 2012.
8. Eskin, G.I., Principles of Ultrasonic Treatment: Application for Light Alloys Melts. *Adv. Perform. Mater.*, 4, 223–232, 1997.

9. Vinod, B., Ramanathan, S., Ananthi, V., Selvakumar, N., Fabrication and Characterization of Organic and In-Organic Reinforced A356 Aluminium Matrix Hybrid Composite by Improved Double-Stir Casting. *Silicon*, 11, 817–829, 2019.
10. Gnanavelbabu, A., Sunu Surendran, K.T., Kumar, S., Influence of ultrasonication power on grain refinement, mechanical properties and wear behaviour of AZ91D/nano-Al2O3 composites. *Mater. Res. Express*, 7, 16544, 2020.
11. Li, Q., Qiu, F., Dong, B., Geng, R., Lv, M., Zhao, Q., Jiang, Q.-C., Fabrication, microstructure refinement and strengthening mechanisms of nanosized SiCP/Al composites assisted ultrasonic vibration. *Mater. Sci. Eng. A*, 735, 310–317, 2018.
12. Prasad Reddy, A., Vamsi Krishna, P., Rao, R.N., Tribological Behaviour of Al6061–2SiC-xGr Hybrid Metal Matrix Nanocomposites Fabricated through Ultrasonically Assisted Stir Casting Technique. *Silicon*, 11, 2853–2871, 2019.
13. Zhou, X., Su, D., Wu, C., Liu, L., Tensile Mechanical Properties and Strengthening Mechanism of Hybrid Carbon Nanotube and Silicon Carbide Nanoparticle-Reinforced Magnesium Alloy Composites. *J. Nanomater.*, 2012, 1–7, 2012.
14. Lan, J., Yang, Y., Li, X., Microstructure and microhardness of SiC nanoparticles reinforced magnesium composites fabricated by ultrasonic method. *Mater. Sci. Eng. A*, 386, 284–290, 2004.
15. Harichandran, R. and Selvakumar, N., Microstructure and mechanical characterization of (B4C+ h-BN)/Al hybrid nanocomposites processed by ultrasound assisted casting. *Int. J. Mech. Sci.*, 144, 814–826, 2018.
16. Srivastava, N. and Chaudhari, G.P., Strengthening in Al alloy nano composites fabricated by ultrasound assisted solidification technique. *Mater. Sci. Eng. A*, 651, 241–247, 2016.
17. Éskin, G.I., Crystallization of Ingots of Magnesium Alloys with Ultrasonic Treatment of the Melt. *Metallurgist*, 47, 265–272, 2003.
18. Chen, Y.-J., Hsu, W.-N., Shih, J.-R., The Effect of Ultrasonic Treatment on Microstructural and Mechanical Properties of Cast Magnesium Alloys. *Mater. Trans.*, 50, 401–408, 2009.
19. Chen, G., Zhou, T., Yan, H., Li, K., Effects of ultrasonic field on microstructures and properties of semi-solid AZ91D magnesium alloy. *J. Wuhan Univ. Technol. Sci. Ed.*, 25, 555–560, 2010.
20. Khosro Aghayani, M. and Niroumand, B., Effects of ultrasonic treatment on microstructure and tensile strength of AZ91 magnesium alloy. *J. Alloys Compd.*, 509, 114–122, 2011.
21. Bhingole, P.P., Patel, B., Chaudhari, G.P., Ultrasonic Processing and Microstructural Analysis of AZ91 Magnesium Alloy. *Mater. Sci. Forum*, 02–703, 975–978, 2012.
22. Bhingole, P. and Chaudhari, G., Effect of Ultrasonic Treatment on the Grain Refinement and Mechanical Properties of AZ91 Magnesium Alloy. *Mater. Sci. Forum*, 710, 463–468, 2012.

23. Katsarou, L., Mounib, M., Lefebvre, W., Vorozhtsov, S., Pavese, M., Badini, C., Molina-Aldareguia, J.M., Jimenez, C.C., Pérez Prado, M.T., Dieringa, H., Microstructure, mechanical properties and creep of magnesium alloy Elektron21 reinforced with AlN nanoparticles by ultrasound-assisted stirring. *Mater. Sci. Eng. A*, 659, 84–92, 2016.
24. Dieringa, H., Katsarou, L., Buzolin, R., Szakács, G., Horstmann, M., Wolff, M., Mendis, C., Vorozhtsov, S., StJohn, D., Ultrasound Assisted Casting of an AM60 Based Metal Matrix Nanocomposite, Its Properties, and Recyclability. *Metals (Basel)*, 7, 388–401, 2017.
25. Mounib, M., Pavese, M., Badini, C., Lefebvre, W., Dieringa, H., Reactivity and Microstructure of Al 2 O 3 -Reinforced Magnesium-Matrix Composites. *Adv. Mater. Sci. Eng.*, 2014, 1–6, 2014.
26. Kumar, H. and Chaudhari, G.P., Creep behavior of AS41 alloy matrix nano-composites. *Mater. Sci. Eng. A*, 607, 435–444, 2014.
27. Liu, S., Gao, F., Zhang, Q., Li, W., Mechanical Properties and Microstrutures of Nano-Sized SiC Particles Reinforced AZ91D Nanocomposites Fabricated by High Intensity Ultrasonic Assisted Casting. *Mater. Sci. Forum - MATER SCI FORUM*, vol. 618–619, pp. 449–452, 2009.
28. Cao, G., Choi, H., Konishi, H., Kou, S., Lakes, R., Li, X., Mg–6Zn/1.5%SiC nanocomposites fabricated by ultrasonic cavitation-based solidification processing. *J. Mater. Sci.*, 43, 16, 5521–5526, 2008.
29. Cao, G., Konishi, H., Li, X., Mechanical properties and microstructure of SiC-reinforced Mg-(2,4)Al-1Si nanocomposites fabricated by ultrasonic cavitation based solidification processing. *Mater. Sci. Eng. A*, 486, 357–362, 2008.
30. Cao, G., Konishi, H., Li, X., Mechanical Properties and Microstructure of Mg/SiC Nanocomposites Fabricated by Ultrasonic Cavitation Based Nanomanufacturing. *J. Manuf. Sci. Eng.*, 130, 1–6, 2008.
31. Nie, K.B., Wang, X.J., Wu, K., Hu, X.S., Zheng, M.Y., Xu, L., Microstructure and tensile properties of micro-SiC particles reinforced magnesium matrix composites produced by semisolid stirring assisted ultrasonic vibration. *Mater. Sci. Eng. A*, 528, 8709–8714, 2011.
32. Jia, X.Y., Liu, S.Y., Gao, F.P., Zhang, Q.Y., Li, W.Z., Magnesium matrix nanocomposites fabricated by ultrasonic assisted casting. *Int. J. Cast Met. Res.*, 22, 196–199, 2009.
33. WANG, Z., WANG, X., ZHAO, Y., DU, W., SiC nanoparticles reinforced magnesium matrix composites fabricated by ultrasonic method. *Trans. Nonferrous Met. Soc China*, 20, s1029–s1032, 2010.
34. De Cicco, M., Konishi, H., Cao, G., Choi, H.S., Turng, L.-S., Perepezko, J.H., Kou, S., Lakes, R., Li, X., Strong, Ductile Magnesium-Zinc Nanocomposites. *Metall. Mater. Trans. A*, 40, 12, 3038–3045, 2009.
35. Chen, L.-Y., Peng, J.-Y., Xu, J.-Q., Choi, H., Li, X.-C., Achieving uniform distribution and dispersion of a high percentage of nanoparticles in metal

matrix nanocomposites by solidification processing. *Scr. Mater.*, 69, 634–637, 2013.
36. Song, S., Zhou, X., Li, L., Ma, W., Numerical simulation and experimental validation of SiC nanoparticle distribution in magnesium melts during ultrasonic cavitation based processing of magnesium matrix nanocomposites. *Ultrason. Sonochem.*, 24, 43–54, 2015.
37. Choi, H., Sun, Y., Slater, B.P., Konishi, H., Li, X., AZ91D/TiB2 Nanocomposites Fabricated by Solidification Nanoprocessing. *Adv. Eng. Mater.*, 14, 291–295, 2012.

13

# Friction Stir Welding of Dissimilar Magnesium Alloys: Analytical Modeling, Simulation, and Experimental Validation

**Biswajit Mallick[1], Jagadeesha T.[1*], Mothilal.T.[2] and Seeniappan Kaliappan[3]**

*[1]Department of Mechanical Engineering, National Institute of Technology NIT Calicut, Kerala, India*
*[2]Department of Automobile Engineering, KCG College of Technology, KCG Nagar, Karapakkam, Chennai, Tamil Nadu, India*
*[3]Department of Mechatronics Engineering, KCG College of Technology, KCG Nagar, Karapakkam, Chennai, Tamil Nadu, India*

## *Abstract*

Friction stir welding (FSW) represents a solid-state welding technique used to join dissimilar materials. This process presents a multiphysics challenge, encompassing heat transfer, mechanical deformation, and phase changes, thus introducing complexity. Given the difficulties in measuring numerous parameters during welding processes, the simulation of the actual welding procedure becomes crucial. In this chapter, an analytical modeling approach is employed to predict heat generation and temperature distributions when two dissimilar plates are joined. Rather than ascribing heat generation to plastic deformation, the focus is on accounting for the heat generated by the frictional aspect of the process. To enhance the computational model, the incorporation of a moving heat source is achieved using the DFLUX subroutine of ABAQUS. The objective is to validate simulations through experiments utilizing an H13 steel tool and optimize parameters to achieve the highest weld quality. Two distinct heat models are considered in these simulations: the surface heat model and the Goldak heat model. Notably, the surface heat model yields results that closely approximate reality compared to the Goldak model. While various techniques for joining metals have emerged, FSW remains the predominant method for solid-state welding, especially when the goal is to

**Corresponding author*: jagdishsg@nitc.ac.in

R. Thanigaivelan, Pradeep Kumar Krishnan, Kamalakanta Muduli and Santosh Kumar Tamang (eds.) New Materials, Processing and Manufacturability: Fabrication and Processing of Advanced Materials, (247–268) 

avoid exceeding the material's melting temperature. In this research, the focus is on joining two dissimilar magnesium alloys, namely, AZ31 and AZ61, *via* experimental approaches. Magnesium alloys have extensive applications, from automotive components to aerospace materials. Conventional welding methods that involve melting two dissimilar magnesium alloys are less desirable due to issues related to poor weldability. In conventional welding, the melting of dissimilar materials leads to problems such as different thermal coefficients, melting points, and mechanical properties like creep, resulting in intermetallic and transgranular defects in intermetallic phases. One significant challenge when welding magnesium alloys conventionally is that magnesium has a self-ignition temperature lower than its melting point. Therefore, caution must be exercised during conventional welding. Friction stir welding offers a means to minimize some of these welding defects, making it a promising technique for joining magnesium alloys.

***Keywords*:** Dissimilar metals, intermetallic compounds, microstructure, residual stress

## 13.1 Introduction

The pursuit of weight reduction in both the automobile and space research sectors has been on the rise, primarily due to its potential to decrease fuel consumption and enhance system efficiency. Among the various welding methods employed in industries, friction stir welding (FSW) stands out as the most commonly used. FSW is a multiphysics problem involving intricate phenomena such as melting, phase changes, the movement of heat sources, and mechanical stresses. Consequently, numerous researchers have directed their efforts toward simulating the FSW process. Biswajith *et al.* conducted an extensive review of FSW applied to magnesium alloys, elucidating its fundamental principles and various facets [1]. Zhu and Chao developed a simulation model that considers the nonlinearity of thermal and mechanical properties, recognizing FSW as a dynamic process. They noted that poor weld quality results from variations in residual stress and unsteady temperature distributions [2].

Sarma *et al.* devised a methodological approach for optimizing tool geometry, considering both mechanical and thermal loading with six different pin shapes. Their findings pointed to triangular-shaped pins as having the least flux and thus being the most effective among the six [3]. Hasan focused on optimizing process parameters to achieve the optimum tensile strength, strain, and hardness. The primary factors under consideration were tool rotational speed, traverse speed, and tilt angle. The stir zone exhibited the highest hardness due to its finer grain size compared to adjacent zones [4]. Zhang *et al.* delved into the role of friction in FSW, recognizing it as a source of heat generation. They employed a modified Gaussian heat source model to represent the friction at the tool-workpiece interface and solid-state deformation near the tool and workpieces. Their analysis revealed that at lower rotational speeds, heat generation was primarily due to friction, while at higher speeds, plastic deformation of the workpiece played a major role in heat generation [5, 6]. Schmidt *et al.* developed a mathematical formulation to express heat production near the tool and workpiece using a cylindrical pin configuration. They assumed that shear stress along the tool-workpiece interface is uniform, though this is not entirely accurate due to the sliding of primary and secondary shear zones during FSW, and they did not consider deformation in the bulk of the workpiece [7].

Yuqing *et al.* investigated the influence of an eccentric tool pin on mechanical properties and microstructure in FSW, employing a 7075 aluminum alloy plate. They observed that an eccentric cylindrical pin produced sound joints with better mechanical properties, along with lower maximum temperatures due to reduced heat generation compared to a cylindrical tool [8]. Essa *et al.* derived an analytical model for heat generation in aluminum alloys using an eccentric cylindrical pin. Their model was in good agreement with the maximum temperature expression, considering a constant beta to account for transverse velocity when using an eccentric pin [9]. Chen *et al.* explored a retractable pin tool FSW, considering different levels of retraction. They found that the total volume of deforming material decreased significantly with increased retraction, and the temperature gradient in the vicinity of the shoulder was nonuniform [10].

Elangovan *et al.* analyzed the impact of tool shoulder diameter and tool pin profile on the formation of the friction stir processed zone [11]. Vijay Shivaji Ghadakh *et al.* presented an analytical model of heat generation in FSW using a tapered cylindrical pin with a flat shoulder, highlighting the importance of considering temperature-dependent friction coefficients and the relationship between taper angle and heat generation [12]. Thomas and Nicholas emphasized the role of eccentricity between the tool shoulder and pin in creating dynamic orbits, which increased friction at the tool-work interface [13, 14]. Hamilton *et al.* introduced the concept of a slip factor, which accounted for increased energy input due to enhanced friction length and effective contact area. This factor was found to be dependent on various process parameters, such as speed, interface temperature, and heat input. They noted that thermal models may struggle to predict temperature distribution at low temperatures due to the unpredictability of solidus temperature but performed well at high temperatures with elevated energy levels stemming from increased interface friction [15–18].

## 13.2 Analytical Modeling

Before one can create a model for heat generation, certain assumptions must be made. These assumptions are as follows: Heat generated is presumed to be primarily influenced by the frictional force and not the plastic deformation in the workpiece material. $\tau_{contact}$ is assumed to be due to purely frictional contact and the contact type is sliding in nature. Sticking contact is not considered for the above analytical model. Therefore, $\tau_{contact} = \mu p$, where μ is the frictional coefficient and p is the pressure applied on the tool shoulder. Table 13.1 gives the mechanical properties of the material used, SDSS 2507.

**Table 13.1** Mechanical properties of SDSS 2507.

| Symbol | Description |
|---|---|
| $Q_1$ | Shoulder heat generation (W) |
| $Q_2$ | Tool probe heat generation (W) |
| $Q_3$ | Probe side heat generation (W) |
| $Q_{total}$ | Overall heat generation (W) |
| $\tau_{contact}$ | Contact shear stress (Pa) |
| $R_{shoulder}$ | Radius of tool shoulder (mm) |
| $R_{probe}$ | Radius of tool probe (mm) |
| $H_{probe}$ | Height of tool probe (mm) |
| $\omega$ | Angular velocity of tool (rad/s) |
| $e$ | Eccentricity (mm) |
| $\alpha$ | Tool shoulder angle (°) |

## 13.3 Heat Input Modeling

In practice, FSW is the result of both friction and plastic deformation occurring between the tool and the workpiece. When solely considering the friction aspect and neglecting plastic deformation, the heat generation in various sections of the tool can be broken down into three distinct components.

The tool shoulder heat production is given by

$$Q_1 = \frac{2}{3}\pi\tau_{contact}\omega\left(R_{shoulder}^3 - R_{probe}^3\right)(1+\tan\alpha)$$

The tool probe side heat production is given by

$$Q_2 = 2\pi\tau_{contact}\omega R_{probe}^2 H_{probe}$$

The tool probe bottom heat production is given by

$$Q_3 = \frac{2}{3}\pi\tau_{contact}\omega R_{probe}^3$$

Therefore, the total heat generated can be summarized as the sum of the above three heat inputs

$$Q_{total} = \frac{2}{3}\pi\tau_{contact}\omega(\left(R_{shoulder}^3 - R_{probe}^3\right)(1+\tan\alpha) + 3R_{probe}^2 H_{probe} + R_{probe}^3)$$

As we have assumed, the cylindrical pin with no tool shoulder angle, i.e., $\alpha = 0$

$$Q_{total} = \frac{2}{3}\pi\tau_{contact}\omega\left(R_{shoulder}^3 + 3R_{probe}^2 H_{probe}\right)$$

We will also assume the case of heat generated in an FSW tool with pin eccentricity as shown in Figure 13.1. The heat generated for the eccentric tool can be calculated in the same way as for a normal tool taking the eccentricity into consideration.

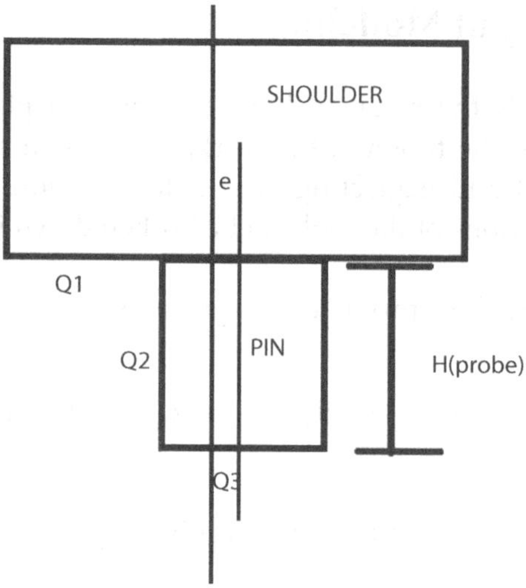


**Figure 13.1** Eccentricity in FSW tool.

The tool shoulder heat production is given by

$$Q_1 = \frac{2}{3}\pi\tau_{contact}\omega\left(\frac{R_{shoulder}^3 - R_{probe}^3}{\cos\alpha}\right)$$

The tool probe side heat generation is given by

$$Q_2 = 2\pi\tau_{contact}\omega\left(R_{probe} + e\right)^2 H_{probe}$$

The tool probe bottom heat generation is given by

$$Q_3 = \frac{2}{3}\pi\tau_{contact}\omega\left((R_{probe} + e)^3 - e^3\right)$$

Therefore, the total heat generated can be summarized as the sum of the above three heat inputs

$$Q_{total} = \frac{2}{3}\pi\tau_{contact}\omega\left[\left(\frac{R_{shoulder}^3 - R_{probe}^3}{\cos\alpha}\right) + 3(R_{probe} + e)^2 H_{probe} + (R_{probe} + e)^3 - e^3\right]$$

Based on certain parameters, different heat generated can be calculated for both the tools and the heat generated can be used as an input parameter for heat flux while modeling in commercial software like ABAQUS.

### 13.3.1 Calculation of Heat Generation

Heat generation is calculated using the abovementioned parameters. The initial assumption of heat being generated is only due to the frictional contact and not due to plastic deformation in the material. Table 13.2 shows heat generated in both normal and eccentric tools.

It can be seen that the heat generated in the eccentric tool is more than that of the normal tool. While the tool shoulder is the major source of heat generated in both cases, the percentage of heat generated by the pin probe and pin bottom is higher in the eccentric tool.

### 13.3.2 Heat Flux Modeling

For the heat flux to be given in the simulation, we have assumed the flux to be of two types, i.e., surface heat model and double elliptical.

i) Surface heat model (Gaussian):
It is assumed that the heat flux, q(r), is linearly distributed in the radial direction of the pin tool shoulder

$$q(r) = \frac{12Qr}{\pi(d_0^3 - d_1^3)}$$

where $r_1 < r < r_2$

**Table 13.2** Heat generated in both normal and eccentric tools.

| Tool angular speed | Cylindrical tool | | | | Eccentric pin tool | | | |
|---|---|---|---|---|---|---|---|---|
| | Q1 | Q2 | Q3 | Q | Q1 | Q2 | Q3 | Q |
| 600 | 362.84 | 46.41 | 12.66 | 421.91 | 362.84 | 99.24 | 32.56 | 494.65 |
| 700 | 423.32 | 54.15 | 14.77 | 492.23 | 423.32 | 115.78 | 37.99 | 577.09 |
| 800 | 483.79 | 61.88 | 16.88 | 562.55 | 483.79 | 132.32 | 43.42 | 659.53 |
| 900 | 544.27 | 69.62 | 18.99 | 632.87 | 544.27 | 148.86 | 48.84 | 741.97 |

ii) Goldak heat source model:

$$q(x,y,z,t) = \frac{6\sqrt{3} f_r Q}{abc_r \pi \sqrt{\pi}} e^{-3\frac{(x-x_c)^2}{a^2}} e^{-3\frac{(y-y_c)^2}{b^2}} e^{-3\frac{(z-z_c)^2}{c^2}}$$

Details are given in Figure 13.2,
where x, y, and z – the points where the temperature to be found
xc, yc, and zc – the coordinates of the heat source
Heat input = Q
η – the heat source efficiency
t – the time
a, b, and c – the semi-major axes of the ellipse along the x, y, and z directions, respectively
ff and fr – the fraction of heat deposited in the front and rear quadrants.

The dimensions for the Goldak model generally take the weld pool dimensions into consideration. Given that we are working on FSW, the dimension will mainly depend on the tool and HAZ. The dimension values are given in Table 13.3.

On the boundary or the surfaces of the workpiece, convection and radiation in heat transfer are responsible for heat loss to room temperature. Such heat convection and radiation on all plate surfaces except for the bottom surface, the heat flux loss is evaluated by

$$q_s = h(T - T_o) + \varepsilon k(T^4 - T_0^4)$$

where
$h =$ the convection coefficient
$\varepsilon =$ the emissivity of the plates
k = Stefan-Boltzmann constant.

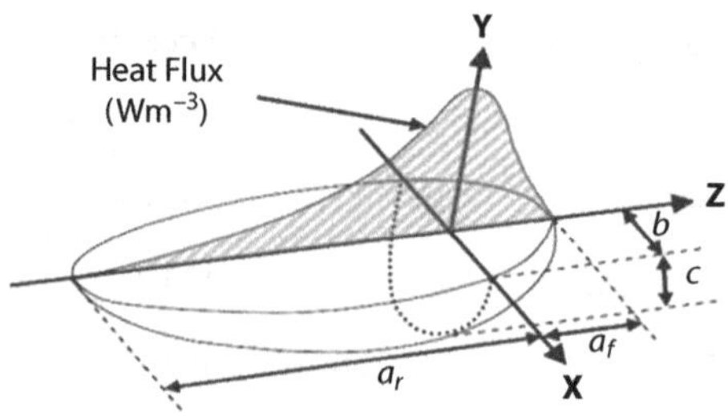


**Figure 13.2** Double ellipsoidal heat source configuration.

**Table 13.3** Values for the Goldak model.

| Dimension | Value |
|---|---|
| b | 0.015 |
| c | 3.8 |
| $a_r$ | 0.009 |
| $a_f$ | 0.009 |
| $f_f$ | 0.7 |
| $f_r$ | 1.3 |

A solid homogenous model is selected for the simulation purpose. The workpiece was made with dimensions of 100 mm × 60 mm × 4 mm. Two different material properties were given on different sections of the workpiece. Material properties such as density, elasticity, specific heat, and thermal conductivity were taken as temperature-dependent properties. The step was taken as coupled temperature displacement for the sake of simulation and step time was taken as per the traverse speed. On the boundary or the surfaces of the workpiece, convection and radiation in heat transfer are responsible for heat loss to room temperature. Heat flux was taken as user-defined and is given through the subroutine. The subroutine variables were changed, and accordingly, the simulation was done for various other parameters. The meshing was done in a gradient-biased manner so that there would be finer mesh along the weld line. Tables 13.4 and 13.5 show the properties of AZ31 and AZ61.

**Table 13.4** Thermal properties of AZ31.

| Temp (°C) | Conductivity (J/kg K) | Specific heat (W/m K) | Density (kg/m$^3$) |
|---|---|---|---|
| 100 | 1005.9 | 90 | 1730.18 |
| 200 | 1031.9 | 95.8 | 1687.97 |
| 300 | 1049.3 | 98.4 | 1656.83 |
| 400 | 1068.9 | 98.6 | 1626.91 |

**Table 13.5** Thermal properties of AZ61.

| Temp (°C) | Conductivity (J/kg K) | Specific heat (W/m K) | Density (kg/m³) |
|---|---|---|---|
| 100 | 1018.8 | 70 | 1732.87 |
| 200 | 1037.2 | 76.9 | 1704.96 |
| 300 | 1076.2 | 81.4 | 1655.28 |
| 400 | 1098.6 | 83.5 | 1610.43 |

In the following simulation, the role of tool material is not taken into account, and the friction coefficient is taken as 0.5 according to the journal papers. Also, the back plate of the workpiece which is in contact with the machine surface is assumed to have a greater heat loss coefficient of 175 $W/m^{2\circ}C$. The following figures show the simulation part that was carried out in ABAQUS CAE 6.14 and show the various stages involved in the simulation: Preparing the part, assigning material properties, step section of the simulation, and interaction step of the simulation. In the interaction step of the simulation, the surrounding temperature/sink temperature is taken as 25°C, the film coefficient is taken as 30 $W/m^{2\circ}C$, and the amplitude of the film coefficient is taken as instantaneous. The above figure shows the interaction step of the simulation. Convection is assumed to be on the top and side surfaces of the workpiece, whereas radiation heat loss is assumed to be whole over the body. Load section of the simulation: The most important part of the simulation is the load section where the degree of freedom of the model is given, i.e., the whole body is encapsulated so that when load is given it will not move in any direction. The heat flux is made to be user-defined and the magnitude is taken as 1, which will be given in the later part of the simulation through the DFLUX subroutine. Biased meshing of the model: A solid homogenous model is made for the simulation purpose. The workpiece was made with dimensions of 100 mm × 120 mm × 4 mm. Two different material properties were given on different sections of the workpiece. A biased meshing was done so as to get finer mesh on the weld line and coarser mesh on the rest of the workpiece. An eight-node thermally coupled brick, trilinear displacement, and temperature (C3D8T) type of element was used for the mesh. Number of nodes: 18480. Number of elements: 11988. Fortran codes were developed for Gaussian heat flux and for the Goldak heat flux model.

The temperature distribution in both the heat models varies from each other along with the maximum temperature achieved. Throughout the simulation, we saw that there is a greater temperature distribution in the Goldak model as compared to the Gaussian one. The accuracy of both models can be determined with a validation of the simulation with experimental results. Temperature distribution for a trial simulation was done with 600, 700, and 800 rpm, and only the heat generated from the shoulder was considered. It was found that the maximum temperature was attained at the end of the simulation process. The temperature distribution for various process parameters is shown in Figures 13.3 to 13.8.

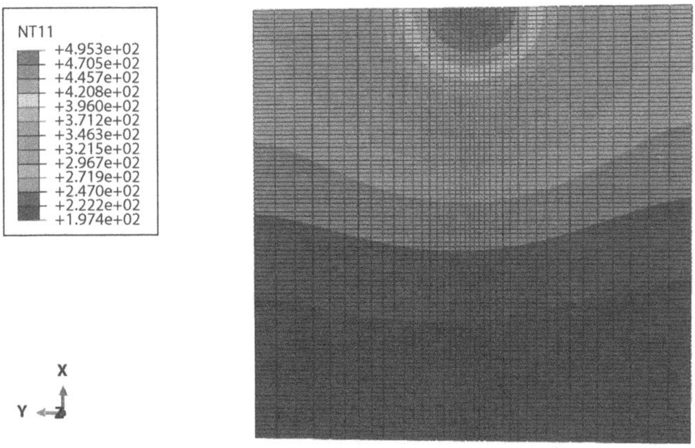


**Figure 13.3** Temperature distributions with 600 rpm, traverse speed of 40 mm/min in the Goldak model.

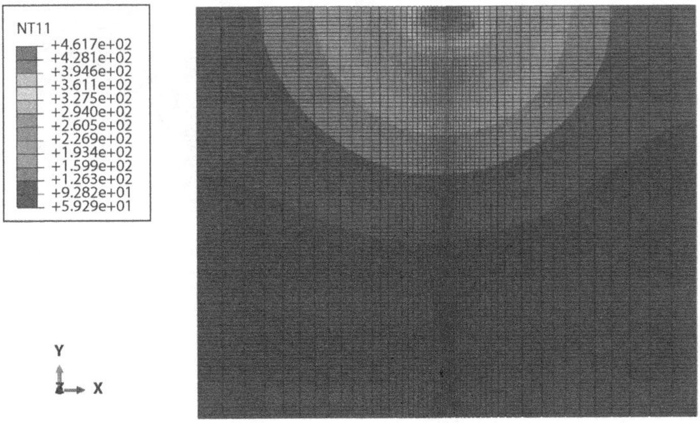


**Figure 13.4** Temperature distribution with 600 rpm, traverse speed of 40 mm/min in Gaussian heat flux model.

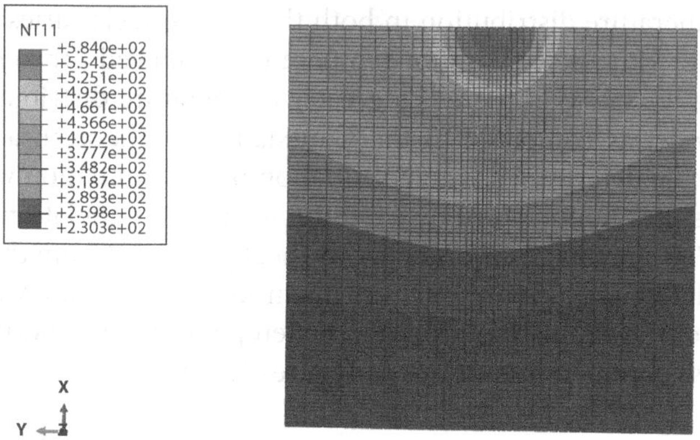


**Figure 13.5** Temperature distribution with 700 rpm, traverse speed of 40 mm/min for the Goldak model.

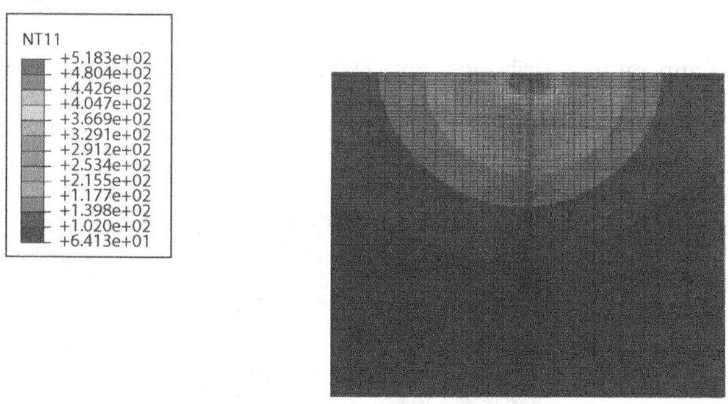


**Figure 13.6** Temperature distributions with 700 rpm, traverse speed of 40 mm/min for Gaussian model.

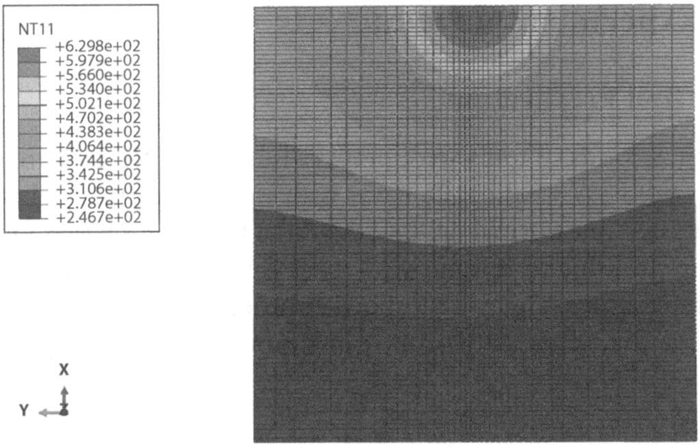


**Figure 13.7** Temperature distribution with 800 rpm, traverse speed of 40mm/min for the Goldak model.

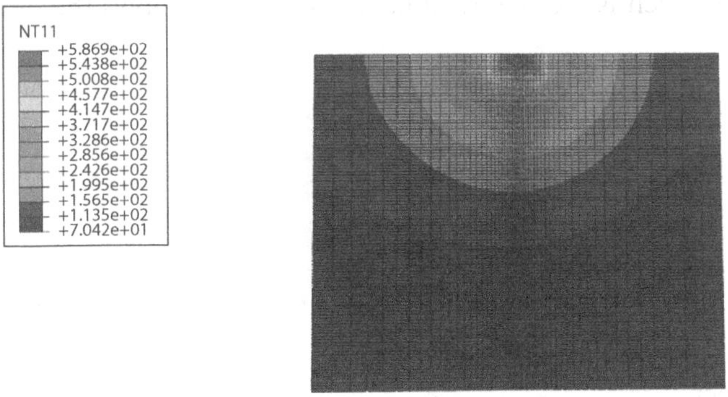


**Figure 13.8** Temperature distribution with 800 rpm, traverse speed of 40 mm/min for Gaussian model.

# 13.4 Experimental Detail

## 13.4.1 Tool Preparation

Two different kinds of tools were prepared using a CNC lathe machine for the purpose of FSW welding. One was with an eccentric pin tool with 1-mm eccentricity and one normal pin tool. Material for tool preparation was taken as H13 die steel due to its ability to resist change in shape and size under higher temperatures. Figure 13.9 shows the tool prepared for this purpose.

Shoulder diameter = 18 mm, pin diameter = 6 mm, pin height = 3.8 mm.

## 13.4.2 Work Material

Eight workpieces of dimensions $100\ \text{mm} \times 50\ \text{mm} \times 4\ \text{mm}$ are cut using a band saw cutter machine and then filed to get a better joint along the weld line. The composition of both materials is shown in Table 13.6.

# 13.5 Experimental Setup

Mainly three parameters were considered for the experimental design, i.e., tool rotational speed, weld speed or traverse speed, and eccentricity of the tool which is a categorical factor. Table 13.7 shows the details of the

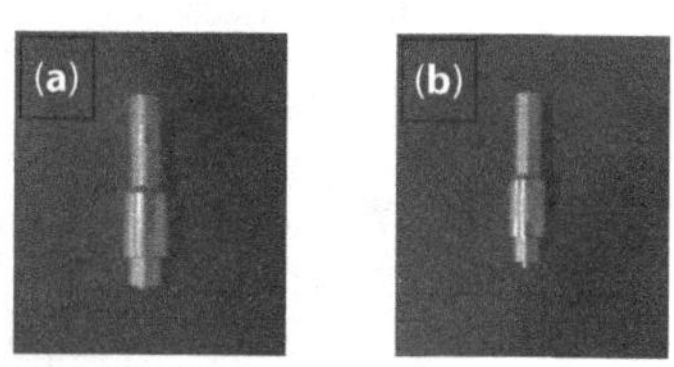


**Figure 13.9** (a) FSW tool with eccentric cylindrical pin and (b) FSW tool with normal cylindrical pin.

**Table 13.6** Composition of AZ31 and AZ61.

| Alloy | Al | Zn | Mn | Si | Fe | Ca | Mg |
|---|---|---|---|---|---|---|---|
| AZ31 | 3.1 | 0.73 | 0.25 | 0.02 | 0.005 | 0.0014 | Balance |
| AZ61 | 6.2 | 0.74 | 0.23 | 0.04 | 0.004 | 0.0013 | Balance |

**Table 13.7** Experimental design of the FSW process.

| Standard | Run | Tool RPM | Traverse speed (mm/min) | Eccentricity |
|---|---|---|---|---|
| 8 | 1 | 1200 | 50 | Centric |
| 3 | 2 | 900 | 50 | Eccentric |
| 7 | 3 | 900 | 50 | Centric |
| 6 | 4 | 1200 | 25 | Centric |
| 2 | 5 | 1200 | 25 | Eccentric |
| 4 | 6 | 1200 | 50 | Eccentric |
| 5 | 7 | 900 | 25 | Centric |
| 1 | 8 | 900 | 25 | Eccentric |

parameters chosen. The total number of experiments that were carried out was eight and the number of replications that were kept was one. ETA linear FSW was used for the welding part. AZ31 was kept on the advancing side for all experiments and AZ61 on the retreating side. Both were clamped tightly on the vice of the machine table. Welding was done at 900 and 1200 rpm with welding speeds of 25 and 50 mm/min. One trial run was done for 100 mm/min and 900 rpm. The flashes were removed with the help of a plier. Figure 13.10 shows the experimental setup and the welding process.

### 13.5.1 Tensile Test

Tensile test specimens were prepared using wire EDM and the dimensions were kept according to ASTM E8 standard for thickness of 4 mm. The feed

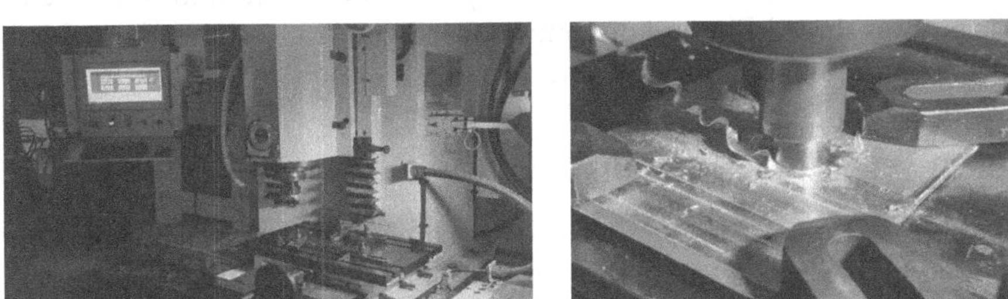

(a) (b)

**Figure 13.10** (a) Linear friction stir welding machine at IIT Bhilai. (b) Workpiece being welded.

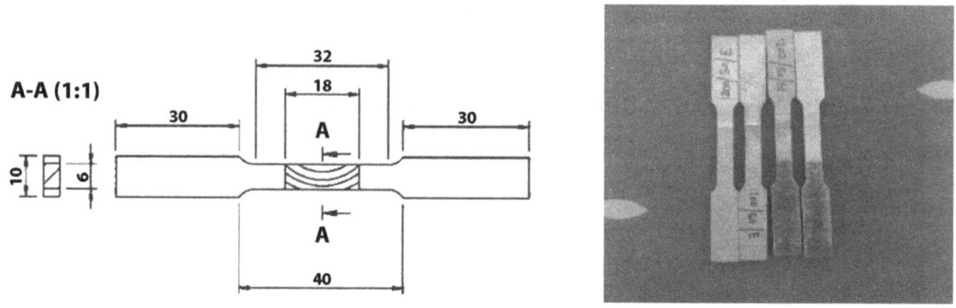


**Figure 13.11** Dimensions of the specimen.

rate for wire EDM was around 3.3 mm/min for the welded plates. A total of eight specimens were prepared for each experiment to be tensile tested and analyzed. Figure 13.11 shows the dimensions of the specimen.

### 13.5.2 Microhardness Test

Only the weld zone was cut and polished using a disc polisher machine to prepare the part for hardness testing. Vicker's hardness testing machine was used to measure the microhardness perpendicular to the weld zone, traversing along the vertical face of the weld. The force was kept at 2 kgf and the dwell time was 10 s for the indentation to happen.

## 13.6 Experimental Results

### 13.6.1 Weldments

The welds looked sound- and defect-free from outer views but a close look showed tunnel defects present in the weld done with an eccentric pin. Normal flash defects were there in every weld but the flash for the eccentric pin was thread-like and discontinuous. Figure 13.12 shows the welded specimen for different process parameters.

### 13.6.2 Tensile Test Results

The tensile strength for the welded specimen done with a normal pin showed higher tensile strength than that of an eccentric pin. The process parameter of 1200 rpm, 50 mm/min with a normal pin showed the highest tensile strength of 183.28 MPa, while the process parameter of 900 rpm, 50mm/min with an eccentric pin showed the least tensile strength. Table 13.8 shows the Tensile test results for different process parameters.

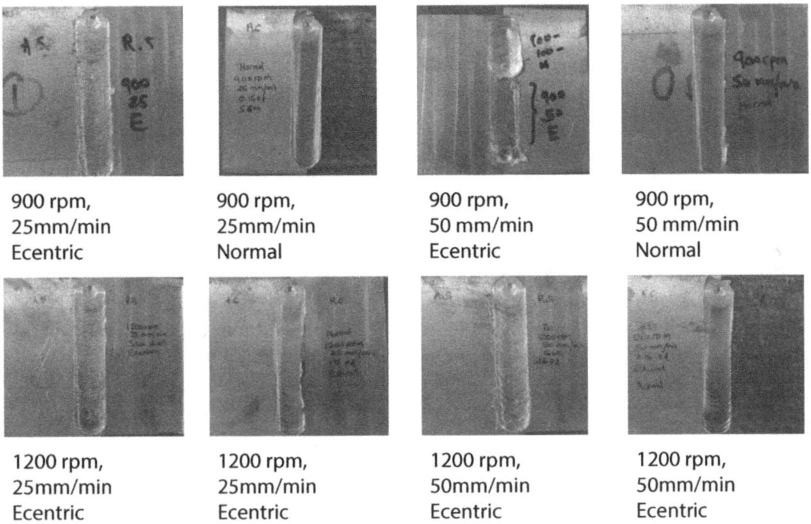


**Figure 13.12** Welded specimen for different process parameters.

**Table 13.8** Tensile test results for different process parameters.

| Standard | Run | Tool RPM | Traverse speed (mm/min) | Eccentricity | Tensile strength (MPa) |
|---|---|---|---|---|---|
| 8 | 1 | 1200 | 50 | Centric | 183.286 |
| 3 | 2 | 900 | 50 | Eccentric | 89.34 |
| 7 | 3 | 900 | 50 | Centric | 168.56 |
| 6 | 4 | 1200 | 25 | Centric | 135.472 |
| 2 | 5 | 1200 | 25 | Eccentric | 108.658 |
| 4 | 6 | 1200 | 50 | Eccentric | 117.680 |
| 5 | 7 | 900 | 25 | Centric | 179.756 |
| 1 | 8 | 900 | 25 | Eccentric | 89.35 |

The tensile strength of the best weld was 30% less than the tensile strength of the base material which suggests that the process parameter of 1200 rpm, 50 mm/min with normal pin gives a very good weld with little to no defects. Analyzing the stress-strain curves also suggests that higher rpm with lower traverse speed may negate the effect of eccentricity and give good weld. Figure 13.13 shows the stress-strain diagram for various cutting and process parameters.

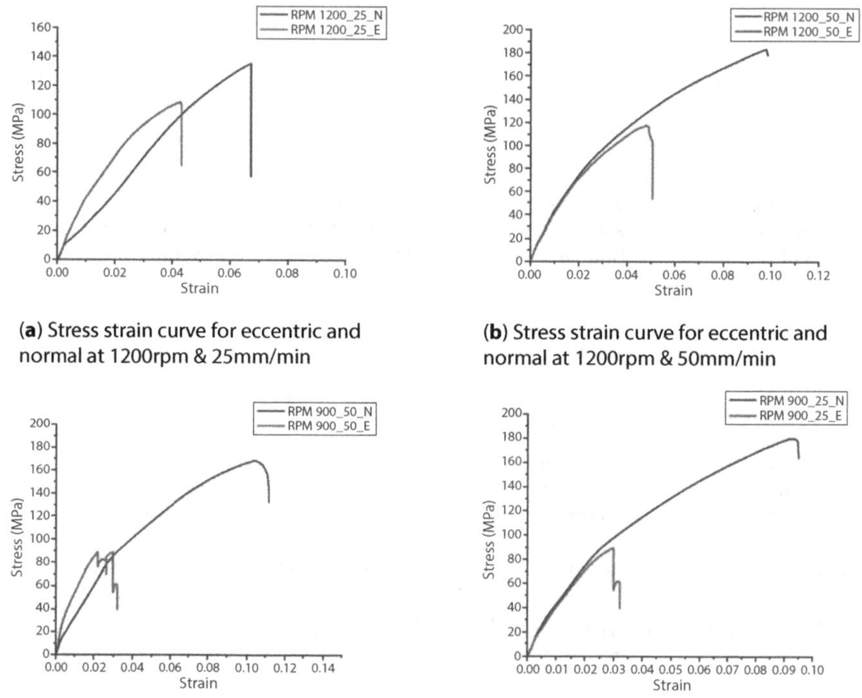


(**a**) Stress strain curve for eccentric and normal at 1200rpm & 25mm/min

(**b**) Stress strain curve for eccentric and normal at 1200rpm & 50mm/min

(**c**) Stress strain curve for eccentric and normal at 900rpm & 50mm/min

(**d**) Stress strain curve for eccentric and normal at 900rpm & 25mm/min

**Figure 13.13** Stress-strain diagram for various process parameters.

## 13.6.3 Hardness Test Results

A hardness test was only done for a welded specimen of the normal pin as the eccentric pin had tunnel defects throughout the weld. The hardness was measured perpendicular to the welding direction, traversing along the

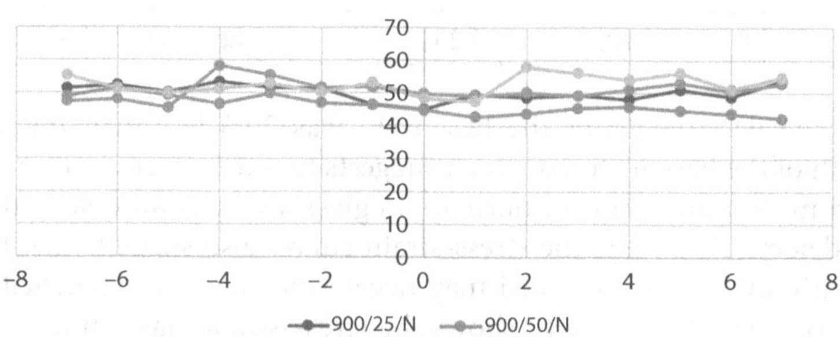


**Figure 13.14** Hardness measured across the weldment.

different weld zones of HAZ, TMAZ, and stir zone. The lower deviation in hardness throughout the weldment may be attributed to the fact that the stirring created a stable joint and no intermetallic compounds were formed throughout the stir zone which is generally the case. Figure 13.14 shows hardness measured across the weldment.

## 13.7 Conclusion

In the study, specimens were subjected to various FSW parameters to join AZ31 and AZ61 Mg alloys. The FSW process parameters were found to significantly influence the weld quality and mechanical properties. One key finding was that the ultimate tensile strength of the welded samples, using a rotational speed of 1200 rpm and a traverse speed of 50 mm/min with a normal cylindrical pin, reached 183 MPa. However, this value was observed to be 30% lower than that of the parent materials, indicating a reduction in strength after welding. The maximum elongation was achieved with a rotational speed of 900 rpm and a traverse speed of 50 mm/min, using a normal pin type, resulting in an elongation of approximately 4.15 mm. The research highlighted the importance of process parameters in avoiding defects during FSW. Increasing the tool's rotational speed led to higher temperatures due to increased relative velocity between the workpiece and the tool, resulting in more friction in the weld zone per unit time. Interestingly, the study also revealed that pin shape and size had less impact on the weld quality compared to pin eccentricity. Although theoretically, an eccentric pin was expected to create a larger stir zone, experimental work showed that poor weld quality was often observed, except at higher rotational speeds. Furthermore, it was noted that there was low variation in hardness, indicating the stability of the joint and the absence of intermetallic compounds, which is typically a concern in dissimilar FSW welding. In conclusion, the study emphasized the significance of FSW process parameters in achieving sound welds and highlighted the need for more comprehensive simulations that consider factors like plastic deformation and material flow for better results.

## References

1. Biswajit, M., Numerical simulation, experimental studies and optimization of FSW of dissimilar magnesium alloys, M. Tech Thesis, National Institute of Technology, Calicut, 2022.

2. Zhu, X.K. and Chao, Y.J., Numerical Simulation of Transient Temperature and Residual Stresses in Friction Stir Welding of 304L Stainless Steel. *J. Mater. Process. Technol.*, 146, 2, 263–272, 2004.
3. Sarma, K.L.N.S., Srikanth, A., Venkateshwarlu, B., Methodological Approach for Best Tool Geometry Determination in Friction Stir Welding Process. *Mater. Today, Proc.*, 46, 7105–7114, 2021.
4. Hasan, M.R., Davies, I.J., Pramanik, A. *et al.* Impact of process parameters on improving the performance of 3D printed recycled polylactic acid (rPLA) components. Int J Adv Manuf Technol 131, 3751–3779, 2024. https://doi.org/10.1007/s00170-024-13140-7.
5. Zhang, S., Chen, G., Liu, Q., Li, H., Zhang, G., Wang, G., Shi, Q., Numerical Analysis and Analytical Modeling of the Spatial Distribution of Heat Flux during Friction Stir Welding. *J. Manuf. Process.*, 33, 245–255, 20182018.
6. Zhu, X.K. and Chao, Y.J., Numerical Simulation of Transient Temperature and Residual Stresses in Friction Stir Welding of 304L Stainless Steel. *J. Mater. Process. Technol.*, 146, 2, 263–272, 2004.
7. Schmidt, H., Hattel, J., Wert, J., An Analytical Model for the Heat Generation in Friction Stir Welding, *Model. Simul. Mater. Sci. Eng.*, 12, 1, 143–157, 2014.
8. Mao, Y., Ke, L., Liu, F., Liu, Q., Huang, C., & Xing, L., Effect of tool pin eccentricity on microstructure and mechanical properties in friction stir welded 7075 aluminum alloy thick plate. *Mater. Des.*, (1980-2015), 62, 334–343, 2014.
9. Essa, A.R.S., Ahmed, M.M.Z., El-Nikhaily, A.E., An analytical model of heat generation for eccentric cylindrical pin in friction stir welding. *J. Mater Res. Technol.*, 5, 3, 234–240, 2016.
10. Chen, K., Liu, X., & Ni, J., A review of friction stir–based processes for joining dissimilar materials. *The International Journal of Advanced Manufacturing Technology*, 104, 1709–1731, 2019.
11. Elangovan, K. and Balasubramanian, V., Influences of tool pin profile and tool shoulder diameter on the formation of friction stir processing zone in AA6061 aluminum alloy. *Mater. Des.*, 29, 362–373, 2008.
12. Gadakh, V. S., & Adepu, K., Heat generation model for taper cylindrical pin profile in FSW. *J. Mater. Res. Technol.*, 2(4), 370–375, 2013.
13. Thomas, W. M., Nicholas, E. D., & Kallee, S. W., Friction based technologies for joining and processing. In *TMS friction stir welding and processing conference*. 1–8, November, 2001.
14. Serindag, H.T. and Kiral, B.G., Friction Stir Welding of AZ31 Magnesium Alloys – A Numerical and Experimental Study. *Lat. Am. J. Solids Struct.*, 14, 1, 113–130, 2017.
15. Hamilton, C., Sommers, A., & Dymek, S., A thermal model of friction stir welding applied to Sc-modified Al–Zn–Mg–Cu alloy extrusions. *Int. J. Mach. Tool Manu.*, 49, 3-4, 230–238, 2009.
16. Kaliappan, S., Shanmugam, A., Johnson, P., Karthick, M., Sekar, S., Patil, P.P., Govindaraajan, V., Impact of AlN-SiC Nanoparticle Reinforcement on the

Mechanical Behavior of Al 6061-Based Hybrid Composite Developed by the Stir Casting Route. *Adv. Mater. Sci. Eng.*, 2022.

17. Suryanarayanan, R., Sridhar, V.G., Natrayan, L., Kaliappan, S., Merneedi, A., Sathish, T., Yeshitla, A., Improvement on mechanical properties of submerged friction stir joining of dissimilar tailor welded aluminum blanks. *Adv. Mater. Sci. Eng.*, 1–6, 2021.
18. Venkatesh, R., Manivannan, S., Kaliappan, S., Socrates, S., Sekar, S., Patil, P.P., Bayu, M.B., Influence of different frequency pulse on weld bead phase ratio in gas tungsten arc welding by ferritic stainless steel AISI-409L. *J. Nanomater.*, 2022.

14

# Advancements in Welding Techniques: Surface and Mechanical Property Insights

**Jagadeesha T.[1*], Manoj Nikam[2], T. Mothilal[3] and M. D. Raj Kamal[4]**

*[1]Department of Mechanical Engineering, National Institute of Technology NIT, Calicut, Kerala, India*
*[2]Department of Mechanical Engineering, Bharati Vidyapeeth College of Engineering, Navi, Mumbai, India*
*[3]Department of Automobile Engineering, KCG College of Technology, KCG Nagar, Karapakkam, Chennai, Tamil Nadu, India*
*[4]Department of Mechanical Engineering, Velammal Institute of Technology, Chennai, Tamil Nadu, India*


### Abstract

The welding industry has evolved to demand highly efficient, precise, and accurate welding processes, driving the need for increased productivity. In recent years, robotics has played a significant role in automating welding operations, vastly improving efficiency compared to manual methods used in the past. This chapter explores various welding processes, their applications, and resulting properties. One particularly notable welding process is friction welding, known for producing welds with exceptional strength and fatigue resistance. This method is versatile and capable of joining dissimilar materials effectively. Additionally, friction welding minimizes distortion and eliminates the need for melting or filler materials, making it an environmentally friendly option. The mechanical properties of friction welds depend on several key factors, including material properties and welding parameters like friction force, forging force, and welding time. Monitoring the weld's temperature and the tool's position is a critical aspect in assessing the quality of friction stir welding, a cutting-edge technology within the realm of friction welding.


**Corresponding author*: jagdishsg@nitc.ac.in

R. Thanigaivelan, Pradeep Kumar Krishnan, Kamalakanta Muduli and Santosh Kumar Tamang (eds.) New Materials, Processing and Manufacturability: Fabrication and Processing of Advanced Materials, (269–286) © 2024 Scrivener Publishing LLC

***Keywords*:** Welding processes, friction welding, welding metallurgy, welding temperature

## 14.1 Introduction

Welding is the joining process in which similar or dissimilar materials are joined with or without the application of heat and pressure. Different kinds of welding processes are available to join the metals or materials [1]. New welding processes are developing to improve safety, efficiency, accuracy, quality of weld, less time-consuming, etc. One of the latest developments in welding is the use of advanced robotic welding systems, which offer several benefits over traditional welding methods [2]. These systems are used as computer-controlled robots to perform welding operations with high precision and consistency, which decreases the risk of human error and improves overall welding quality [3]. Another new welding process has developed laser welding. Some welding processes such as arc welding, resistance welding, friction welding, gas welding, brazing, and soldering have been going on for many years [4]. Friction welding is a solid-state type of welding process in which welding is completed by applying pressure after gaining the desired heat at the interface of metals where we have to weld. In this friction welding, heat is generated by the relative motion of two metals with respect to each other [5].

The two main factors that affect the weld quality are the temperature of the weld and tool positioning during friction stir welding. The new method for measuring process parameters in real time using an infrared thermal imaging camera and a high-speed camera along with a new algorithm for analyzing the data collected from these measurements helps in improving the surface characteristics of weld [6]. The choice of process parameters is important in controlling the microstructures and mechanical properties of the welds. The post-weld heat treatment also helps in achieving the desired microstructure and mechanical properties. Welding dissimilar materials using friction welding poses several challenges due to differences in their physical and metallurgical properties [7].

### 14.1.1 Arc Welding Processes

Arc welding is a cost-effective and versatile welding method widely utilized in various applications. Its affordability and broad range of uses make it a preferred choice in many industries, including motor vehicle manufacturing, shipbuilding, and power plants. In arc welding, a coated electrode is held by an electrode holder.

The welding process begins with the generation of an arc between the wire electrode and the workpiece. This arc provides the necessary energy to melt the flux, which then fuses with the liquid metal of the workpiece and settles into the groove. Due to its lower density, the flux rises to the top surface of the metals, creating a protective slag coating.

The area surrounding the weld, where the temperature exceeds the recrystallization point, is referred to as the heat-affected zone (HAZ). Arc welding offers various processes suitable for welding similar or dissimilar materials, allowing for the selection of the most appropriate method based on job requirements and economic advantages.

### 14.1.2 Tungsten Inert Gas Welding

In this welding process, an arc is generated between a nonconsumable tungsten electrode and the workpiece within a protective atmosphere of inert gas, typically argon. The external filler material is necessary to complete the weld due to the nonconsumable nature of the electrode. Both AC and DC machines can be employed for this process, but AC is commonly used. The tungsten electrode serves as the negative pole, while the workpiece acts as the positive pole. This arrangement is preferred because the positive pole has a lower melting temperature, making it more prone to melting, while the negative pole has a higher melting temperature. Tungsten inert gas (TIG) welding is well-suited for materials like aluminum, magnesium, and other reactive metals, especially when they are in the form of thin plates. However, it is considered a costly welding process. Its versatility extends to underwater welding, thanks to its ability to maintain a protective inert medium in challenging environments.

*Advantages of TIG Welding:*

- High-quality welds: TIG welding produces high-quality, precise, and clean welds. The process allows for fine control over the heat input, arc length, and weld pool, resulting in welds with minimal distortion, porosity, or spatter.
- No flux: Unlike other welding processes, TIG welding does not require any flux or filler metal, which reduces the risk of contamination and eliminates the need for post-weld cleaning.

- Low heat input: TIG welding has a low heat input, which minimizes the risk of warping or distorting the workpiece.
- Low fumes: TIG welding produces minimal fumes and smoke, making it a safer and healthier option for the operator and surrounding environment.
- Aesthetics: TIG welding produces clean and aesthetically pleasing welds. The process allows for precise control over the weld pool, resulting in a smooth and even bead appearance.

*Disadvantages of TIG Welding:*

- Susceptible to contamination: TIG welding requires a clean work environment, as even small amounts of dirt, oil, or other contaminants can affect the quality of the weld.
- Cost: TIG welding equipment is expensive compared to other types of welding equipment, such as metal inert gas (MIG) or stick welding. TIG welding also requires the use of expensive tungsten electrodes and shielding gas.
- Limited to thin materials: TIG welding is typically used for thinner materials, such as aluminum, stainless steel, and copper, as it produces a precise and clean weld. It may not be as effective for thicker materials, which may require a more robust welding process.

*Applications of TIG Welding:*

- Automotive industry: TIG welding is often used to repair exhaust systems, as well as to weld thin body panels and other automotive components.
- Aerospace industry: TIG welding is used extensively in the aerospace industry, particularly for welding thin sections of aircraft parts and components.
- Medical industry: TIG welding is used in the production of medical devices and equipment, such as surgical instruments, as it produces clean and precise welds.
- Art and sculpture: TIG welding is often used by artists and sculptors to create intricate metalwork designs, as it allows for precise control and produces high-quality welds.

### 14.1.3 Metal Inert Gas Welding

In MIG welding, an arc is established between a consumable electrode and the workpiece. Unlike in TIG welding, there is no need for an external filler material, as the consumable electrode serves this purpose. MIG welding finds widespread application in sheet metal industries and can be effectively used for thicknesses ranging from 5 to 30 mm, making it suitable for thicker materials compared to TIG welding.

The automation of MIG welding processes is on the rise, with robotics and other automated machinery becoming increasingly prevalent. This shift is driven by the desire for enhanced accuracy, consistency, and overall efficiency in the welding process. An emerging technique in MIG welding, known as synergic welding, employs a unique power supply system that automatically adjusts the voltage and wire-feed rate to maintain a stable arc. This innovative approach aims to minimize welding defects and improve the overall quality of the welds.

*Advantages of MIG Welding:*

- Strong welds: MIG welding produces strong and durable welds that can withstand high levels of stress and strain.
- Less cleaning required: MIG welding produces little to no slag, which means less cleaning is required after welding.
- High welding speed: MIG welding is a fast process compared to other types of welding. The welding wire is continuously fed into the weld, allowing for a faster welding speed and higher productivity.

*Disadvantages of MIG Welding:*

- High heat output: MIG welding produces a lot of heat, which can cause warping or distortion in the metal being welded.
- Shielding gas required: MIG welding requires a shielding gas to protect the weld from contamination, which can add to the cost of the process.
- Limited mobility: MIG welding requires a power source and a gas cylinder, making it less mobile than other welding processes, such as stick welding.

*Application of MIG Welding:*

- Manufacturing industry: MIG welding is used in the manufacturing industry to join metal parts together, such as in the production of appliances, furniture, and machinery
- Construction industry: MIG welding is used in the construction industry to join metal parts together, such as in the production of structural steel frames and supports.
- Shipbuilding industry: MIG welding is used extensively in the shipbuilding industry to join metal parts together, such as in the production of hulls and decks.
- Automotive industry: MIG welding is widely used in the automotive industry to join thin sheets of metal together, such as in the production of car bodies and exhaust systems.

### 14.1.4 Submerged Arc Welding

Submerged arc welding employs a fusible granular flux to facilitate the welding process. This method is commonly employed in the fabrication of pressure vessels, LPG cylinders, bridges, and shipbuilding applications. Typically, this welding technique is conducted in the horizontal position, enabling the welding of thick plates in a single pass. Both AC and DC power sources find application in this welding process. Automated submerged arc welding systems incorporate advanced sensing and control technologies, such as laser vision systems and adaptive welding control, to enhance the welding process and ensure high-quality welds.

*Advantages of SAW Welding:*

- High productivity: SAW is a high-speed welding process that can be used to weld long, continuous joints with a high deposition rate. This makes it well-suited for industrial applications where productivity is key.
- Minimal spatter: SAW produces minimal spatter, which reduces the need for post-weld cleaning and makes the process more efficient.
- Can be automated: SAW is a highly automated process that can be easily integrated into robotic welding systems. This makes it ideal for high-volume welding applications.
- High-quality welds: SAW produces high-quality welds with minimal porosity and excellent fusion. The flux layer also

helps to protect the weld from contamination, resulting in strong and durable welds.

*Disadvantages of SAW Welding:*

- Flux disposal: SAW requires a large amount of flux, which can create waste disposal issues. The used flux must be properly disposed of, which can add to the cost and complexity of the process.
- Complex setup: SAW requires a complex setup with multiple components, including the welding head, the power source, and the flux delivery system.
- Limited mobility: SAW is a stationary process that requires the workpiece to be positioned beneath the welding head. This limits its mobility and makes it less suitable for welding in tight or hard-to-reach spaces.

*Application of SAW Welding:*

- Pipeline industry: SAW is used to weld pipelines in the oil and gas industry, as well as in other pipeline applications such as water and sewage pipelines.
- Pressure vessel industry: SAW is used to weld large pressure vessels, such as boilers and storage tanks.
- Wind turbine industry: SAW is used in the fabrication of wind turbine towers, as well as in the production of other wind turbine components.
- Heavy equipment industry: SAW is used in the manufacturing of heavy equipment, such as bulldozers, excavators, and cranes.

### 14.1.5 Flux-Cored Arc Welding

Flux-cored welding constitutes an arc welding variant that employs a wire electrode continually fed with flux to establish the arc and shield the weld pool from atmospheric impurities. This flux offers several advantages, including the reduction of spatter, acceleration of welding speed, and enhancement of bead appearance. The application of this welding technique finds relevance in the automotive industry, aerospace sector, and shipbuilding industry.

*Advantages of Flux-Cored Arc Welding:*

- Shielding gas not always required: Some types of flux-cored welding wire are self-shielding and do not require an external shielding gas, which can save on costs and simplify the welding process.
- Better penetration: Flux-cored welding can produce deeper penetration than other welding processes, which makes it useful for welding thick materials.
- All-position welding: Flux-cored welding can be used in all positions, including vertical and overhead welding, which makes it more versatile than some other welding processes.
- High productivity: Flux-cored welding is a high-speed welding process that can produce a high deposition rate, making it ideal for applications where high productivity is important.

*Disadvantages of Flux-Cored Arc Welding:*

- Porosity: Flux-cored welding can be more prone to porosity than other welding processes, particularly if the welding conditions are not properly controlled.
- Limited joint accessibility: Flux-cored welding is not as versatile as some other welding processes when it comes to joint accessibility, particularly for tight or hard-to-reach areas.
- Slag removal: Flux-cored welding produces a slag coating on the weld, which must be removed after welding, adding to the cleanup time and cost.
- Equipment costs: Flux-cored welding requires a different type of welding machine and gun than traditional MIG welding, which can increase equipment costs.

*Application of Flux-Cored Arc Welding:*

- Construction: Flux-cored arc welding (FCAW) is widely used in the construction industry for welding structural steel, beams, columns, and other components of buildings and bridges.
- Oil and gas industry: FCAW is used in the oil and gas industry for welding pipelines, storage tanks, and other equipment.

- Repair and maintenance: FCAW is often used for repair and maintenance work, such as welding broken parts, repairing machinery, and fixing damaged equipment.
- Heavy equipment: FCAW is used in the manufacturing of heavy equipment such as bulldozers, cranes, and excavators.

## 14.2 Resistance Welding

It is a high-current and low-voltage process. Kind of resistance welding processes are as follows.

### 14.2.1 Spot Welding Processes

In spot welding, a set of plates or sheets is positioned between copper electrodes. An electric current flows through these sheets, generating heat due to their high resistance. Once the desired level of heat is reached, pressure is applied to the electrodes, effectively welding the sheets together and creating a small, rounded weld known as a nugget. The integration of robotics into spot welding processes enhances efficiency, and precision and reduces the likelihood of errors, resulting in improved welding outcomes.

*Advantages of Spot Welding:*

- Reduced material waste: Spot welding requires minimal filler material, resulting in less material waste and cost savings.
- Fast and efficient: Spot welding is a fast and efficient welding process that can produce a high volume of welds quickly, making it ideal for mass production applications.
- Minimal material distortion: Since spot welding only heats a small area, it produces minimal distortion or deformation of the materials being welded.

*Disadvantages of Spot Welding:*

- Limited material thickness: Spot welding is typically limited to materials that are 3 mm or less in thickness, which may not be sufficient for some applications.
- Lack of strength in certain applications: Spot welding may not be suitable for high-stress applications, as the welded joint may not have sufficient strength to withstand the load.

- Heat control issues: If the welding parameters are not properly controlled, spot welding can produce excessive heat, leading to warping or deformation of the materials being welded.

*Application of Spot Welding:*

- Automotive manufacturing: Spot welding is commonly used in the automotive industry for welding sheet metal components such as body panels, doors, and hoods.
- Aerospace manufacturing: Spot welding is used in the aerospace industry for joining thin sheet metal components, such as fuselage and wing skins.
- Packaging: Spot welding is used in the packaging industry for joining metal cans and containers.
- Jewelry manufacturing: Spot welding is used in the jewelry industry for joining small metal components, such as chains and clasps.

### 14.2.2 Projection Welding

In a single setup using the same heat source employed for spot welding, multiple spot welds are performed. This welding technique is evolving with a focus on advanced sensing and control systems. These innovations aim to minimize welding defects while enhancing the precision, accuracy, and overall productivity of the welding process.

*Advantages of Projection Welding:*

- Accurate management: Projection welding offers meticulous management of welding parameters, including heat, pressure, and duration, ensuring dependable and consistent welds.
- Enhanced efficiency: This welding method is known for its rapid and efficient performance, enabling the creation of top-notch welds swiftly, making it particularly well-suited for high-volume production tasks.

*Disadvantages of Projection Welding:*

- Thickness constraints: Projection welding is generally constrained to materials with thicknesses of 3 mm or less, which may be insufficient for certain applications.

- Joint strength limitation: In certain instances, the joint strength achieved through projection welding may be inferior to that attainable with alternative welding techniques, like resistance spot welding.
- Electrode wear challenge: The electrodes employed in projection welding often experience rapid wear, necessitating frequent replacement and leading to heightened maintenance expenses.

*Application of Projection Welding:*

- Aerospace production: Projection welding finds application in the aerospace sector, where it is employed to weld crucial aircraft elements such as landing gear, engine components, and structural parts.
- Medical equipment fabrication: In the manufacturing of medical devices like surgical instruments, implantable devices, and diagnostic equipment, projection welding plays a pivotal role.
- Automotive industry: Projection welding is a prevalent technique within the automotive manufacturing realm, utilized for welding various components including brake parts, suspension elements, and structural components

### 14.2.3 Seam Welding

Seam welding relies on ohmic resistance as its fundamental welding mechanism. This process can be described as continuous spot welding, wherein an overlapping region is effectively joined. To achieve this, a pair of sheets is positioned between copper roller electrodes. Once the desired pressure is attained by applying electrical current, the roller moves along the overlapped area of the sheets, resulting in the completion of the welding process.

*Advantages of Seam Welding:*

- Leak-proof: Seam welding gives leak-proof joints such as tanks.
- Reduced material waste: Seam welding requires minimal filler material, resulting in less material waste and cost savings.

*Disadvantages of Seam Welding:*

- Limited material thickness: Seam welding is typically limited to materials that are 6 mm or less in thickness.
- Limited joint strength: In some cases, the strength of the joint created by seam welding may be lower than that of other welding processes, such as TIG welding or MIG welding.

*Application of Seam Welding:*

- Automotive manufacturing: Seam welding is commonly used in the automotive industry for welding fuel tanks.
- Electronics manufacturing: Seam welding is used in the production of electronic components, such as battery cells, capacitor cases, and other electronic enclosures.
- Food and beverage manufacturing: Seam welding is used to create airtight seals on food and beverage packaging, such as cans and containers.
- Construction: Seam welding can be used in the construction industry for welding pipes.

### 14.2.4 Butt Welding

In this welding procedure, the initial step involves bringing the surfaces to be welded into contact with each other. Subsequently, an electric current is conducted through the workpiece. Once the desired level of heat is achieved, pressure is applied from both sides of the circular workpiece. It is essential to note certain limitations in this process, including the requirement for both circular cross-section workpieces to have the same melting temperature. Additionally, it is crucial to ensure that the surface finish of the cross section is properly prepared before commencing the welding process.

*Advantages of Butt Welding:*

- No filler metal required: In butt welding, no additional filler material is required, which makes it a cost-effective process, especially for high-volume production.

*Disadvantages of Butt Welding:*

- Potential for warping: During the welding process, the heat generated can cause the metal to warp, particularly in thin materials. This can affect the fit and finish of the final product.
- High equipment cost: Butt welding requires specialized equipment such as welding machines, clamps, and fixturing, which can be expensive to purchase and maintain.
- Requires precise preparation: Butt welding requires precise preparation of the edges of the two pieces of metal to be joined, which can be time-consuming and requires skilled labor.

*Application of Butt Welding:*

- Construction industry: Butt welding is used in the construction industry for joining steel beams, pipes, and other structural components.
- Oil and gas industry: Butt welding is used in the oil and gas industry for joining pipelines and other components used in the transmission of oil and gas.

## 14.3 Hybrid Welding

This cutting-edge welding technique combines the principles of both arc welding and laser welding, simultaneously applied to the same location within the weld pool. This fusion results in a host of advantageous properties, including exceptionally high energy density and minimal heat input, enabling rapid welding, deep penetration, and enhanced tensile strength. Additionally, it offers excellent gap-bridging capabilities, the option to introduce filler metal, and the ability to manipulate the metal's microstructure. Furthermore, this method allows for the welding of thicker materials in a single pass, delivering significant cost benefits in the present and foreseeable future. Recently, a wide range of industries have embraced this technology to boost welding productivity and efficiency, with a predominant focus on steel welding applications.

## 14.4 Rotary Friction Welding

The source of friction welding is mechanical work. In friction welding, first, we have to keep in contact with two circular metals and then give relative motion. Due to relative motion, friction causes heat to be generated. When a sufficient amount of heat is generated, then axial pressure is applied to complete the welding. The details are shown in Figure 14.1.

## 14.5 Friction Stir Welding

Wayne Thomas has pioneered an innovative welding technique that involves the insertion of a welding probe into the joint of the workpiece, followed by the initiation of tool rotation. During this rotation, friction generates heat at the interface between the tool and the workpiece. Once a sufficient amount of heat is generated, a vertical force is applied to the tool to finalize the welding process, as illustrated in Figure 14.2 accompanying this description.

The effectiveness of the friction stir welding process hinges primarily on two crucial factors: maintaining the appropriate weld temperature and precise tool positioning. Several methods are employed to achieve this, including the attachment of thermocouples to either the workpiece or the tool, the use of thermal cameras, the application of numerical simulations, and the utilization of pyrometers.

*Advantages of Friction Stir Welding:*

- High-strength joints: Stir friction welding produces joints that are typically stronger than those produced by other

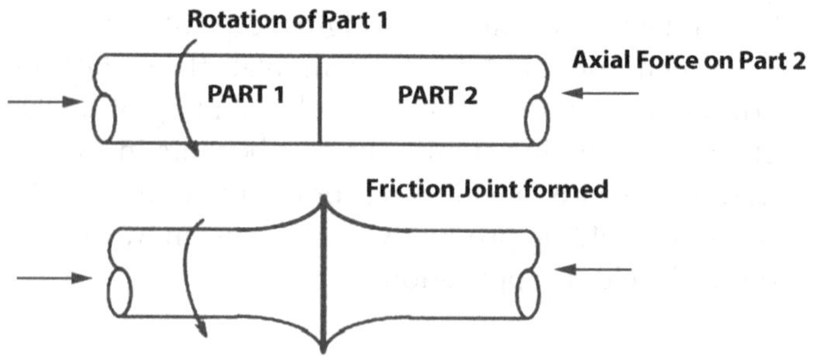


**Figure 14.1** Schematic diagram of rotary friction welding.

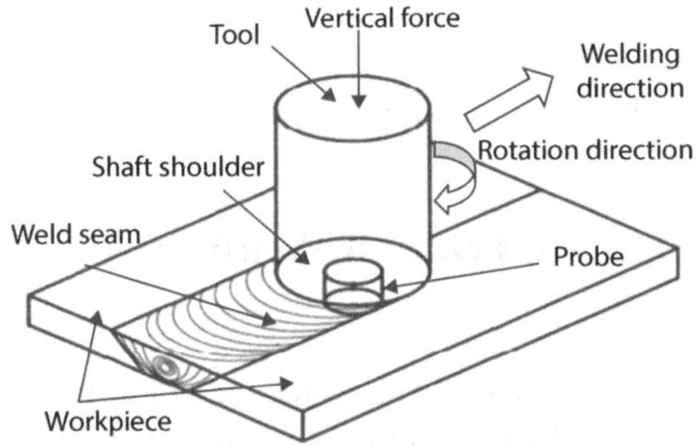


**Figure 14.2** Schematic diagram of friction stir welding.

welding processes, such as fusion welding. This is because the process creates a solid-state bond between the two metals, without melting the base material, resulting in a joint that is free from porosity, voids, and cracks.

- Low heat input: Stir friction welding generates heat through friction between the two metals, without the need for external heating. This results in a low heat input to the base material, reducing the risk of distortion, cracking, or other heat-related defects.

*Disadvantages of Friction Stir Welding:*

- Limited joint configuration: FSW is best suited for linear joints or joints with relatively simple geometries. Welding more complex joint configurations can be difficult and may require additional equipment or modifications to the welding process.

*Application of Friction Stir Welding:*

- Aerospace: FSW is commonly used in the aerospace industry to join aluminum and titanium alloys for aircraft structures, such as wings and fuselages. It is also used for joining rocket fuel tanks.
- Railways: FSW is used in the railway industry to join aluminum and magnesium alloys for lightweight components, such as railcar bodies, frames, and floors.

- Defense: FSW is used in the defense industry for joining armor plates, vehicle structures, and other components where high strength and durability are required.

## 14.6 Properties of Friction Welding

- **Tensile Strength**

If we evaluate the efficiency of the weld joint of copper and aluminum which had welded by friction welding, then we need to do the tensile test, after doing this test, we determine the tensile strength of the joint. The tensile strength of the weld joint is increasing as increasing of forging pressure. The strength of the joint directly relates to forging pressure and friction time. The time of friction depends on some factors such as material composition and workpiece rotation speed.

- **Impact Strength**

The result of the impact test tells how much effect temperature has on welded joints of aluminum and copper. The toughness is measured by the impact test. The toughness of the weld may increase or decrease when the temperature varies. The impact strength increases with increasing forging pressure and decreases with increasing friction pressure, i.e., temperature.

- **Microhardness**

To define the mechanical properties of welded joints, hardness is one of the most important factors to decide it. Hardness is affected by temperature at the interface of aluminum and copper due to friction.

- **Fractured Surface**

Fractured workpieces were observed by SEM after the tensile and impact test to find the location of fracture and type of failure. Tensile fracture in most of the joints is obtained near to aluminum interface

## 14.7 Future Research Opportunities and Conclusions

The performance of friction welding is highly dependent on process parameters such as rotational speed, axial force, and welding time. Future research could focus on optimizing these parameters to improve the quality and mechanical properties of the resulting weld. Future research could focus on using advanced microscopy techniques to investigate the microstructure of friction-welded joints and how it relates to their mechanical properties. Future research could focus on applying friction welding to new materials, such as composites or high-temperature alloys, and investigating the resulting properties of the joints. While this chapter does not encompass all the latest advancements, it is important to note that the discussed technologies are already in use to some extent within industrial applications. Achieving high-quality joints necessitates the development of highly efficient and easily implementable welding techniques. This requires a meticulous examination of welding equipment, methods, and materials. There is an evident shift toward increased reliability and heightened quality standards in welding. Furthermore, the trend toward automating welding processes will gain momentum. Welding techniques such as friction stir welding, laser welding, and laser arc hybrid welding will become increasingly significant, owing to the rapid transfer of knowledge from research to practical applications.

Automation in welding is particularly well-suited for use in hazardous environments and various positions, as well as for mass production purposes. Designers and companies can now analyze parameters and simulate welding processes using finite element analysis models before setting up physical test setups. It is worth noting that friction welds exhibit higher tensile strength than their base metals. Additionally, there is an increase in microhardness toward the weld joint interface due to the effects of strain hardening.

## References

1. Kah, P. and Martikainen, J., Current trends in welding processes and materials: improve in effectiveness. *Rev. Adv. Mater. Sci.*, 30, 189–200, 2012.
2. Ghosh, A. and Mallick, A.K., *Manufacturing Science*, Second edition, East West Press Pvt Ltd., New Delhi, 2018.
3. Mi., S.T., C. Zhang., Y.-, Fu., X., Yan., F., Shen., J., A measurement method of weld temperature and tool positioning during friction stir welding process. *Weld. World*, 65, 202–230, 2021.
4. Muralimohan, C., Haribabu, S., Hariprasada reddy, Y., Muthupandi, V., Sivaprasad, Evaluation of microstructures and mechanical properties of dissimilar materials by friction welding,Advance in manufacturing and material engineering. *Proc. Mater. Sci.*, 5, 1107–1113, 2014.
5. Cheepu, M., Muthupandi, V., Che., W.S., Improving Mechanical Properties of Dissimilar Material Friction Welds. *Appl. Mech. Mater.*, 877, 157–162, 2017.

6. W.M. Thomas, E.D. Nicholas, J.C. Needham, M.G. Murch, P. Temple- Smith, C.J. Dawes, Friction Stir Butt Welding. GB Patent Application No. 9125978. US Patent No. 5460317, 1991.
7. Tannous, Q., Ghnatios, C., Barasinski, A., A new friction welding process for soft packaging applications. *Int. J. Mater. Form.*, 15, 19–26, 2022.

15

# Advanced 3D Printing for Industrial Components: Welded Joint Analysis and Strength Assessment

**Jagadeesha T.[1]*, Amith Kumar Gajakosh[2], Avinash Malladi[3] and Natrayan L.[4]**

*[1]Department of Mechanical Engineering, National Institute of Technology NIT, Calicut, Kerala, India*
*[2]Amith Kumar Gajakosh, Department of Mechanical Engineering, BTL Institute of Technology and Management, Bangalore, Karnataka, India*
*[3]Department of Mechatronics Engineering, Faculty of Science and Technology, IcfaiTech, Icfai Foundation for Higher Education, Hyderabad, Telangana, India*
*[4]Department of Mechanical Engineering, Saveetha School of Engineering, SIMATS, Chennai, Tamil Nadu, India*

## *Abstract*
Additive manufacturing plays a pivotal role in the production of industrial components, particularly in the context of welded joints. This process involves the meticulous analysis of 3D printed welded joints and a thorough assessment of their tensile properties. To gain a comprehensive understanding of this procedure, we will embark on a detailed exploration, commencing with an overview of the current state of 3D printing technology as it pertains to welded joints and its manifold applications in contemporary industrial manufacturing. Within this context, it becomes evident that several critical factors contribute to the overall strength and reliability of 3D printed components with welded joints. These factors include judicious material selection, meticulous design considerations, and the attainment of impeccable welding quality. These elements, in concert, hold the key to determining the robustness and durability of the resulting components. One cannot overstate the importance of subjecting 3D printed components with welded joints to a rigorous strength analysis. This is an indispensable step in ensuring their dependability and optimal performance. This paper delves into

**Corresponding author*: jagdishsg@nitc.ac.in

R. Thanigaivelan, Pradeep Kumar Krishnan, Kamalakanta Muduli and Santosh Kumar Tamang (eds.) New Materials, Processing and Manufacturability: Fabrication and Processing of Advanced Materials, (287–298) © 2024 Scrivener Publishing LLC

the imperative nature of such analyses, shedding light on their significance in the quality assurance process. A pivotal aspect of our exploration is the elucidation of various strength analysis techniques. These methodologies, each with its own set of advantages and limitations, are instrumental in evaluating the tensile characteristics of 3D printed components. Prominently featured among these techniques are finite element analysis, mechanical testing, and nondestructive testing. A comprehensive discussion of their pros and cons provides valuable insights into their suitability for different applications and contexts. However, it is essential to recognize that the examination of welded joints in 3D printed components is not without its challenges. This paper takes a deep dive into these challenges, considering factors such as structural complexity, material heterogeneity, and the need for precision in testing. Addressing these issues effectively is paramount in ensuring the integrity and safety of industrial components produced through additive manufacturing with welded joints.

***Keywords*****:** Welded joint, 3D printing, laser or electron beam

## 15.1 Introduction

A fast-developing technique that is receiving a lot of attention in the additive manufacturing (AM) community is the use of 3D printing with welded joints for the production of industrial components [1]. Using a laser or an electron beam to fuse together layers of metal powder, this procedure involves layer-by-layer deposition to produce delicate and complicated metal objects [2]. It is now possible to produce practical and long-lasting industrial parts using 3D printing technology that has welded joints, which results in less material waste and greater design flexibility [3]. Analysis of the tensile qualities of 3D printed welded joints for industrial components is a crucial step. The performance, dependability, and safety of a 3D printed component in practical applications are greatly influenced by its strength [4]. To ensure that the 3D printed components satisfy the necessary requirements and can resist the operational stresses and pressures they will experience, a thorough strength study is necessary [5–8].

The goal is to offer a thorough examination of the 3D printing of industrial components having welded joints, with an emphasis on strength analysis in particular. This process' many facets will be covered here, including the state of welded joint 3D printing technology at present, the importance of conducting strength analysis, the techniques used for strength analysis, and the difficulties and factors to be considered when analyzing welded joint 3D printed components [9]. This research intends to provide helpful insights and suggestions for successfully analyzing the strength qualities of welded joint 3D printed industrial materials.

## 15.2 Material Used for SLM Technology

For the experiment, Renishaw-provided gas-atomized AISI316L powder is used. It is a very low-carbon austenitic stainless steel that is nonmagnetic. Glow discharge optical emission spectrometry (GDOES) analysis was performed using an optical emission spectrometer from Spectrums Analytic to determine the chemical composition (Table 15.1). The bulk analysis at 700 V and 35 mA of stimulation was used to determine the material's average chemical composition. The average size of the powder particles was 26.33 m, and their average d50 diameter was 25.09 m. On the KYENCE VHX-5000 instrument, grain size was assessed using an optical technique. The composition of AISI316L is shown in Table 15.1

## 15.3 SLM Technology 3D Component

The selective laser melting (SLM) AM technique, also known as three-dimensional printing, utilizes layers of powdered material that are melted and fused together to create intricate and precise three-dimensional objects. This process begins with a digital model of the desired product, which is divided into multiple layers. A bed of powdered material, typically composed of metal or alloy, is spread out on a build platform. The laser is then employed to selectively melt and fuse the powdered material in accordance with the design specifications, solidifying it into a single layer. Subsequently, the build platform is lowered, and this process is repeated layer by layer until the entire object is fully formed. The laser is directed once again to melt and fuse each new layer onto the previous one, resulting in the production of the final item.

As the molten material cools down, it rapidly solidifies, creating a solid bond between the layers. To prevent oxidation and unwanted reactions, this process is conducted in a controlled environment, typically in the presence

**Table 15.1** AISI 316L powder analysis.

| C | Mn | Si | P | S | Cr | Ni | Mo | W | Cu | Ti | Nb | Al | B |
|---|---|---|---|---|---|---|---|---|---|---|---|---|---|
| 0.016 | 1.17 | 0.22 | 0.023 | 0.0067 | 17.72 | 14.24 | 2.73 | 0.19 | 0.077 | 0.0003 | 0.013 | 0.01 | 0.002 |

of inert gas. Various post-processing techniques such as brushing, blowing, or vacuuming are employed to eliminate any remaining powder from the final product once it is removed from the powder bed. SLM offers numerous advantages in this regard. In this work, circular samples were fabricated using an AM400 3D printer, capitalizing on the technology's ability to produce intricate geometries and structures that are often challenging or impossible to achieve using traditional methods. The AM400 3D printer utilized a laser with a maximum nominal power output of 400 W. The focus size was calibrated accordingly, and argon gas with high purity was used.

The researchers opted to employ an inert shielding gas with a composition of 5.0 throughout the entire sample production process, up to the device's technological limit of 70 m. Prior to initiation, the chamber was pressurized, and the inert gas continued to purge any remaining air from the chamber, ensuring that the oxygen content remained below 1000 ppm during the entire construction process. This setup effectively eliminated metal fumes through gas flow and prevented powder oxidation during the build. Renishaw's quant AM software was used to generate the build. The samples had an outer diameter of 20 mm and a wall thickness of 2 mm.

## 15.4 TIG Welding Method for Joining

All samples were heated in the Classic O816 VAK vacuum furnace before any treatment. Using an annealing temperature for 360 min and cooling in a furnace for 12 hours in a vacuum, the internal stress was lowered to 550°C. The end welds on the tubes were heat-treated, and then the edges were chamfered at a 45° angle to prepare them for welding. Figure 15.1 depicts welding completed after stitching. All samples were put together using the welding method 141 (TIG) using a WLa15 electrode with a 1.6-mm diameter.

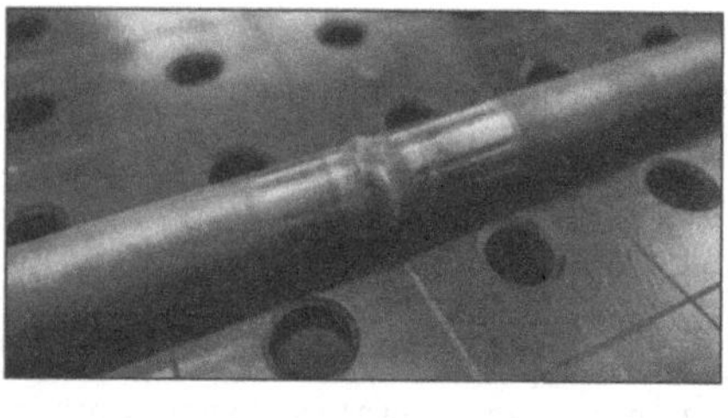

**Figure 15.1** Welded joints.

## 15.5 ABS and PLA 3D Component Material

The thermoplastic polymer known as acrylonitrile butadiene styrene, or ABS, is recognized for its durability, resilience, and strength. Because it melts at a greater temperature than polylactic acid (PLA), frequently between 210°C and 250°C, it can be utilized for applications where parts must withstand high temperatures or mechanical forces. ABS is suitable for use in difficult circumstances because of its great chemical resistance. ABS is recognized for its ability to withstand post-processing techniques like painting, acetone vapor smoothing, and sanding, which allows for further enhancing the surface polish and appearance of printed goods. While ABS is known for its ease of use, PLA (polylactic acid) is a bio-based thermoplastic polymer made from renewable sources like sugarcane or cornmeal. PLA is a well-liked option for sustainable and ecologically sensitive applications because of its biodegradability and environmental friendliness. Strong and long-lasting items can be created using PLA thanks to its outstanding printability.

### 15.5.1 Fused Deposition Modeling

The most popular AM technique for ABS and PLA materials is fused deposition modeling (FDM); however, these materials can be utilized in a variety of AM techniques. The AM process known as FDM uses thermoplastic filament, such as ABS and PLA, as the raw material. The filament is melted and extruded in FDM to produce successive layers that solidify and join to form a 3D printed item using computer-aided design (CAD) data to precisely control a heated nozzle. FDM is well renowned for its flexibility, usability, and accessibility, making it popular in a number of applications from rapid prototyping to the production of functional parts.

The FDM, ABS, and PLA filaments are gathered by the 3D printer's material feeding system and fed into the heated nozzle, where they melt and are deposited layer by layer to create the finished object. The unique requirements of the printed product, such as its mechanical properties, environmental considerations, and intended application, will determine whether ABS or PLA is selected. Due to the compatibility of the technology and the accessibility of filaments for FDM printers, ABS and PLA are typically utilized in FDM procedures even if there are materials for alternative AM techniques like selective laser sintering (SLS) and stereolithography (SLA).

## 15.6 Welding Method Used for ABS and PLA

Being thermoplastic materials, ABS and PLA are often not joined using heat-based welding processes, in contrast to conventional welding techniques. There are alternative ways to "weld" or connect ABS and PLA 3D printed components together, though.

### 15.6.1 Solvent Welding

In the process of solvent welding, the surface of the thermoplastic material is dissolved with a solvent, which then evaporates to form a bond. Acetone is frequently utilized as a solvent for solvent-welding ABS. Acetone vapor or brushing with acetone causes the surface of ABS components to liquefy and fuse together, forming a solid connection. Contrarily, PLA does not dissolve in acetone, making acetone solvent welding ineffective for PLA. Other solvents, such as isobutyl acetate or methyl ethyl ketone, may be used for solvent welding with PLA. Details are given below:

Step 1: Shape programming with solvent
Step 2: Solvent-induced transesterifications
Step 3: Weldable material from step 2
Step 4: Topological arrangement
Step 5: Reconfigurable with solvent from step 4
Step 6: Healable with solvent from step 4.

### 15.6.2 Mechanical Fastening

Mechanical fastening, which involves utilizing mechanical devices like screws, nuts, bolts, or other hardware to keep the parts together, is another way to link ABS and PLA 3D printed components. This approach, which depends on physical connections rather than chemical or thermal processes, can be successful for both ABS and PLA.

### 15.6.3 Adhesive Bonding

Adhesive bonding is another method for fusing ABS and PLA components together. There are numerous adhesives that can be used to bind thermoplastic materials together, including cyanoacrylate (CA) and epoxy-based adhesives. These adhesives provide a strong bind between the ABS or PLA material's surfaces by chemically adhering to them.

## 15.7 Characterisation Techniques

### 15.7.1 Examining the Base Material's Porosity

The Archimedes approach was used to determine the overall density based on the observations and analysis. The printed base metal has an average density of 7.849 g/cm$_3$. For reference, the homogeneous cast and rolled material of AISI316L steel has a total density of 7.999 g/cm$_3$. Equation 15.1 is used to compare the measured substance's density to that of a predetermined reference material in order to estimate the relative density.

$$\rho_{relative} = {}^{\rho}/{\rho_{ref}} \times 100 \qquad (15.1)$$

### 15.7.2 Analysis of Mechanical Properties

The lower hardness values in the heat-affected zone (HAZ) compared to the base material are attributed to the base material's surface being harder than the weld metal. This discrepancy occurs due to the precipitation of carbides in the base material, followed by partial dissolution of carbides in the HAZ as a result of the tungsten inert gas (TIG) welding process. This dissolution of carbides contributes to a decrease in material hardness. In contrast, the rupture in the other samples occurred outside the HAZ and was primarily caused by an increase in porosity during the fabrication of the SLM samples. When fractures did occur within the HAZ zone, only one sample proved relevant for the tensile test analysis. It is worth noting that the fracture surfaces exhibited variations in their characteristics.

Several factors can lead to premature rupture, resulting in the breaking of a material. This rupture can be characterized in different ways, depending on the circumstances. On one hand, it can be caused by the presence of large interconnected pores and non-molten powder particles within the fracture surface. In such cases, a transcrystalline ductile fracture pattern is observed, typically affecting the ridges on the fracture surface. This type of fracture occurs when the material loses its strength over time and eventually cracks. When it comes to welding ABS and PLA materials, the outcomes can vary significantly based on various factors, including the choice of solvent or adhesive, surface preparation methods, and the welding technique employed. Here are some general observations:

Formation of Pores: Pores can be observed in the welded area, and their formation can be attributed to inadequate melting of particles or surface turbulence that traps gasses. Fusion porosity, resulting from insufficient melting, often forms along the boundaries of layers and is characterized by irregular shapes extending along the X and Y planes. Process variables such as laser power, layer thickness, scanning speed, and hatching distance can influence the size and occurrence of these pores. The lack of uniform energy density distribution across the layer's surface is the primary cause of this porosity, as it disrupts the cohesion between the layers. One solution to this issue is the adoption of a re-melting procedure, where scanning each layer twice can nearly achieve 100% density.

Gas Trapping: Another potential cause of porosity is the trapping of gas by the turbulent surface flow of molten metal in the melting bath. This gas may originate from the evaporation of the material or be a protective atmospheric gas, or it could result from a combination of both processes.

In summary, premature rupture in materials can result from various factors, and the outcomes of welding ABS and PLA materials depend on factors such as the choice of solvent, surface preparation, and welding technique. Porosity formation can occur due to insufficient melting and gas trapping, and addressing these issues may involve re-melting procedures and controlling the welding environment.

## 15.8 Challenges Faced in the Development Process

### 15.8.1 Due to the Solvent Welding

- Weak joints: The joint may be weak and fail under stress if the solvent is not administered appropriately or if the bonding surfaces are improperly prepared.
- Cracking: The material may crack or craze if the solvent used is incompatible with the thermoplastic substance being bonded.
- Warping: If the solvent is not applied evenly or if the material is not maintained in place during the bonding process, solvent welding may cause the bonded materials to warp or distort
- Toxic fumes can be harmful to the health of the person performing the welding since they are released by the solvents used in solvent welding.

### 15.8.2 Due to Mechanical Fastening

Materials can be joined using either welding or mechanical fastening. While welding involves melting and fusing the materials together at the joint, mechanical fastening involves joining two or more materials together mechanically using tools like screws, bolts, rivets, or clips. To make a stronger junction, mechanical fastening and welding can occasionally be combined. A hybrid joint is what this is. A hybrid junction is one in which the components are first mechanically joined and then strengthened by welding. Using a hybrid joint has a number of benefits. First, compared to mechanical fastening or welding alone, it can offer a stronger and more resilient union. Second, it can lessen the chance that stress concentration and surface deterioration will happen while welding. Last but not least, it can offer an alternative to welding that is more effective and economical, especially for large or complex components.

### 15.8.3 Due to Adhesive Bonding

Adhesives can be incompatible with some materials, which results in the degradation or weakening of those materials over time. Cure time, the time needed for the adhesive to cure during adhesive bonding, might cause a delay in the production process.

### 15.8.4 Method to Avoid Issue

Solvent welding requires strict adherence to the manufacturer's recommendations concerning both the solvent and the specific thermoplastic materials to be bonded. When engaging in solvent welding, it is imperative to prioritize safety by wearing appropriate protective gear, such as gloves and a respirator. For mechanical fastening to ensure a durable and enduring bond, a thorough evaluation of the materials to be joined, the choice of fasteners, and the welding technique is essential. Safety measures while welding must be observed, including wearing the necessary protective equipment and ensuring proper ventilation. In adhesive bonding, the selection of the correct adhesive for the materials in question and adhering to the manufacturer's application guidelines are paramount in preventing potential issues. Prior to applying the adhesive, meticulous surface preparation is crucial to ensure cleanliness and the absence of contaminants. Additionally, it is vital to consider the environmental impact of the adhesive used and take appropriate precautions to minimize any potential harm.

## 15.9 Conclusion

This chapter contributes to the enhancement of our knowledge and the advancement of techniques for connecting components manufactured through 3D printing, particularly using SLM with AISI316L material. The welding process employed in this study was the TIG technique, with a total of eight welded joints created. The selection of welded samples was done randomly, enabling a comprehensive assessment through various tests including microhardness, tensile testing, and porosity evaluation.

The SLM approach utilized in this context resulted in components with an impressive 98.12% relative density. However, the mechanical properties of the welded joints were notably influenced by defects in the base material, primarily in the form of pores. This issue led to premature fracture occurrences either before or slightly above the yield threshold. Material failure in the base material region was typically observed near the HAZ. It is important to note that the microhardness in the HAZ zone was approximately 30% lower than that of the base material.

It is worth mentioning that various techniques are available for connecting components, including mechanical, solvent-based, and heat-based methods. However, when it comes to joining ABS (acrylonitrile butadiene styrene) and PLA (polylactic acid), challenges may arise due to their distinct material properties and chemical compositions. While mechanical techniques like using fasteners or adhesives can be employed to join ABS and PLA, it is important to acknowledge that the joint's strength might be lower compared to heat-based or solvent-based methods.

Keep in mind that the specific types of ABS and PLA materials used, the chosen welding technique, and the intended application of the welded components all play crucial roles in determining the success of the ABS-PLA joint. To identify the optimal welding process and settings for a particular ABS-PLA joint and to ensure safety when working with these materials, conducting comprehensive testing and experimentation is strongly recommended.

In summary, welding ABS and PLA together can present challenges, but with careful consideration of tools, methods, and safety precautions, effective bonding can be achieved.

## References

1. Gurupranes, S.V., Kaliappan, S., Natrayan, L., Pater., P.B., Sekar, Jayaraman, P., Aravinda Pandian, C., Esakkiraj, Investigation of Physicochemical Properties and Characterization of Leaf Stalk Fibres Extracted from the Caribbean Royal Palm Tree. *Int. J. Chem. Eng.*, 10, 421–442, 2022.

2. Eckel, Z.C., Zhou, C., Martin, J.H., Jacobsen, A.J., Carter, W.B., Schaedler, T.A., Additive manufacturing of polymer-derived ceramics. 351, 58–62, Science2016.AU: Please provide journal title.
3. Zhang, C., Bao, Y., Zhu, H., Nie, X., Zhang, W., Zhang, S., Zeng, X., A comparison between laser and TIG welding of selective laser melted AlSi10Mg. *Opt. Laser Technol.*, 120, 105696, 2019.
4. Liverani, E., Toschi, S., Ceschini, L., Fortunato, A., Effect of selective laser melting (SLM) process parameters on microstructure and mechanical properties of 316L austenitic stainless steel. *J. Mater. Process. Technol.*, 249, 255–263, 2017.
5. Löber, L., Flache, C., Petters, R., Kühn, U., Eckert, J., Comparison of different post processing technologies for SLM generated 316l steel parts. *Rapid Prototyping J.*, 19, 173–179, 2013.
6. Kumar, R., Singh, R., Ahuja, I.P.S., Amendola, A., Penna, R., Friction welding for the manufacturing of PA6 and ABS structures reinforced with Fe particles. *Compos. B. Eng.*, 132, 244–257, 2018.
7. Tiwary, V.K., Ravi., N., Arunkumar, P., Sivakumar, S., Deshpande, A.S., Malik., V.R., Investigations on friction stir joining of 3D printed parts to overcome bed size limitation and enhance joint quality for unmanned aircraft systems. *Proc. Inst. Mech. Eng. C. J.*, 234, 24, 4857–4871, 2020.
8. Lin, C.B., Wu, L.C., Chen, Y.Y., Friction welding of similar materials: polypropylene, high-density polyethylene, and nylon-6. *J. Appl. Polym. Sci.*, 91, 5, 2771–2780, 2020.
9. Hasegawa, M., Asada, T., Ozawa, Y., Study of friction welding of polyethylene. *Weld. Int.*, 16, 4, 537–543, 2002.

16

# Advancements in Electrochemical Surface Coatings: Innovations, Applications, and Future Prospects

**Jagadeesha T.**[1], **Soundarrajan M.**[2*], **Seeniappan Kaliappan**[3] **and M. D. Raj Kamal**[4]

[1]*Department of Mechanical Engineering, National Institute of Technology NIT, Calicut, Kerala, India*
[2]*Muthayammal Engineering College, Rasipuram, Tamil Nadu, India*
[3]*Department of Mechatronics Engineering, KCG College of Technology, KCG Nagar, Karapakkam, Chennai, Tamil Nadu, India*
[4]*Department of Mechanical Engineering, Velammal Institute of Technology, Chennai, Tamil Nadu, India*

## *Abstract*

In recent years, electrochemical surface coating processes have garnered considerable attention due to their remarkable ability to confer numerous advantages upon the materials they are applied to. These processes involve the controlled application of electrical currents to deposit metallic or ceramic coatings onto various substrates, offering precise control over factors such as coating thickness and composition. As a result, these coatings yield surfaces that are more uniform and enduring, leading to enhancements in corrosion resistance, wear resistance, and surface hardness. Various techniques are employed in electrochemical surface coating, including electroplating, electroless plating, electrophoretic deposition, and anodizing. A standout feature of these methods is their capacity to meticulously regulate coating thickness and composition, thereby elevating the properties of the underlying substrates. Additionally, these techniques are relatively straightforward and cost-effective, rendering them attractive choices for a wide array of industries. Nevertheless, it is worth noting that electrochemical surface coatings demand specialized equipment and a high degree of technical expertise.

**Corresponding author*: soundarrajan05@gmail.com

R. Thanigaivelan, Pradeep Kumar Krishnan, Kamalakanta Muduli and Santosh Kumar Tamang (eds.) New Materials, Processing and Manufacturability: Fabrication and Processing of Advanced Materials, (299–316) © 2024 Scrivener Publishing LLC

In recent times, significant strides have been made in the field of electrochemical surface coatings. These advancements encompass the development of novel materials and coatings, as well as refinements in processing techniques. These breakthroughs have expanded the potential applications of electrochemical surface coatings, particularly in the aerospace, biomedical, and electronics sectors. Consequently, these coatings offer the prospect of delivering consistent and enduring layers with meticulous control over thickness and composition, offering substantial benefits across diverse industries. Looking ahead, it is anticipated that ongoing research and development efforts in this domain will continue to unearth fresh possibilities and further refine electrochemical surface coating applications. This will likely lead to the emergence of innovative and improved coating technologies, with far-reaching implications for a variety of industries. In this term paper, we will delve into the intricate details of the associated technologies, processing methods, and characterization techniques pertinent to electrochemical surface coatings.

***Keywords***: Electrodeposition, anodization, electrophoretic deposition, electroless plating, surface coating, electroplating

## 16.1 Introduction

Surfaces are subject to various environmental factors such as corrosion, wear and tear, weathering, and chemical exposure that can degrade their performance and life span [1]. Surface coating is an effective solution to address these issues and ensure that surfaces maintain their functionality and aesthetic appeal. Surface coating involves the application of a protective layer onto a surface to provide a barrier against these factors, thereby improving the surface's durability, resistance, and longevity. Surface coating is essential for many industries, as it provides protection and enhances the performance of surfaces in harsh environments [2]. Surface coating by electrochemical methods is a process that involves the application of a thin layer of material onto the surface of a substrate using electrochemical reactions. The process involves immersing the substrate (e.g., metal, plastic, or ceramic) into an electrolyte solution that contains dissolved ions of the coating material [3]. The substrate is then connected to the negative terminal of a power supply, while a conductive anode made of the coating material is connected to the positive terminal [4]. When an electric current is passed through the system, the ions in the electrolyte are attracted to the substrate and are reduced to form a thin coating layer on its surface. The thickness and properties of the coating can be controlled by adjusting the current density, the composition of the electrolyte, and other process parameters.

The benefits of electrochemical surface coating include improved corrosion resistance, wear resistance, and aesthetic appearance [5]. There are different types of electrochemical methods and the surfaces are to be prepared for coating by various techniques. The surface characteristics of coating can be studied by appropriate characterization methods [6]. Electrochemical surface coating methods are widely used in various industries, including automotive, aerospace, electronics, and biomedical engineering [7]. Examples of coatings that can be applied using electrochemical methods include metals, metal oxides, polymers, and ceramics. The advancement in coating methods has resulted in the possibility of this technique in dental implants. However, surface coating has some limitations like cost, threat during toxic coating, and peeling of coating over time that need to be considered [8].

This chapter efforts to study various electrochemical deposition methods and surface preparation procedures. The role of electrochemical coating in various industries is also discussed. The aim of this chapter is to show the significance of coating in surface protection. Although coating offers a lot of merits, it faces some challenges also. With the advancement in materials and coating methods, the application of surface coating in various fields will be improved.

## 16.2 Fundamentals of Electrochemical Process

Electrochemical processes involve the transformation of chemical reactions into electrical energy. These processes are rooted in the fundamental principles of electrochemistry, which explores the intricate relationship between electrical energy and chemical reactions [9]. One significant application of these principles is seen in electrochemical surface coating. This technique involves immersing a metal surface into an electrolytic solution containing ions of the coating material. Subsequently, both the solution and the metal surface are subjected to a direct electrical current. The outcome of this process is the deposition of the coating material onto the metal surface through an oxidation-reduction reaction.

In this electrochemical surface coating process, the metal surface attracts anions and cations from the electrolytic solution due to their opposing charges. When a direct current is applied, anions migrate toward the anode (the positive electrode), while cations move toward the cathode (the negative electrode). At the cathode, the cations interact with the electrons provided by the direct current, resulting in the formation of a metal layer on the metal surface. By adjusting factors such as voltage, current, and the composition of the electrolytic solution, it becomes possible to control both the thickness and composition of the coating [10]. The fundamental setup for such electrochemical processes involves an electrochemical cell. This cell consists of anode and cathode electrodes immersed in an electrolyte solution. Within this cell, ions in the electrolyte solution can undergo reduction or oxidation processes. When a voltage is applied across the electrodes, a chemical process takes place, allowing electrons to flow from the anode to the cathode.

Electrochemical energy is a key outcome of these processes, as it can be harnessed for various applications, including powering devices such as batteries and fuel cells [11]. Batteries, for instance, rely on chemical reactions to generate electrical energy that can be used as a power source. Electrolysis is another important application of electrochemistry, where an electric current is employed to drive non-spontaneous chemical reactions. In this process, the anode and cathode are connected to an external power supply, providing the necessary energy to facilitate the reaction [12]. At the heart of electrochemistry are redox (reduction-oxidation) reactions. In an electrochemical cell, oxidation takes place at the anode, where the species losing electrons becomes a cation. Conversely, reduction occurs at the cathode, where the species gaining electrons becomes an anion [13]. These redox reactions are central to understanding and harnessing electrochemical processes.

### 16.2.1 Electrode Kinetics

Electrode kinetics, in the context of electrochemical surface coating, pertain to the phenomena that transpire at the boundary between an electrode and a solution. Numerous factors exert influence over this intricate process, encompassing the properties of the electrolyte solution, the nature of the electrode and coating material, the magnitude of the applied voltage, and the density of the current applied [14]. Within this dynamic interplay, several key variables play pivotal roles in dictating the course of electrode kinetics. These encompass the velocity at which ions migrate from the bulk electrolyte solution to the electrode's surface, the rate of electron transfer occurring between the electrode and ions, and the pace of nucleation and subsequent growth of the coating on the electrode's surface. These intricate processes, in turn, are subject to modulation by factors such as the concentration of the coating material within the electrolyte solution, the pH level of the solution, the applied voltage, the current density, and the prevailing temperature conditions [15].

### 16.2.2 Faraday's Laws of Electrolysis

Faraday's laws of electrolysis are a set of quantitative laws that describe the relationship between the amount of substance produced or consumed during an electrolytic reaction and the amount of electric charge passed through the electrolyte [16].

Faraday's First Law of Electrolysis:
According to Faraday's first law of electrolysis, the amount of electricity passed through the electrolyte is directly proportional to the mass of any substance deposited or liberated at an electrode.

$$m \propto Q$$

$$m = ZQ$$

where Z is the Constant of proportionality, known as the electrochemical equivalent. It is the mass of the substance deposited or liberated per unit charge.

Faraday's Second Law of Electrolysis:
According to this law, when the same amount of electricity is passed through a series of electrolytes, the electrochemical equivalence is proportional to the masses of the substance liberated at the electrode.

$$m \propto E$$

## 16.3 Types of Electrochemical Surface Coatings

Electrochemical surface coatings are divided into electrodeposition, electroless plating, anodizing, and electrophoretic deposition based on the deposition method.

### 16.3.1 Electrodeposition

This is a process in which a thin metal layer is applied to the surface of another metal to alter its surface characteristics. It relies on the principles of electrolysis, wherein positively charged ions of the desired material are diminished through the application of an electric current and are subsequently deposited as a thin film onto a conductive substrate surface.

Through this technique, specific enhancements such as improved electrical conductivity, corrosion resistance, heightened heat tolerance, reduced wear and friction, and the creation of decorative surface coatings can be achieved [17].

### 16.3.2 Anodization

This electrochemical technique is employed to enhance the corrosion resistance, wear resistance, and surface hardness of metals like aluminum and titanium by creating a protective oxide layer on their surfaces. Anodization transforms the metal surface into a metal oxide surface, safeguarding the underlying metal from oxidation. During this process, a material typically crafted from lead or stainless steel serves as the cathode, while the metal substrate, immersed in an electrolyte solution, acts as the anode. As an electric current flows through, oxygen ions are released at the anode and react with the metal, forming the oxide layer. By adjusting factors such as current density, voltage, and anodization duration, one can control the thickness and properties of the resulting oxide layer [18].

### 16.3.3 Electrophoretic Deposition

Electrophoretic deposition is a method that involves the application of an electric field to deposit charged particles onto a substrate. In this process, the substrate functions as the cathode, while a solution containing charged particles is brought into contact with it. The particles, suspended within a colloid solution, are attracted and gathered onto the substrate through a two-step mechanism. The substrate in need of coating serves as one of the electrodes within an electrochemical cell where the deposition occurs. When voltage is applied, the suspended coating particles become polarized and are drawn toward the substrate, resulting in a loose coating. At this stage, the coating primarily consists of a powder agglomeration, necessitating additional processing like sintering to enhance the bond between the particles and the substrate [19].

For electrophoretic deposition to be successful, the polymers used must either be polarizable or have the capability to support induced dipoles. This method offers several advantages, including its ability to generate uniform coatings on three-dimensional substrates and its efficiency in coating the entire exposed surface, simplifying the application of coatings to both the inner and outer surfaces of cylindrical objects.

### 16.3.4 Electroless Plating

This method is alternatively referred to as autocatalytic plating or chemical plating, as it does not necessitate an electrical current for coating formation. Instead, it relies on a suitable reducing agent to generate electrons, which subsequently facilitate the reduction of metal ions. The primary components of electroless plating encompass metal ion salt, pH adjuster, complexing agent, and reducing agent.

The advantages of electroless plating are the following:

- It deposits layers of uniform thickness even on small cavities and complex-shaped parts.
- Electrical contact is not required so it can be used for plating nonconductors like plastics.
- The deposits will have low porosity and unique mechanical, chemical, and magnetic properties.

The low deposition rate and high cost of chemical components in mass production are the limitations of this process.

The widely used electroless plating techniques are the following:

- Immersion plating: Displacement reactions deposit a very thin layer of surface coating.
- Autocatalytic plating: The metal coating is done by redox reactions in the presence of a reducing agent. This process deposits a thicker and uniform layer with more quality.

## 16.4 Overview of Surface Preparation Methods

In the electrochemical surface coating process, surface pre-treatment stands as a crucial step. The quality of the surface preparation significantly impacts the adherence, uniformity, and longevity of the coating. The choice of substrate, the coating material, and the desired coating characteristics collectively determine the specific surface preparation method employed. Among the procedures involved in surface preparation are cleaning, degreasing, surface activation, and surface roughening [20–22].

### 16.4.1 Cleaning

The initial phase of surface preparation involves thorough cleaning of the substrate to eliminate any contaminants such as dirt, grease, oil, and rust. Several methods can be employed for this purpose:

Solvent Cleaning: Solvent cleaning entails the application of a solvent, either by wiping or spraying, onto the substrate to dissolve and remove impurities. Commonly used solvents include alcohol, acetone, and chlorinated solvents.
Detergent Cleaning: Detergent cleaning involves the use of a detergent solution to emulsify and eliminate impurities. This method is particularly effective in removing organic pollutants like oils and greases.
Abrasive Cleaning: Abrasive cleaning relies on physically removing impurities with the use of abrasive tools such as sandpaper, grit blasting, or wire brushing.
Ultrasonic Cleaning: In the ultrasonic cleaning method, the substrate is immersed in an ultrasonic cleaning solution and exposed to high-frequency sound waves. These sound waves create high-pressure bubbles that dislodge and remove pollutants from the surface.

### 16.4.2 Degreasing

Following the initial cleaning process, it is essential to degrease the substrate to ensure the complete removal of any residual oils and impurities. The choice of degreasing method depends on the type of substrate and the specific pollutants that need to be eliminated. Three common degreasing methods are employed:

**Solvent Degreasing**: Solvent degreasing involves the application of a solvent that can dissolve and remove oils and greases. This solvent is either sprayed or wiped onto the substrate, and typical examples include alcohol, acetone, and chlorinated solvents.
**Alkaline Degreasing**: Alkaline degreasing utilizes an alkaline cleanser to emulsify and eliminate oils and greases. This cleanser can be applied through spraying, immersion, or ultrasonic cleaning and typically has a pH level ranging from 9 to 14.
**Acidic Degreasing**: Acidic degreasing relies on an acidic cleanser to dissolve and remove oils and greases. The cleaner can be either sprayed on or used in an immersion process, with its pH typically falling between 1 and 6.

### 16.4.3 Surface Activation

An essential step in enhancing the adhesion of a coating to a substrate involves surface activation, which generates a reactive surface. Surface activation can be achieved through various techniques, including the following:

**Mechanical Surface Preparation**: This method involves roughening the surface through processes such as sandblasting, shot peening, or grinding. The roughened surface provides increased surface area for the coating to adhere to, enhancing adhesion.
**Chemical Surface Preparation**: Chemical surface preparation entails the application of acids or other chemicals to the surface to render it reactive and conducive to adhesion. Examples of chemical treatments include acid etching, chromate conversion coating, and phosphating.
**Electrochemical Surface Preparation**: In this approach, the substrate's surface is modified using electrochemical methods to facilitate adhesion. An example is anodic oxidation, where the substrate is anodized to create a thin oxide layer that promotes adhesion. Other examples include electrochemical etching and electroplating.

### 16.4.4 Surface Roughening

Surface roughening is a method employed to impart a textured, roughened quality to the surface of a substrate, thereby enhancing the adhesion of coatings to that surface. This can be achieved through either mechanical or chemical means.

**Mechanical Surface Roughening**: Various techniques such as sandblasting, shot peening, or grinding are utilized. Sandblasting entails directing high-pressure streams of abrasive particles onto the substrate's surface, effectively removing material and creating a roughened texture. These abrasive particles not only increase the surface area but also provide additional bonding sites for the coating. Shot peening, on the other hand, employs tiny metallic beads that leave minute dents in the substrate when they come into contact with it, thereby enhancing surface area and strengthening coating adhesion. Grinding involves scraping material off the substrate's surface using an abrasive wheel, resulting in a profiled surface that improves coating adhesion.
**Chemical Surface Roughening**: Chemical surface roughening, on the other hand, involves modifying the substrate's surface chemically to create

the desired rough texture. Several chemical methods can be employed, including acid etching, chromate conversion coating, and phosphating. In acid etching, a small layer of material is dissolved from the substrate's surface using an acidic solution, yielding a rough texture. The depth of this profile can be controlled by adjusting the concentration of the acid and the duration of exposure. Chromate conversion coating involves applying a solution containing chromium ions to the substrate's surface while phosphating entails applying a solution containing phosphate ions to achieve the desired surface texture.

## 16.5 Characterization of Electrochemically Coated Surface

A coated surface exhibits its desired characteristics when the process is executed without any errors. To assess the coated surface, diverse methods are employed to analyze its morphology, crystal structure, composition, evenness, thickness, and more. SEM can be used to study the morphology and microstructure of surface coatings, including their thickness, uniformity, and adhesion to the underlying substrate. EDS can provide information on the chemical composition of a surface coating. Information on the thickness, homogeneity, and roughness of surface coatings can be obtained using surface profilometry. It can also be used to test how effectively a coating adheres to the substrate beneath it and to find coating flaws like fractures or delamination. XRD can be used to characterize the crystal structure of the coating material.

## 16.6 Applications of Electrochemical Surface Coating in Various Fields

The advancement of new materials, coatings, and manufacturing techniques has opened up the possibilities for employing electrochemical surface coatings across a wide range of applications, including aerospace, automotive, biomedical, and electronics industries.

### 16.6.1 Automotive

Electrochemical surface coating plays a vital role in the automotive industry, offering several benefits.

- **Corrosion Resistance**: It effectively prevents corrosion in critical parts like underbodies, suspensions, and engine components, using durable coatings like zinc and zinc nickel.
- **Wear Resistance**: It creates tough and durable coatings for engine components, reducing wear and extending their life span, often using nickel and chrome coatings.
- **Aesthetic Enhancement**: Electrochemical coatings provide a range of finishes and colors, enhancing the visual appeal of vehicle parts like wheels, trim, and emblems.
- **Cost-Effective**: This method is cost-effective, delivering high-quality coatings at lower costs compared to other techniques, with minimal material waste.

Overall, electrochemical surface coating is a versatile and efficient solution for various automotive applications.

### 16.6.2 Aerospace

Electrochemical surface coating is a crucial technology in the aerospace industry, offering various benefits:

**Corrosion Protection**: It shields aircraft parts from harsh environments using coatings made from metals like aluminum, titanium, and chromium.
**Thermal Barrier Coatings**: These coatings provide insulation and safeguard against high temperatures in aerospace applications, particularly in jet engines, using materials like ceramics.
**Wear Resistance**: Nickel and chrome coatings are employed for wear resistance in aviation, enhancing the durability of parts exposed to high temperatures and stress.
**Electrical Conductivity**: Electrochemical surface coating can create conductive coatings, essential for specific aircraft applications, using materials like copper and silver.
**Aesthetic Enhancement**: It also enhances the visual appeal of aircraft components, suitable for substrates such as brass, stainless steel, and aluminum.

In summary, electrochemical surface coating plays a vital role in aerospace, enhancing the performance, durability, and aesthetics of critical components.

### 16.6.3 Biomedical Application

Electrochemical surface coatings find diverse applications in biomedicine:

**Targeted Drug Delivery**: They enable controlled drug release triggered by factors like pH or temperature, delivering medications precisely where needed.
**Enhanced Implants**: These coatings improve biocompatibility, reducing infection risk, and can include antibacterial properties or tissue growth stimulation.
**Advanced Biosensors**: Electrochemical coatings create sensitive and specific biosensors for precise chemical detection in biological samples.
**Tissue Engineering**: They enhance biomaterial properties, promoting cell adhesion and tissue integration for better regeneration.

### 16.6.4 Application in Dental Implants

Electrochemical surface coatings offer promising enhancements to dental implants:

**Improved Osseointegration**: By applying electrochemical techniques, we can augment the osseointegration capabilities of dental implants. Hydroxyapatite deposition on the implant surface, achieved through electrochemical methods, reinforces the bond with surrounding bone tissue, enhancing stability and durability.
**Antibacterial Defense**: Electrochemical coatings can mitigate the risk of bacterial infection, a common concern with dental implants. Coating implant surfaces with materials like silver nanoparticles creates a hostile environment for bacteria, reducing the chance of implant failure due to microbial colonization.
**Enhanced Surface Roughness**: Surface texture influences how well an implant integrates with bone tissue. Electrochemical surface coatings can modify the implant's roughness, promoting better osseointegration. Techniques like anodic oxidation can create a rougher surface conducive to bone tissue formation.
**Optimized Biocompatibility**: Ensuring dental implant biocompatibility is crucial to tissue health. Electrochemical coatings can enhance biocompatibility by adding a layer of materials like polyethylene glycol to the implant surface, reducing the risk of inflammation and rejection.

### 16.6.5 Microelectronics

In the realm of microelectronics research and development, the electrochemical surface coating offers a wide range of valuable applications, including surface passivation, surface functionalization, electroplating, and surface modification for sensing purposes. These processes allow precise control over the surface properties of electronic devices, leading to performance enhancements.

**Surface Passivation**: Electrochemical surface coating protects electronic device surfaces from environmental threats like oxidation, corrosion, and contamination. For instance, it can deposit a protective oxide or nitride layer, safeguarding against damage.
**Surface Functionalization**: This technique enables the introduction of functional groups on device surfaces, facilitating specific chemical or biological interactions. For example, it can create biosensors by depositing a conductive polymer layer for immobilizing biological molecules.
**Electroplating**: Electrochemical surface coating optimizes metal layer deposition, enhancing its quality and adherence. It provides control over surface characteristics, critical for improving device performance.
**Surface Modification for Sensing**: This approach tailors device surfaces for chemical or biological sensing. For instance, it can modify a field-effect transistor with recognition molecules like antibodies, enabling the detection of specific analytes, such as proteins or nucleic acids, in sensing applications.

## 16.7 Challenges and Limitations of Electrochemical Surface Coating Methods

Electrochemical surface coating techniques have advantages but face challenges:

**Cost**: Expensive equipment and materials, along with time-intensive processes, raise costs.
**Adhesion**: Achieving strong coating-substrate adherence is difficult, affecting efficiency.
**Scalability**: Scaling up for large-scale production is challenging, especially for complex geometries.
**Safety**: Hazardous chemicals used require strict safety measures for workers and environmental protection.

**Composition Control**: Managing coating composition and structure is complicated.
**Electrolyte**: The stability and composition of electrolytes can affect coating quality.
**Coating Types**: Effective mostly for metallic coatings, not nonmetallic ones.
**Substrates**: Limited to conductive substrates, requiring treatments for nonconductive materials.
**Process Parameters**: Precise control of pH, temperature, and current density is crucial to avoid quality issues.

## 16.8 Conclusions

Electrochemical surface coating is a versatile method that enhances substrate properties through electrochemical reactions. Various techniques exist:

**Electroplating**: It deposits metal ions onto a substrate by passing an electric current through a metal salt solution.
**Electroless Plating**: This chemical process coats a substrate with metal without using an electric current.
**Electrophoretic Deposition**: Charged particles in a colloidal suspension move via an electric field to coat an oppositely charged conductive substrate.
**Anodizing**: An electric current creates an oxide layer on metal to improve wear and corrosion resistance.

These techniques require specific pre-treatment of the surface. Coating thickness is controlled by factors like deposition duration and electrolyte choice. Various characterization methods analyze coating surface properties.

Electrochemical surface coating is widely used in industries like automotive, aerospace, electronics, and biomedical due to its versatile benefits. However, it demands specialized equipment and expertise. Ongoing developments focus on creating new materials, coatings, and improved processing methods.

## 16.9 Future Trends

The future of electrochemical surface coating is poised for significant advancements driven by various key factors. These developments will encompass the exploration of novel and sophisticated materials, the adoption of cutting-edge technologies like nanostructured coatings, a strong emphasis on sustainability, and the optimization of processes through automation, robotics, digitalization, and data analytics. These emerging trends are expected to foster innovation and meet the increasing demands of diverse industries seeking high-performance coatings.

**Advanced Materials**: One pivotal aspect shaping the future of electrochemical surface coating involves the continuous development of advanced materials. This includes the utilization of composite materials, which blend two or more substances to impart special properties such as increased strength, durability, and wear resistance. Researchers are also delving into advanced alloys and metal-matrix composites, which offer superior corrosion resistance and other desirable characteristics. The creation of more advanced materials is expected to open up a broader range of applications for electrochemical surface coating.

**Nanostructured Coatings**: Another key facet is the evolution of nanostructured coatings, a process in which materials with layer thickness ranging from a few nanometers to a few microns are employed. These coatings can enhance properties such as hardness, adhesion, and longevity.

**Robotics and Automation**: The future of electrochemical surface coating is set to embrace extensive use of robotics and automation. This encompasses robotic handling and precise positioning of workpieces, as well as automated systems for monitoring and managing the coating process. The integration of robotics and automation is expected to ensure consistent coating quality, reduce waste, and enhance process efficiency.

**Digitalization and Data Analytics**: The fusion of digital technology and data analytics is poised to exert a significant influence on the future of electrochemical surface coating. This involves the deployment of sensors for monitoring process parameters, leveraging data analytics for optimizing these parameters and reducing defects, and utilizing artificial intelligence for quality control and predictive maintenance. The synergy of digital technology and data analytics is anticipated to lead to more productive and efficient coating processes.

**Hybrid Micromachining and Microfabrication Techniques**: Future research opportunities in this field include addressing the challenges associated with hybrid micromachining and microfabrication techniques, which require the simultaneous use of various types of energy in the same fabrication area. Fundamental issues related to material removal capabilities and the mechanisms of hybrid systems are key research objectives. Additionally, there is a need for simulation and modeling of hybrid micromachining and microfabrication procedures, with approaches such as molecular simulation and multiscale modeling being explored. These simulations and models contribute to a deeper understanding of the fundamental aspects of hybrid processes. To support these techniques, ultraprecision machines with high stiffness, precise control mechanisms, accurate feed drives, and the ability to compensate for dynamic and static alignment errors are essential. Further improvements in accuracy are achieved by minimizing tool faults through reduced tool holding and repositioning.

## References


1. Kim., K.-H. and Ramaswamy., N., Electrochemical surface modification of titanium in dentistry. *Dent. Mater. J.*, 28, 20–36, 2008.
2. Bi, P., Li, H., Zhao, G., Ran, M., Cao, L., Guo, H., Xue, Y., Robust Super-Hydrophobic Coating Prepared by Electrochemical Surface Engineering for Corrosion Protection. *Coatings*, 9, 452–467, 2019.
3. Yu., S., Guo., D., Han., J., Sun., L., Zhu., H., Yu., Z., Dargusch., M., Wang., G., Enhancing Antibacterial Performance and Biocompatibility of Pure Titanium by a Two-Step Electrochemical Surface Coating. *ACS Appl. Mater. Interfaces*, 12, 40, 433–446, 2020.
4. Frank Walsh, C. and de, León., C.P., Versatile electrochemical coatings and surface layers from aqueous methanesulfonic acid. *Surf. Coat. Technol.*, 259, 676–697, 2014.
5. Qu., X., Huang., H., Wan., T., Hu., L., Yu., Z., Liu., Y., Dou., A., Zhou., Y., Su., M., Peng., X., Wu., H.-H., Wu., T., Chu., D., An integrated surface coating strategy to enhance the electrochemical performance of nickel-rich layered cathodes. *Nano Energy*, 91, 106665, 2022.
6. Zhuang., J. and Yu, G.-R., Effects of surface coatings on electrochemical properties and contaminant sorption of clay minerals. *Chemosphere*, 49, 619–628, 2002.
7. Park, K.S., Son, J.T., Chung, H.T., Kim, S.J., Lee, C.H., Kang, K.T., Kim, H.G., Surface modification by silver coating for improving electrochemical properties of LiFePO4. *Solid State Commun.*, 129, 5, 311–314, 2004.

8. Sinha, P.K. and Feser, R., Phosphate coating on steel surfaces by an electrochemical method. *Surf. Coat. Technol.*, 161, 158–168, 2002.
9. Bian., X., Fu., Q., Bie., X., Yang., P., Qiu., H., Pang., Q., Chen., G., Du., F., Wei., Y., Improved Electrochemical Performance and Thermal Stability of Li-excess $Li_{1.18}Co_{0.15}Ni_{0.15}Mn_{0.52}O_2$ Cathode Material by $Li_3PO_4$ Surface Coating. *Electrochim. Acta*, 174, 875–884, 2015.
10. Chen, X.-B., Easton, M., Birbilis, N., Yang, H.Y., Abbott, Trevor., Corrosion-resistant electrochemical plating of magnesium (Mg) alloys. *Surf. Eng.*, 1, 315–346, 2013.
11. Li, L., Zhang., S.L., Zhang, J., Guo, P., Zheng., J., Zhao, X.S., Enhancement of Electrochemical Performance of Macroporous Carbon by Surface Coating of Polyaniline. *Chem. Mater.*, 22, 3, 1195–1202, 2010.
12. Haïdopoulos, M., Turgeon, S., Sarra-Bournet, C., Development of an optimized electrochemical process for subsequent coating of 316 stainless steel for stent applications. *J. Mater. Sci. Mater. Med.*, 17, 647–657, 2006.
13. Singh Raman, R.K., Chakraborty Banerjee, P., Lobo., D., Gullapalli, H., Sumandasa, M., Kumar, A., Choudhary, L., Tkacz, R., Ajayan, P., Majumder, M., Protecting copper from electrochemical degradation by graphene coating. *Carbon*, 50, 4040–4045.0, 2012.
14. JR, V.K. and Mythili, T., Recent Trends in Non-Traditional Machining of Alloys and Composites, in: *Advanced Manufacturing Techniques for Engineering and Engineered Materials*, pp. 61–81, IGI Global, USA, 2022.
15. Sivashankar, N. and Thanigaivelan, R., Electrochemical micromachining of magnesium AZ31 alloy using minimum quantity electrolyte. *Mater. Manuf. Processes*, *38*, 11, 1406–1415, 2023.
16. Maniraj, S., Thanigaivelan, R., Gunasekaran, K., Saravanan, K.G., Optimization of Process Parameters in Electrochemical Micromachining of AMCs by Using Different Techniques of Weight Evaluation. *Adv. Mater. Sci. Eng.*, 1–9, 2023.
17. Thanigaivelan, R., Saravanan, K.G., Makesh, M., Surface Modification Techniques for Improving the Material Performance, in: *Advanced Manufacturing Techniques for Engineering and Engineered Materials*, pp. 1–19, IGI Global, USA, 2022.
18. Thanigaivelan, R., Prakash, S., Maniraj, S., Surface Modification Techniques for Bio-Materials: An Overview, in: *Advanced Manufacturing Techniques for Engineering and Engineered Materials*, pp. 42–60, 2022.
19. Rajan, N., Thanigaivelan, R.R., Muthurajan, K.G., Effenct of Electrochemical Machining Process Parameters on Anisotropic Property of Metal Matrix Composites Al7075. *J. New Mater. Electrochem. Syst.*, *21*, 4, 239–242, 2018.
20. Li, J., Zhou, P., Attarilar, S., Shi, H., Innovative surface modification procedures to achieve micro/nano-graded Ti-based biomedical alloys and implants. *Coatings*, *11*, 6, 647, 2021.

21. Sharan, J., Lale, S.V., Koul, V., Mishra, M., Kharbanda, O.P., An Overview of Surface Modifications of Titanium and its Alloys for Biomedical Applications. *Trends Biomater. Artif. Organs*, 29, 2, 176–187, 2015.
22. Mansfeld, F. and Wang, Y., Corrosion protection of high copper aluminium alloys by surface modification. *Br. Corros. J.*, *29*, 3, 194–200, 1994.

17

# Process Variable Impact Analysis in Unconventional Machining: Enhancing MRR, Accuracy, and Surface Quality

**Jagadeesha T.[1]*, Manoj Nikam[2], Seeniappan Kaliappan[3] and Natrayan L.[4]**

[1]*Department of Mechanical Engineering, National Institute of Technology NIT, Calicut, Kerala, India*

[2]*Department of Mechanical Engineering, Bharati Vidyapeeth College of Engineering, Navi Mumbai, Maharashtra, India*

[3]*Department of Mechatronics Engineering, KCG College of Technology, KCG Nagar, Karapakkam, Chennai, Tamil Nadu, India*

[4]*Department of Mechanical Engineering, Saveetha School of Engineering, SIMATS, Chennai, Tamil Nadu, India*

***Abstract***

This chapter delves into an exploration of four unconventional machining techniques: electro-discharge machining (EDM), electrochemical machining (ECM), abrasive jet machining (AJM), and ultrasonic machining (USM). The aim is to analyze how various process variables impact surface quality, accuracy, and material removal rate (MRR). In EDM, the process variables under scrutiny are pulse-on time and peak current. For ECM, the focus shifts to electrolyte concentration, current density, and electrolyte temperature. In AJM, attention is given to abrasive particle size, abrasive flow rate, and standoff distance. Lastly, in USM, the key variables are tool vibration frequency and amplitude. These experiments are carried out on workpieces made of titanium and stainless steel. The results reveal that in EDM, increasing pulse-on time and peak current boosts MRR but comes at the cost of decreased accuracy and surface quality. In ECM, higher electrolyte concentration, current density, and electrolyte temperature lead to increased MRR, while lower current density and higher electrolyte temperature result in improved surface finish. In AJM, elevating the abrasive flow rate enhances MRR, but smaller

**Corresponding author*: jagdishsg@nitc.ac.in

R. Thanigaivelan, Pradeep Kumar Krishnan, Kamalakanta Muduli and Santosh Kumar Tamang (eds.) New Materials, Processing and Manufacturability: Fabrication and Processing of Advanced Materials, (317–334) © 2024 Scrivener Publishing LLC

abrasive particle size and greater standoff distance yield a superior surface finish. In USM, greater vibration frequency and amplitude yield higher MRR but diminish accuracy and surface quality. In conclusion, this study underscores the paramount importance of selecting the appropriate process parameters to attain the desired outcomes in terms of MRR, accuracy, and surface quality in unconventional machining processes.

***Keywords:*** Surface quality, unconventional machining, process variables, material removal rate (MRR), accuracy

## 17.1 Introduction

Modern manufacturing industries frequently utilize unconventional machining techniques such as electrical discharge machining (EDM) and abrasive jet machining (AJM) to address the challenges posed by difficult-to-machine materials [1]. The efficiency and effectiveness of these methods are heavily dependent on various process parameters. In AJM, these parameters encompass pulse duration, pulse current, voltage, abrasive concentration, and nozzle diameter, while in EDM, they encompass pulse duration, current, voltage, electrode material, and dielectric fluid.

Optimizing these process parameters is imperative to achieve the desired material removal rate (MRR), accuracy, and surface quality. MRR measures the rate at which material is removed from the workpiece and is influenced by parameters like voltage, current, and pulse duration. Accuracy refers to how closely the machined workpiece aligns with the specified shape and dimensions and can be influenced by variables such as electrode material, gap distance, and dielectric fluid. Surface quality assesses the polish and texture of the machined surface and can be impacted by pulse duration, voltage, and electrode material.

Understanding how these process variables affect MRR, accuracy, and surface quality is pivotal in optimizing the machining process to achieve the desired outcomes. For example, in AJM, enhancing abrasive concentration can enhance MRR but may compromise surface quality due to increased surface roughness. In EDM, the formation of a recast layer can lead to poor surface quality, even if a longer pulse duration enhances MRR [2].

In summary, process parameters wield significant influence in nontraditional machining (NTM) processes, necessitating their careful optimization to attain the desired MRR, accuracy, and surface quality. A comprehensive grasp of the impact of these parameters can enhance process control and bolster machining efficiency.

## 17.2 Nontraditional Machining Processes

A machining process is classified as nontraditional when its method of material removal fundamentally differs from traditional processes. In NTM, an alternative form of energy, distinct from the conventional use of force applied by a tool in direct contact with the workpiece, is employed to eliminate excess material from the work surface or to partition the workpiece into smaller components.

### 17.2.1 Electro-Discharge Machining

The most widely adopted unconventional machining technique employs a thermoelectric approach involving a series of electric sparks generated between an electrode and the workpiece. In this method, the hardness of the workpieces does not play a crucial role [3]. Instead, both the tool and the workpiece must remain immersed in a dielectric material throughout the machining process. Electrical sparks are used in the unconventional machining technique known as EDM to remove material from a workpiece. The working principle of EDM involves the following steps:

1. Workpiece and Tool: A small gap between a workpiece and a tool (referred to as an electrode) is maintained in a dielectric fluid (such as oil or deionized water), and this space is referred to as the spark gap.
2. Material Removal: The molten and vaporized dielectric fluid and the gas produced by the spark then remove material from the workpiece's surface.
3. Repetition: The process is repeated many times, with the tool gradually moving closer to the workpiece, until the workpiece's size and required shape are obtained.

The EDM procedure can be managed by adjusting various parameters, such as the electrical voltage, pulse duration, frequency, and polarity, as well as the dielectric fluid and the material of the electrode. Hard and challenging-to-machine materials like hardened steel, titanium, and tungsten carbide can all be effectively machined with EDM [3].

Overall, the working principle of EDM involves the use of high-frequency electrical sparks to generate heat and pressure that erodes and removes material from the workpiece surface, making it an important process for the creation of intricate forms and precise parts [4].

### 17.2.2 Parameters Affecting the MRR in EDM

**Peak current:**
The pulse current is the amount of electrical current that flows between the tool and the workpiece during the electrical discharge. Increasing the pulse current not only can increase the MRR but can also result in higher electrode wear and lower surface finish.

**Voltage:**
The voltage is the potential difference that exists during the electrical discharge between the tool and the workpiece. The MRR may not only rise with higher voltage but may also result in a worse surface polish and more electrode wear.

**Pulse-on Time:**
The pulse-on time is the period of time during which an electrical discharge is applied between the tool and the workpiece. While lengthening the pulses can boost MRR, doing so can potentially harm electrode surface quality and speed up electrode degradation.

**Pulse-off Time:**
As $T_{off}$ rises, MRR declines. This $T_{off}$ permits a period of time during which material can harden and be removed with the aid of dielectric fluid.

**Dielectric Fluid:**
The eroded material is removed using the dielectric fluid during the EDM procedure. The MRR can be significantly influenced by the nature and characteristics of the dielectric fluid. Using a dielectric fluid with good flushing properties and low electrical conductivity can result in higher MRR.

### 17.2.3 Parameters Affecting the Surface Quality and Accuracy in EDM

**Voltage:**
When investigating the influence of voltage on surface roughness while keeping a constant current, pulse duration, and pulse interval, it becomes

evident that higher voltage levels lead to improved surface quality. Elevating the voltage results in longer discharge times, ultimately leading to a substantial gap in discharge timing [5]. Consequently, this leads to enhanced discharge conditions, accompanied by a reduction in the number of discharge cycles within a specified time frame. As a result, stable machining ensues, leading to a decrease in surface roughness values. Simultaneously, increased voltage levels also contribute to improved machining accuracy.

**Current (I):**
As the current increases, the surface quality deteriorates. As the input parameter's current increases, the surface roughness rises. At peak current, the discharge energy will be greater depending on the current value (I). This causes the surface roughness to rise and a reduction in surface quality. As the current rises, machining accuracy declines.

**Pulse-on Time ($T_{on}$):**
As $T_{on}$ increases, surface roughness increases in value, and surface quality is impacted. The quantity of discharge energy (E), which rises with an increase in $T_{on}$, has the biggest impact on surface quality. The size of the spark crater affects the roughness of the surface [6]. A shallow crater with a larger diameter creates a rougher surface on the workpiece. Deeper erosion craters will form on the surface as a result of a huge amount of discharging energy producing intense sparks. Due to the effect of discharge energy on the accuracy, surface quality and accuracy both decline as pulse-on time increases.

**Pulse-off Time ($T_{off}$):**
As the duration of the pulse interval ($T_{off}$) increases, there is a corresponding decrease in the discharge energy (E), illustrating a reduction in surface roughness and an enhancement in surface quality [7]. Furthermore, when the discharge energy decreases, there is a concurrent reduction in roughness, resulting in a more precise workpiece.

## 17.3 Electrochemical Machining

Electrochemical machining (ECM) uses electrochemistry to remove material from a workpiece. Voltage is applied between a tool (cathode) and the workpiece (anode) submerged in an electrolyte. Material dissolves due to the current, with the tool often customized for shape, typically made of copper or brass. ECM is known for precision, ideal for tough materials, intricate shapes, and industries like aerospace, medical, and automotive [8]. It is slower than conventional methods, requiring special tools and training. ECM immerses the workpiece in an electrolyte like NaCl, $NaNO_3$, or NaOH. The negative power supply terminal connects to the tool electrode, the positive to the workpiece. As current flows, metal ions dissolve from the workpiece's surface in contact with the tool electrode, typically made of inert materials like copper or graphite. Dissolved ions are carried away by the electrolyte, and the tool electrode continuously wears away. The cavity's shape depends on the tool electrode's geometry. ECM excels in precision, favored for intricate parts in the aerospace, automotive, and medical industries.

### 17.3.1 Parameters Affecting the MRR in ECM

**Electrolyte Concentration:**
When the concentration of an electrolyte is elevated, the rate of ion transformation from the anode to the electrolyte solution also accelerates, leading to an increase in the MRR.

**Interelectrode Gap:**
As the interelectrode gap between a tool (cathode) and the workpiece (anode) increases, the flow rate from anode to cathode firstly decreases and then increases the MRR. For more MRR, the interelectrode gap between electrodes is kept less.

**Applied Voltage (V):**
As voltage is applied between the electrodes in a system, initially, the MRR tends to decrease, followed by a gradual increase. However, beyond a certain point, as we continue to raise the applied voltage, the MRR will start to increase at a constant rate.

**Electrolyte Flow Rate:**
Increasing the flow rate of an electrolyte in an ECM system, facilitated by a pump, results in a reduction of the MRR from the workpiece (anode). Therefore, as the electrolyte flow rate increases, the MRR in an ECM operation decreases.

### 17.3.2 Parameters Affecting the Surface Quality and Accuracy in ECM

**Electrolyte Concentration:**
The surface quality initially deteriorates as the electrolyte content increases but then experiences a sudden improvement. This phenomenon is attributed

to the higher concentration of electrolytes and the accumulation of multiple generations of machined products, which modify the properties of the electrolyte [9, 10]. This alteration leads to increased surface roughness and decreased surface quality. Therefore, as the concentration of the electrolyte rises, surface quality and accuracy will deteriorate initially, but beyond a certain concentration threshold, they will progressively improve.

**Applied Voltage (V):**
Surface roughness increases and surface quality degrades gradually with an increase in the value of voltage. An increase in voltage causes the surface roughness (Ra) to rise and surface quality to decrease, consequently decreasing the accuracy of the workpiece operation [11, 12].

## 17.4 Ultrasonic Machining

Ultrasonic machining (USM) is an NTM process that utilizes abrasive particles to impact the workpiece and remove material. The duration of the USM operation is influenced by several factors, including the frequency of the tool's vibrations, the size of the abrasive particles suspended in the slurry, and the viscosity of the fluid used [13]. Typically, abrasive grains such as BCN or SiC are preferred for their enhanced rigidity. If the viscosity of the fluid is low, it becomes possible to wash out previously used abrasives.

### 17.4.1 Parameters Affecting the MRR in USM

**Amplitude:**
MRR increases with an increase in amplitude with increasing frequency. Amplitude is an end-to-end motion, or an expansion or a contraction, of a tool of USM. Therefore, the motion (velocity) of a tool increases and, consecutively, MRR will increase.

**Mean Grain Diameter:**
MRR of a USM process increases with an increase in mean grain diameter. Slurry size has more effect on MRR than its concentration. Therefore, the slurry size is directly proportional to the MRR of the USM process.

**Feed Force:**
Feed force has a better effect on the MRR of a USM process. When the feed force rises, the MRR of a USM process also rises up to a certain point before it begins to decline when the feed force rises beyond that point. As a result, MRR initially rises and then falls as feed force increases.

**Work/Tool Hardness:**
MRR of a USM decreases with an increase in work/tool hardness. As the hardness of the workpiece increases, it will take more time to remove the material from the workpiece. Therefore, the rate at which material leaves the workpiece will decrease. This consequently decreases the MRR of a USM technique.

### 17.4.2 Parameters Affecting the Surface Quality and Accuracy in USM

**Mean Grain Diameter:**
As the grain diameter of an abrasive used in a USM increases, then the surface quality of a workpiece decreases which means the surface roughness of the workpiece increases. The accuracy of a system becomes less and tolerance limit can be exceeded. In that way, accuracy gets lowered with an increase in mean grain diameter.

**Amplitude of Vibration:**
The amplitude of vibration is the maximum displacement of the tool tip during the machining process. A higher amplitude of vibration results in a higher removal rate but can also cause surface damage and decrease accuracy.

**Frequency of Vibration:**
The frequency of vibration is the number of vibrations per second. Higher frequencies not only result in a smoother surface finish and higher accuracy but also require more energy and can cause tool wear.

**Tool Material:**
The tool material affects the surface finish and accuracy in USM. The tool must be able to withstand high-frequency vibrations and maintain its shape to ensure accuracy. Workpiece material: The workpiece material affects the surface finish and accuracy in USM. Harder materials require higher amplitude and frequency of vibration to achieve the desired result.

**Tool Geometry:**
The geometry of the tool affects the surface finish and accuracy in USM. The tool must be designed to efficiently transfer the vibrations to the workpiece and remove material in a controlled manner.

**Machining Time:**
The length of time the tool is in contact with the workpiece affects the surface finish and accuracy. Longer machining times can result in a smoother surface finish but may also result in a decrease in accuracy. Overall, achieving the desired surface quality and accuracy in USM requires careful selection and optimization of these parameters.

### 17.4.3 Abrasive Jet Machining (AJM)

AJM is an unconventional machining technique that employs a high-speed stream of abrasive particles to remove material from a workpiece. These abrasive particles are typically composed of materials like aluminum oxide, silicon carbide, or even diamond. In AJM, these abrasive particles are carried within a rapidly moving gas stream, which is then directed toward the workpiece. Common choices for the gas stream include compressed air, nitrogen, or gas mixtures. The gas stream imparts high velocities to the abrasive particles, causing them to impact the workpiece and erode the material. One significant advantage of AJM is its ability to process materials that are notoriously difficult to machine using conventional methods, such as ceramics, glass, and composite materials. Moreover, it is a relatively cost-effective process, requiring less complex equipment compared to other machining techniques. However, AJM does come with its limitations. It is a more labor-intensive process compared to alternative machining methods, and the resulting surface finish is generally not as smooth. Additionally, the abrasive particles can cause wear and tear on the nozzle used to direct the gas stream, leading to increased operational expenses. In summary, AJM serves as a valuable machining technique for applications involving the fabrication of tough, brittle materials, particularly in industries like aerospace, electronics, and medical devices.

The AJM process follows these key steps:

**Mixing of Abrasive Particles:** Abrasive particles are introduced into a mixing chamber along with a carrier gas, typically compressed air. Here, they become entrained within the gas stream.

**Nozzle:** The mixture of gas and abrasive particles is then channeled through a small nozzle, where the particles are accelerated to high velocities, typically in the range of 200 to 400 m/s.
**Impingement on Workpiece:** The high-velocity abrasive particles are directed toward the workpiece, which is securely held in place. As these particles strike the workpiece's surface, they remove material through erosion.
**Waste Material Removal:** Another device, such as a conveyor belt or vacuum system, is used to clear the workpiece of any waste materials generated during the machining process.

Adjusting parameters such as the number of abrasive particles in the gas stream, the size and shape of the nozzle, the distance between the nozzle and the workpiece, and the pressure of the carrier gas allows for precise control over the machining operation. Increasing the concentration of abrasive particles in the gas stream can expedite material removal. Ultimately, AJM is a reliable method for eroding material from workpieces, particularly when dealing with hard, brittle materials that are challenging to machine using traditional methods.

### 17.4.4 Parameters Affecting the MRR in AJM

In AJM, among other things, the following factors influence the MRR:

**Abrasive Characteristics:**
The MRR can be strongly impacted by the characteristics of the abrasive employed in AJM, such as its hardness, shape, size, and density. Harder and sharper abrasives can remove material more quickly, resulting in a higher MRR.

**Nozzle Diameter:**
The size of the nozzle used in AJM can also affect the MRR. A larger nozzle can deliver a higher flow rate of abrasive particles, resulting in a higher MRR.

**Pressure of Abrasive Stream:**
The pressure of the abrasive stream is another critical factor that affects the MRR. Increasing the pressure of the abrasive stream can increase the speed at which material is removed, resulting in a higher MRR.

**Distance Between Nozzle and Workpiece:**
The distance between the nozzle and the workpiece also plays a role in determining the MRR. A closer distance between the nozzle and the workpiece can result in a higher MRR.

**Traverse Rate:**
The speed at which the nozzle is moved across the workpiece, known as the traverse rate, can also affect the MRR. A higher traverse rate can result in a lower MRR, while a slower traverse rate can result in a higher MRR.

**Workpiece Properties:**
The properties of the workpiece being machined, such as its hardness, toughness, and surface finish, can also affect the MRR. Softer materials are generally easier to remove, resulting in a higher MRR, while harder materials may require more time to remove material, resulting in a lower MRR.

Overall, a combination of these factors, as well as other parameters, such as the type and concentration of abrasive used, the angle of the nozzle, and the type of carrier gas, can all affect the MRR in AJM.

### 17.4.5 Parameters Affecting the Surface Roughness and Accuracy in AJM

**Abrasive Particle Size:**
The size of the abrasive particles used in the jet stream can have a significant impact on the surface roughness. Generally, smaller abrasive particles produce smoother surfaces than larger ones.

**Jet Pressure:**
The pressure of the abrasive jet stream can also affect the surface roughness. Higher pressures typically result in rougher surfaces, as the abrasive particles are more likely to create deeper grooves in the workpiece.

**Standoff Distance:**
The distance between the nozzle and the workpiece, known as the standoff distance, can also influence the surface roughness. Generally, shorter standoff distances result in rougher surfaces, as the abrasive particles have less time to impact and smooth out the surface.

**Abrasive Flow Rate:**
The rate at which the abrasive particles are fed into the jet stream can also affect the surface roughness. Higher flow rates typically result in rougher surfaces, as more abrasive particles are being used to remove material.

**Workpiece Material:**
The type of material being machined can also play a role in the surface roughness. Softer materials tend to produce smoother surfaces, while harder materials can result in rougher surfaces due to the increased difficulty in removing material.

**Nozzle Diameter:**
The diameter of the nozzle can also affect the surface roughness. Smaller nozzles typically produce smoother surfaces, as the jet stream is more focused and precise, while larger nozzles may result in rougher surfaces due to their wider spray pattern.

**Abrasive Material:**
The type of abrasive material used can also impact the surface roughness. Softer abrasives, such as glass beads, tend to produce smoother surfaces, while harder abrasives, such as diamond, can result in rougher surfaces due to their ability to remove material more aggressively.

## 17.5 Laser Beam Machining (LBM)

Laser beam machining (LBM) is an unconventional manufacturing technique that employs a high-energy laser beam to erode material from a workpiece. Typically, a laser source generates the beam, which is then guided toward the workpiece through a system of mirrors and lenses. When the laser beam contacts the workpiece's surface, it generates intense heat, causing the material to undergo processes such as melting, vaporization, or burning, thus shaping the material as desired. This material removal process is influenced by various factors, including the laser beam's intensity, exposure duration, and the characteristics of the workpiece material. The laser beam's focus is adjusted using an array of mirrors and lenses, allowing for control over both the precision and the amount of material removed during machining [14]. Consequently, operators can achieve precise and intricate shapes.

Laser beam manipulation techniques vary, with options like raster scanning, where the laser beam traverses the workpiece in parallel lines, and vector scanning, which targets specific points on the workpiece to create the desired shape. In essence, LBM relies on the controlled application of high-energy laser beams to selectively remove material from a workpiece, enabling the creation of intricate and precise shapes.

### 17.5.1 Parameters Affecting the MRR in LBM

The following parameters can affect the MRR in LBM [15, 16]:

**Laser Power:**
The laser power is one of the most critical parameters affecting the MRR in LBM. Increasing the laser power not only leads to higher MRR but also increases the risk of thermal damage to the material being machined.

**Scanning Speed:**
The MRR is also impacted by the laser beam's scanning speed. A lower MRR results from a faster scanning rate since the laser spends less time at each location on the surface.

**Focal Length:**
The focal length of the laser beam affects the spot size and intensity at the workpiece surface. A shorter focal length leads to a smaller spot size and higher intensity, which can result in a higher MRR.

**Material Properties:**
The material being machined also affects the MRR in LBM. Harder and more reflective materials require higher laser power and longer exposure times, resulting in a lower MRR.

Assist gas: The MRR may be impacted by the use of an aid gas during LBM. The assist gas makes the laser more effective by assisting in the removal of molten material from the cutting zone. The choice of assist gas and its flow rate can affect the MRR.

### 17.5.2 Parameters Affecting the Surface Roughness and Accuracy in LBM

**Laser Power:**
Both the rate of material removal and the surface quality are impacted by laser power. Increasing the laser power can result in a rougher surface due to the formation of surface irregularities caused by thermal damage to the

material. A lower laser power not only can result in a smoother surface but it can also decrease the MRR.

**Spot Size:**
The resolution and accuracy of the machined surface are significantly influenced by the laser beam's spot size. A smaller spot size not only can result in a higher accuracy and a smoother surface but it can also increase the processing time.

**Material Properties:**
The material being machined can also affect the surface roughness and accuracy. Harder materials tend to produce rougher surfaces, while softer materials tend to produce smoother surfaces. The material's thermal conductivity, absorption coefficient, and reflectivity also play a role in determining the surface quality.

**Gas Assist:**
The use of an assist gas can also affect the surface roughness and accuracy. The assist gas helps to remove molten material from the cutting zone and can help to reduce the formation of surface irregularities. The choice of assist gas and its flow rate can affect the surface quality.

## 17.6 Conclusion

Nonconventional machining techniques comprise a range of manufacturing methods that do not depend on traditional cutting tools. These methods include AJM, LBM, USM, ECM, and EDM. Within these techniques, various process variables play a pivotal role in determining MRRs, precision, and surface quality. ECM, USM, LBM, AJM, and EDM are all affected by a variety of process parameters that collectively influence the speed of material removal, precision, and the quality of the resultant surface finish

In NTM processes, the MRR serves as a critical performance indicator, indicating the amount of material that can be removed within a specific time frame. Several process variables influence MRR, including the power of the machining source, feed rate, and the type and size of abrasive particles used. In AJM, MRRs are relatively low compared to other nontraditional methods, typically falling within the range of 0.5 to 20 $mm^3/min$. During AJM, a high-speed stream of abrasive particles is directed at the workpiece to gradually remove material. USM employs a vibrating tool that strikes the workpiece with abrasive particles. Consequently, USM generally achieves higher MRR values than AJM, typically ranging from 10 to 300 mm$^3$/min. For EDM, the MRR depends on the material being machined. However, in general, EDM can yield MRR values ranging from 1 to 100 mm$^3$/min. Therefore, in summary, the sequence of MRRs in NTM processes from lowest to highest is AJM (0.5 to 20 mm$^3$/min), EDM (1 to 100 mm$^3$/min), and USM (10 to 300 mm$^3$/min): LBM < USM < AJM < EDM < ECM.

In NTM, accuracy is a critical factor that reflects the ability of the process to produce components that meet the required specifications. This precision is affected by several key parameters, including the resolution of the machining equipment, the stability of the machining configuration, and the expertise of the operator. The order of importance for achieving accuracy in NTM can be summarized as follows: AJM< USM < EDM < ECM < LBM.

Surface quality plays a pivotal role in NTM, representing the level of smoothness and consistency achieved on the machined part's surface. This critical parameter is influenced by various factors, such as the type and dimensions of abrasive particles used, the feed rate, and the machining parameters. In NTM, surface quality is typically categorized as follows: USM < AJM < EDM < ECM < LBM.

To sum up, NTM is significantly influenced by process parameters that affect MRRs, precision, and surface quality. It is imperative to carefully choose and fine-tune these parameters to attain the desired performance outcomes in NTM procedures.

## 17.7 Future Research Opportunities

NTM has become a crucial method for handling challenging materials like ceramics, composites, and hardened metals. However, there is still significant potential for growth and enhancement in this field. Here are some areas of prospective research and development in NTM:

Integration of AI and Machine Learning: The incorporation of AI and machine learning (ML) techniques can significantly boost the control, optimization, and productivity of NTM processes. ML can be employed to analyze real-time data and enhance machining performance, while AI algorithms can predict the optimal process parameters for specific materials.

Hybrid NTM Processes: The development of hybrid NTM processes that combine two or more NTM methods, such as laser-assisted ECM, has the potential to enhance the efficiency, effectiveness, and quality of machining.

These hybrid techniques can also enable the precise machining of complex, multi-material components.

In summary, NTM is an evolving field with ample opportunities for future research and improvement. Advances in AI, ML, hybrid processes, materials science, and sustainability are expected to lead to innovative NTM techniques that can meet the growing demand for precision and high-quality machining across various industries.

## References

1. Dewan, P., Lepcha, L.P., Khaling, A., Prasad, N., Rai, S., Experimental Analysis and Optimization of EDM Process Parameters. *IOP Conf. Ser.: Mater. Sci. Eng.*, 377, 012220, 20182019.
2. Chandramouli, S. and Eswaraiah, K., Optimization of EDM Process parameters in Machining of 17-4 PH Steel using Taguchi Method. *Mater. Today Proc.*, 4, 2, Part A, 2040–2047, 2017.
3. Kumar, S., Ghoshal, S., Arora, P., Optimization of Process Parameters in Electric Discharge Machining for SS420 Using Taguchi Technique. *Adv. Electr. Electron. Eng.*, 2, 997–1004, 2021.
4. Hernández-Castillo, I., Sánchez-López, O., Lancho-Romero, G.A., Castañeda-Roldán, C.H., An experimental study of surface roughness in electrical discharge machining of AISI 304 stainless steel. *Ing. e Investig.*, 38, 90–96, 2017.
5. Janardhanan, J., Ramnath, V., Akilesh, S., Kumar, R., Optimization of process parameters on EN24 Tool steel using Taguchi technique in Electro-Discharge Machining (EDM). *IOP Conf. Ser.: Mater. Sci. Eng.*, 149, 012022, 2016.
6. Faisal., N. and Kumar., K., Optimization of Machine Process Parameters in EDM for EN 31 Using Evolutionary Optimization Techniques. *Technologies*, 6, 2, 54–65, 2018.
7. Srinivasan, V.P., Duraithilagar, S., Dhanasekhar, S., Abhishek, A., Sundaravadive, A.P., Optimization of electrical discharge machining process parameters of titanium alloy using Taguchi methodology. *AIP Conf. Proc.*, 2527, 020002, 2022.
8. Kumar, A. and Pabla, B.S., Review on optimized process parameters of ele trochemical machining and its variants. *Mater. Proc.*, 46, Part 20, 10854–10860, 2021.
9. Munda, J. and Bhattacharyya, B., Investigation into electrochemical micromachining (EMM) through response surface methodology-based approach. *Int. J. Adv. Manuf. Technol.*, 7, 821–832, 2008.
10. Das, M.K., Barman, T.K., Kumar, K., Sahoo, P., Effect of Process Parameters on MRR and Surface Roughness in ECM of EN 31 Tool Steel Using WPCA. *Int. J. Mater. Form.*, 4, 2, 19–28, 2017.

11. Soundarrajan, M. and Thanigaivelan, R., Investigation of electrochemical micromachining process using ultrasonic heated electrolyte, in: *Advances in Micro and Nano Manufacturing and Surface Engineering: Proceedings of AIMTDR 2018*, Springer, Singapore, pp. 423–434, 2019.
12. Thanigaivelan, R., Senthilkumar, R., Arunachalam, R.M., Natarajan, N., Impact of the shape of electrode-tool on radical overcut of microhole in electrochemical micromachining. *Surf. Eng. Appl. Electrochem.*, *53*, 486–492, 2017.
13. Rajathesh, C., Vinay, G., Nanjundeswaraswamy, Akki, S., Process Parameters of Ultrasonic Machining. *Int. J. Recent Technol. Eng.*, 8, 8949–8955, 2022.
14. Nguyen., V., Altarazi., F., Tran., T., Optimization of Process Parameters for Laser Cutting Process of Stainless Steel 304: A Comparative Analysis and Estimation with Taguchi Method and Response Surface Methodology. *Math. Probl. Eng.*, 2022, 14–23, 2022.
15. Ganesan, S.K. and Rajasekaran, T., Optimization of Laser Parameters and Dimple Geometry Using PCA-Coupled GRG. *Stroj. Vestn. J. Mech. E.*, 67(10), 525–533, 2021.
16. Thanigaivelan, R., Arunachalam, R.M., Nithish, A., Venkatesh, S., Naveenkumar, P., Selvaganapathy, S., ARAVIND, A.S., Optimization of Laser and Electrochemical Process Parameters for Surface Modification of Hardness and Hydrophobicity on 316L Steel. *Lasers Eng. (Old City Publishing)*, *45*, 2020.

# 18

# Pulsed Power Supplies for Surface Coating Applications: Methods, Materials, and Case Studies

**Jagadeesha T.[1*], Amith Kumar Gajakosh[2], Seeniappan Kaliappan[3] and S. Socrates[4]**

[1]*Department of Mechanical Engineering, National Institute of Technology NIT, Calicut, Kerala, India*

[2]*Amith Kumar Gajakosh, Department of Mechanical Engineering, BTL Institute of Technology and Management, Bangalore, Karnataka, India*

[3]*Department of Mechatronics Engineering, KCG College of Technology, KCG Nagar, Karapakkam, Chennai, Tamil Nadu, India*

[4]*Department of Mechanical Engineering, Velammal Institute of Technology, Chennai, Tamil Nadu, India*

## *Abstract*

A pulse power supply is a device designed to generate brief, high-power bursts of energy at the micro- or nanosecond level. This process involves the conversion of low-power electrical energy from a standard wall outlet into stored energy within capacitors, which is then discharged as a concentrated pulse of energy. This technology finds extensive applications in food processing, water treatment, and various medical processes. This chapter discusses various steps involved in surface coating using pulsed power systems and performance evaluation criteria for such coatings. Different coating materials along with coating methods such as balanced and unbalanced sputtering processes are discussed. Case studies on alumina ceramic coating and plasma electrolysis oxidation with a pulsed power source are demonstrated in this chapter.

***Keywords*:** Surface coating, pulse power supply, magnetron sputtering, plasma electrolysis

**Corresponding author*: jagdishsg@nitc.ac.in

R. Thanigaivelan, Pradeep Kumar Krishnan, Kamalakanta Muduli and Santosh Kumar Tamang (eds.) New Materials, Processing and Manufacturability: Fabrication and Processing of Advanced Materials, (335–350) 

## 18.1 Introduction

Surface coating is an engineering procedure that involves the addition of an extra layer of metal onto a solid material, typically in the form of powder or film. The choice between powder or film depends on the specific application and characteristics desired for the coated material. Surface coating serves various purposes, including aesthetic improvement and alteration of the thermal properties of the base material. To gain a comprehensive understanding of surface coating technology, in-depth study and research are essential, spanning multiple disciplines such as mechanical engineering, electrical engineering, chemical engineering, and material science. When selecting the appropriate coating material, two crucial considerations come into play:

a) Examining the relationship between the composition, structure, and properties of the coating material.
b) Determining how to construct the structure and design of the coated material to achieve a predefined set of properties.

These considerations are pivotal in the successful implementation of surface coating processes. Various steps of the surface coating process are as follows:

a) Functional requirement of the surface
b) Identification of the surface properties
c) Selection of proper materials and understand the physics of coating
d) Selection of the coating process
e) Selection of process parameters

Tool failure can occur due to various limitations associated with the tool, such as the rapid wear and heavy metal buildup on the tool's face, leading to changes in its geometry [1]. These issues are particularly noticeable during dry machining. However, these problems can be mitigated through the application of coatings to the tool [2]. For instance, in the case of dry machining with a carbide tool, failure can result from multiple factors. Yet, by applying a titanium nitride coating with a thickness of 5 μm, the tool's life span can be significantly extended. Likewise, a diamond coating can be utilized to enhance the tool's resistance to welding when working with aluminum [6]. The factors on which the performance of the coating depends are the following:

a) Process used for the coating
b) Structure of the base material and coating material
c) Properties of the base material and coating material
d) Performance of the coating depending on the process parameters chosen

Finally, the performance of the coated materials can be tested by checking the following parameters:

1. Force
2. Temperature
3. Chip
4. Tool wear
5. Workpiece surface quality

Over the past decade, numerous researchers have delved into the realm of surface coating using pulse power supplies. The arc method is a preferred choice, especially for high-insulating materials, as it leads to the formation of droplets. This is achieved by converting the discharge pulse into a frequency ranging from 10 to 200 Hz. When it comes to aluminum oxide coating, the most effective approach involves the use of reactive unbalanced magnetron sputtering with a DC (direct current) power supply [3]. The shape and duration of current pulses play a pivotal role in influencing various properties of the oxide layer [4]. Researchers commonly employ electrochemical impedance spectroscopy and potentiodynamic polarization methods to determine the optimal shape and duration of current pulses. In spectroscopy, the voltage is modified, and the pulse waveform of the source can be captured during plasma electrolytic oxidation (PEO). It is noteworthy that the charging process typically occurs at the onset of the pulse [5]. Micro-arc oxidation (MAO) finds its application primarily with nonferrous metals rather than insulating metals [6]. When coating AZ91D, a magnesium alloy, achieving an optimal surface coating entails fine-tuning the supplied power in pulse form, discharge loop current (DLC), and power parameters [7]. For the creation of a ceramic-like aluminum oxidation coating on stainless steels, an electrolytic oxidation process in an aqueous solution of NaAlO2 can be effectively employed [8].

## 18.2 Different Coating Materials

To select a surface coating process, the following points should be taken into consideration:

1. Economics of process
2. Consideration of the environment
3. Compatibility of the coating process with coating material and substrate material

The choice of coating materials is a crucial factor to consider. It is important to ensure compatibility between the coating material and the type of coating needed to achieve durability and enhance the material's properties. Below are examples of coating materials based on the specific coating requirements.

1. Metal coating: Ti, Cr, W. Mo. Ni. Cu, Ag, Au, and Al
2. Hard coating: TiC, TiN, $Al_2O_3$, CrN, and AlN
3. Super hard coating: Diamond and CBN (carbon boron nitride)
4. Soft coating: $MoS_2$, $WS_2$, and $TaS_2$
5. Composite: $MoS_2$+TiN
6. Multilayer coating: TiC+TiN+$Al_2O_3$
7. Graded coating: TiC+TiCN+TiN

## 18.3 Magnetron Sputtering

Sputtering is a process that involves the transfer of momentum from particles, typically ions of neutral gasses, to the surface of a target material. The efficiency of this process is influenced by various parameters, including the energy, angles, and masses of the incident particles, as well as the binding energy between atoms within the material. When ions collide with the target surface, they can either be absorbed by it or reflected away. As the energy of these ions increases, they can penetrate deeper into the atomic structure of the target material, leading to the degradation of its surface [9]. Once the energy reaches a certain threshold, atoms from the surface can escape into the surrounding environment.

The primary energy, which is needed to eject an atom from the surface of the target material, typically exceeds the binding energy of the target atoms by a factor of approximately 3 to 4. This primary energy is generated by establishing a potential difference between the cathode (where the target material is located) and the anode (comprising the chamber wall and the electrically grounded substrate). This voltage difference ionizes the neutral gas present in the chamber, typically argon, and triggers the ignition of plasma [10].

The process involves positively charged ions being accelerated toward the cathode positioned at a negative potential. If the energy transferred from these incident ions to the target material's surface, located on the cathode, surpasses their initial energy, the atoms removed from the material collide with other atoms, leading to an energy cascade. After several collisions, these atoms, now lacking some of their initial energy, reach the substrate surface. Additionally, during target ion bombardment, secondary electrons are generated, and these play a crucial role in maintaining plasma stability. However, the traditional sputtering method has some drawbacks, including a low deposition rate, inefficient ionization in the plasma, and substrate heating. These limitations are largely mitigated by adopting the magnetron sputtering technique and more recently, the unbalanced magnetron sputtering method.

When all magnets possess equal magnetic power, they generate a configuration known as balanced magnetron sputtering. In this setup, all the magnetic field lines originating from the external magnets pass through the central magnet. Consequently, these magnetic field lines form a closed loop, preventing electrons from escaping the magnetic field. The plasma effectively extends to a region approximately 60 mm away from the target surface. In this region, the structure and properties of the substrate are significantly influenced by the simultaneous bombardment of ions [11].

When the central magnet is weaker in comparison to the ring magnets, it results in an unbalanced magnetron sputtering configuration. In this setup, not all magnetic field lines are effectively captured by the central magnet. As a consequence, the shape of the resultant magnetic field changes, allowing some electrons to easily escape from the plasma toward the substrate. The presence of these escaping electrons in the vicinity of the substrate leads to the ionization of the inert gas and the infiltration of the plasma into this region. This, in turn, results in ions bombarding the substrate, which enhances the quality of the growing thin film.

## 18.4 Ceramic Coatings

Ceramic coatings come in various types, each with distinct compositions and mechanical and metallurgical properties. These coatings often outperform most other materials when used for coating purposes. The method of application is also crucial, as it significantly influences the material's properties. When ceramic coatings are applied using thermal spray techniques such as flame spraying, electric arc spraying, vacuum or plasma-assisted thermal spraying, or high-velocity oxy-fuel (HVOF) spraying, they typically exhibit excellent adhesion to the base material. This is due to the high temperatures or high velocities involved in these processes, which promote strong diffusion and bonding [12]. Furthermore, these methods yield dense coatings while minimizing the waste of coating material. The choice of a ceramic coating hinges on the specific performance characteristics required. Various ceramic coatings offer a range of distinct properties, as outlined below. These properties guide the selection of the most suitable coating material. Additionally, an essential factor in the selection process is the budget and economic considerations associated with the coating application. Characteristics of ceramic coatings are the following:

1. Exceptional longevity
2. Outstanding hardness
3. Superior resistance to wear, scratches, and abrasion
4. Effective oxidation resistance and high-temperature-resistant oxides
5. Excellent UV protection
6. Strong corrosion resistance, owing to the development of a protective oxide layer
7. Defense against chemical attacks
8. Minimal porosity

### 18.4.1 Specific Types of Ceramic Coatings

1. Chrome oxide ($Cr_2O_3$) coatings are renowned for their impressive wear and corrosion resistance capabilities.
2. Zirconia ($ZrO_2$), on the other hand, stands out for its exceptional thermal barrier properties, owing to the presence of thermally stable oxides. This makes it particularly advantageous for high-temperature applications and significantly enhances the fatigue and creep resistance of components, thereby extending their service life.
3. Alumina ($Al_2O_3$) coatings, sometimes referred to as quartz coatings, offer a cost-effective solution and can be applied to various materials to provide thermal or electrical insulation.

They exhibit a favorable combination of wear and corrosion resistance.

4. Titania ($TiO_2$) coatings excel in wear resistance, offering a cost-effective and chemically stable solution. They find widespread use in applications where durable surfaces and resistance to wear are critical.
5. Alumina-titania ($Al_2O_3/TiO_2$) composite coatings, typically comprising a small proportion of titania (approximately 3%), offer a versatile solution with excellent corrosion and chemical resistance properties. These coatings are suitable for a wide range of applications, including those in underwater and marine environments, and the addition of titania significantly enhances their toughness.

## 18.5 Plasma Electrolysis Oxidation

Plasma electrolytic oxide (PEO) coatings are described as highly durable, firmly bonded oxide layers applied to metals such as aluminum and magnesium. This process is sometimes referred to as MAO or spark discharge anodizing [13]. It involves the alteration of a conventional anodized oxide film by applying an electric field stronger than the oxide's dielectric breakdown field. This leads to discharges and plasma-chemical reactions that contribute to the growth of the coating. Additionally, localized conditions of heat and pressure result in the sintering and annealing of the coating. Rapid cooling further modifies the oxide, resulting in a complex combination of amorphous material and nanocrystalline phases [14].

It is noted that this method can be applied to various valve metals, which include aluminum, magnesium, and titanium, along with a wide range of their alloys. The characteristics of the coating are said to depend on factors such as the base alloy used, the type of electrolyte employed, and various parameters within the electrical system. For instance, on aluminum, it is stated that dense alumina coatings with approximately 3% porosity can be achieved, and these coatings can be as thick as 130 μm. It is further mentioned that by utilizing different electrolytes, it is possible to create coatings that are more porous and can reach thicknesses of up to 600 μm [15]. Typical alumina coatings, as explained, comprise a relatively dense polycrystalline alpha-alumina layer with a softer, more porous gamma-alumina layer situated on top.

### 18.5.1 Properties of Plasma Electrolytic Oxide

1. Hardness: Typical hardness of 1400 to 1700 HV is possible
2. Adhesion: Good adhesion to substrate material
3. Heat resistance: Can withstand temperatures up to 2000°C to 2500°C
4. Thermal insulation
5. Wear resistance: Better than hard anodizing
6. Dimensional tolerance: Uniform and controllable coating possible
7. Friction: Coefficient of friction is 0.5 for dry and 0.1 for lubricated
8. Electrical insulation: Gives very high electrical insulation 10 to 20 V/μm.
9. Corrosion resistance: 6000 to 7000 hours in a salt-fog test chamber
10. Impregnation: Teflon can be used to modify t properties

In discussing PEO coatings, it was noted that one of their primary mechanical characteristics is their low stiffness, which, akin to other conventional ceramic coating systems produced through thermal spraying, allows these coatings to endure substantial mechanical strain without experiencing proportionally significant stresses. It was mentioned that elastic modulus values of up to 25 to 40 GPa are achievable [16]. Additionally, it was pointed out that the exceptional wear resistance of PEO coatings stems from their high hardness (with typical values on aluminum ranging from 1500 to 2000 HV). Another noteworthy outcome of this low stiffness property is its ability to enable the coatings to withstand substantial thermally induced strains without undergoing spallation [17, 18]. During the early stages of investigating PEO coatings, it was observed that these coatings exhibited a notable degree of sub-micrometer scale porosity. This particular porosity had gone unnoticed since it was not visible through conventional polished cross-sectional analysis. The importance of this inherent porosity was emphasized as it possesses the remarkable capability to readily absorb lubricants, thus enhancing the performance of lubricated wear. Additionally, it was mentioned that it can effectively be infused with polymers or other substances to ensure strong adhesion for topcoats, such as polytetrafluoroethylene (PTFE), when developing composite surface layers.

The microstructural characteristics of PEO coatings in magnesium alloys were stated to be influenced by operational conditions, with the thickness of these coatings varying from 5 to 200 µm. It was pointed out that, typically, the interface of the coating exhibits a zigzag pattern, creating an integrated appearance with the substrate. The PEO process was described as involving a plasma discharge procedure around a metal immersed in an environmentally friendly electrolyte, and it was noted that these discharge procedures have a significant impact on the structure of the coatings and, consequently, their physical and mechanical properties. It was further explained that during the PEO treatment, three processes occur simultaneously: electrochemical reactions, plasma chemical reactions, and oxygen thermal diffusion. Commonly held beliefs suggest that PEO coatings on magnesium substrates have a three-layer structure. It was mentioned that the surface of PEO coatings typically features pancake-shaped structures with a discharge channel in the center of each pancake. It was noted that molten substances flow through these discharge channels and solidify when exposed to the cold electrolyte, thereby creating distinct boundaries that separate the pancakes. It was stated that the size of these pores depends on the density of the discharge and the process time, with reported pore diameters in magnesium alloys ranging from 0.5 to 50 µm. It was mentioned that discharge channels are often surrounded by molten substances and local microcracks, which may form due to the rapid cooling of molten substances when they come into contact with the cold electrolyte. Additionally, it was noted that high-energy discharges can lead to the fusion of significant portions of the oxide film and substrate, and eventually, molten substances flow toward the surface, forming large ceramic particles as they reach the cold electrolyte. It was also observed that the presence of molten oxides around the pores indicates that the temperature inside these micro-voids can briefly reach several thousand degrees.

## 18.6 Case Study 1: Aluminum Oxide Coatings by Unbalanced Sputtering

The challenges linked to reactive DC sputtering of highly insulating materials like alumina are well-documented. This process typically results in low deposition rates and the accumulation of an insulating layer on the target surface, leading to arcing issues. These arc events disrupt the stable operation and can cause material droplets to be ejected from the target, potentially introducing defects in the growing film. However, research has demonstrated that the occurrence of arcs can be significantly reduced by applying pulsed magnetron discharge within the frequency range of 10 to 200 kHz [19].

Reactive unbalanced magnetron sputtering can be employed in two different approaches for alumina: one involved a DC power supply in conjunction with a fixed frequency pulse unit, and the other utilized a variable frequency power supply with a maximum frequency. Such processes give dense, stoichiometric films with a microhardness of more than 2500 $kg/mm^2$.

Complex objects benefit from the deposition of high-grade metal layers via unbalanced magnetron sputtering, which effectively applies various ceramic coatings, such as titanium nitride [8]. This process coats metal targets at a rate of micrometers per second. However, reactive sputtering significantly reduces the coating rate, dropping to micrometers per hour. The drawback here is the prolonged time required, making it economically unappealing. To address this issue, the pulse approach can be employed, substantially reducing flaws in the thin film.

In unipolar pulsed sputtering, a single magnetron is pulsed, limiting the pulse-on duration to prevent breakdown and subsequent arcing before insulating layers charge to a critical level. During the pulse-off period, the discharge is eliminated through the plasma. Bipolar pulsed sputtering is used when two magnetrons are linked to the same pulse source, with each magnetron source alternately supplying the anode and cathode of a discharge [22]. Periodic pole switching prevents arcing and encourages the discharge of insulating layers. This approach aims to achieve the quick deposition of defect-free ceramic coatings on intricate components, increasing its commercial appeal. Consequently, the PMS (pulsed magnetron sputtering) procedure is gaining more attention. Researchers at Salford University employed the PMS approach to evaluate the deposition of alumina coatings in a closed-field unbalanced magnetron system.

Reactive DC sputtering of aluminum oxide coatings posed challenges. Even with turndown signals at 60% of the pure metal signal, indicating low levels of target poisoning, arcing persisted from the target throughout the runs, leading to process instability. The resulting coating exhibited a granular, porous structure. The composition was sub-stoichiometric, and the microhardness was notably low. After improvements, both of the considered pulse units operated stably with minimal arcing at the target. They also provided better control over the reactive sputtering process, enabling management of the coating's composition and, consequently, its characteristics.

## 18.7 Case Study 2: Power Supply Mode in Plasma Electrolysis Oxidation (Micro-Arc Oxidation Coating) in Magnesium Alloy

A distinctive technique for creating ceramic coatings on a range of nonferrous metals, including Mg, Al, Ti, and their alloys, is known as MAO, which also goes by names such as anodic spark deposition and micro-arc discharge oxidation [14]. MAO is highly effective in enhancing surface properties like wear and corrosion resistance, electrical insulation, and thermal characteristics, primarily due to its ability to produce exceptionally adherent ceramic coatings. It is worth noting that MAO technology is known for its manageability and environmental safety, although the process itself can be quite challenging, involving electrochemical oxidation and high-voltage sparks [20].

In this study, MAO was applied to the magnesium alloy AZ91D using both pulsed DC and pulsed bipolar current (BC) power supply modes. A unique MAO power supply, known as the DLC mode, was developed for the AZ91D magnesium alloy after examining load characteristics under various waveforms [21]. The research investigated the influence of these power supply modes on coating thickness, surface morphology, and corrosion resistance. Furthermore, the study explored the effects of loading parameters on surface morphology and the efficiency of MAO coatings, aiming to enhance efficiency, reduce energy consumption, and prevent localized large arc ablation phenomena [23].

### 18.7.1 Process Parameters

Process parameters are the characteristics of the manufacturing process like temperature. For surface coating, here are the process parameters:

1. Substrate material
2. Size and shape of the substrate
3. Coating material
4. Coating architecture
5. Thickness
6. Uniformity of thickness
7. Roughness of coating
8. Phase purity
9. Stoichiometry
10. Adhesion with substrate

11. Residual stresses
12. Growth rate

### 18.7.2 Advantages of the Process

1. Less droplets of residue on the surface.
2. No heavy metal buildup edges.
3. Can be used in dry machining also.
4. Tools will have a high tool life.
5. The traditional arc method is heavy, and pulse power supply equipment is compact and lightweight in nature. Therefore, easy to plan and execute the process.
6. No effect of changes in the electric network or load. Therefore, it is highly precise.
7. The processing of output waveform is of low cost and also of good quality.
8. The output given by the system is much more stable and highly effective as the pulse power system uses a high-frequency transformer.
9. Consumes less power as it is not continuous in nature.

### 18.7.3 Disadvantages of Process

1. The circuit and equipment are complex.
2. There is a chance of overlap of the current and voltage waveform.
3. Disturbance due to voltage spike.
4. There's a good chance of high-frequency noises.

## 18.8 Conclusions

The above study demonstrates that using a pulse power supply, fully dense alumina coatings can be deposited at a rate 40% higher than traditional methods, and it can reduce the occurrence of pores and loose deposits. The formation of gas bubbles within the coating is attributed to the extremely small equivalent capacitance within the circuit, and these capacitance and resistance values are determined through decay curves. When it comes to unipolar pulse power supply, it is most effective in aggressive environments with elevated temperatures. To minimize coating porosity, the PEO process is employed, utilizing microsecond current pulses. In terms of thickness and corrosion resistance, the DLC outperforms the BC pulse power supply. Additionally, when the voltage increment is increased, system efficiency improves, resulting in reduced energy consumption.

## References


1. Curran, J.A. and Clyne, T.W., Thermo-physical properties of plasma electrolytic oxide coatings on aluminium. *Surf. Coat. Technol.*, 199, 168–176, 2005.
2. Tan, J.C., Tsipas, S.A., Golosnoy, I.O., Curran, J.A., Paul, S., Clyne, T.W., A steady-state bi-substrate technique for measurement of the thermal conductivity of ceramic coatings. *Surf. Coat. Technol.*, 201, 1414–1420, 2006.
3. Wheeler, J.M., Collier, C.A., Paillard, J., Curran, J.A., Evaluation of Micromechanical Behaviour of Plasma Electrolytic Oxide (PEO) Coatings on Ti-6Al-4V. *Surf. Coat. Technol.*, 204, 3399–3409, 2010.
4. Martini, C., Ceschini, L., Tarterini, L., Paillard, J., Curran, J.A., PEO layers obtained from mixed aluminate-phosphate baths on Ti-6Al-4V: Dry sliding behaviour and influence of a PTFE Topcoat. *Wear*, 269, 747–756, 2010.
5. Troughton, S.C., Nominé, A., Nominé, A.V., Henrion, G., Clyne, T.W., Synchronised electrical monitoring and high speed video of bubble growth associated with individual discharges during plasma electrolytic oxidation. *Appl. Surf. Sci.*, 359, 405–411, 2015.
6. Dean, J., Gu, T., Clyne, T.W., Evaluation of residual stress levels in plasma electrolytic oxidation coatings using a curvature method. *Surf. Coat. Technol.*, 269, 47–53, 2015.
7. Dunleavy, C.S., Curran, J.A., Clyne, T.W., Time dependent statistics of plasma discharge parameters during bulk AC plasma electrolytic oxidation of aluminium. *Appl. Surf. Sci.*, 268, 397–409, 2013.
8. Curran, J.A. and Clyne, T.W., Porosity in plasma electrolytic oxide coatings. *Acta Mater.*, 54, 7, 1985–1993, 2006.
9. Curran, J.A. and Clyne, T.W., The thermal conductivity of plasma electr lytic oxide coatings on aluminium and magnesium. *Surf. Coat. Technol.*, 199, 177–183, 2005.
10. Krysmann, W., Kurze, P., Dittrich, K.-H., Schneider, H.G., Process Characteristics and Parameters of Anodic Oxidation. *Cryst. Res. Technol.*, 19, 973–979, 1984.

11. Lu, X., Mohedano, M., Blawert, C., Matykina, E., Arrabal, R., Kainer, K.U., Zheludkevich, M.L., Plasma electrolytic oxidation coatings with particle additions—A review. *Surf. Coat. Technol.*, 307, 1165–1182, 2016.
12. Fattah-Alhosseini, A., Chaharmahali, R., Babaei, K., Effect of particles addition to solution of plasma electrolytic oxidation (PEO) on the properties of PEO coatings formed on magnesium and its alloys. *J. Magnes. Alloys*, 8, 3, 799–818, 2020.
13. Sharifil, H., Aliofkhazraei, M., Darband, G.B., Shrestha, S., A review on adhesion strength of PEO coatings by scratch test method. *Surf. Rev. Lett.*, 25, 183–195, 2018.
14. Gnedenkov, S.V., Sinebryukhov, S.L., Sergienko, V., Composite PEO-Coatings as Defence Against Corrosion and Wear: A Review. *Corros. Sci. Tech.*, 18, 212–219, 2019.
15. Clyne, T.W. and Troughton, S.C., A review of recent work on discharge characteristics during plasma electrolytic oxidation of various metals. *Int. Mater. Rev.*, 64, 127–162, 2018.
16. Tsai, D.S. and Chou, C.C., Review of the Soft Sparking Issues in Plasma Electrolytic Oxidation. *Metals*, 8, 105, 2018.
17. Walsh, F.C., Low, C.T.J., Wood, R.J.K., Stevens, K., Archer, J., Poeton, A.R., Ryder, A., Plasma electrolytic oxidation (PEO) for production of anodised coatings on lightweight metal (Al, Mg, Ti) alloys. *Trans. Inst. Met. Finish.*, 87, 122–135, 2009.
18. Yerokhin, A.L., Nie, X., Leyland, A., Matthews, A., Dowey, S.J., Plasma electrolysis for surface engineering. *Surf. Coat. Technol.*, 122, 73–93, 1999.
19. Sieber, M., Simchen, F., Morgenstern, R., Scharf, I., Lampke, T., Plasma Electrolytic Oxidation of High-Strength Aluminium Alloys—Substrate Effect on Wear and Corrosion Performance. *Metals*, 5, 365–398, 2018.
20. Yerokhin, A., Leyland, A., Matthews, A., Kinetic aspects of aluminium titanate layer formation on titanium alloys by plasma electrolytic oxidation. *Appl. Surf. Sci.*, 200, 172–184, 2002.
21. Monaghan, D.P., Teer, D.G., Logan, P.A., Efeoglu, I., Arnell, R.D., Deposition of wear resistant coatings based on diamond like carbon by unbalanced magnetron sputtering. *Surf. Coat. Technol.*, 60, 1–3, 525–530, 1993.
22. Frach, P., Heisig, U., Gottfried, Walde, H., Aspects and results of long- term stable deposition of Al2O3 with high rate pulsed reactive magnetron sputtering. *Surf. Coat. Technol.*, 59, 1–3, 177–182, 1993.

23. Jiang, B.L. and Ge, Y.F., Micro-arc oxidation (MAO) to improve the corrosion resistance of magnesium (Mg) alloys, Editor(s): Guang-Ling Song, in: *Woodhead Publishing Series in Metals and Surface Engineering, Corrosion Prevention of Magnesium Alloys*, pp. 163–196, Woodhead Publishing, U.K., 2013.

# 19

# Fabrication and Water Absorption Testing of Banana Fiber–Reinforced Composite

**Larikynti Massar, Rohan Sarkar, Nako Komo and Nabam Teyi***

*Department of Mechanical Engineering, North Eastern Regional Institute of Science and Technology (NERIST), Nirjuli, Arunachal Pradesh, India*

***Abstract***

The use of composite materials in the engineering field is increasing day by day. A composite material is a material system composed of a combination of two or more micro or macro constituents that differ in form and chemical composition and are essentially insoluble in each other. The composite material's two main components are the matrix, which surrounds and binds the reinforcements. Fibers have been mostly used as reinforcement. Many traditional materials have become obsolete as prospective reinforcements owing to the widespread use of fibers in everyday life. Even among fibers, because of their exceptional characteristics, natural fibers have seen a major surge in use in recent years. Natural fibers are eco-friendly since the parts that are no longer useful can be recycled. However, one of the disadvantages of natural fiber is its high moisture absorption. Hence, this paper aims to fabricate a natural fiber–reinforced composite and test its water absorption percentage. The natural fiber used for fabrication is banana fiber. Banana fiber is a lignocellulose fiber obtained from the pseudo-stem of the banana plant, which has relatively good mechanical properties. The banana fiber is biodegradable and renewable. Its growth rate is short, and it is available in abundance. Furthermore, the fiber extraction process is simple. The matrix used is HSC-1100, which is an unmodified medium-viscosity liquid epoxy resin derived from epichlorohydrin and bisphenol A. The hardener used is triethylenetetramine (TETA), an organic compound with the formula $[CH_2NHCH_2CH_2NH_2]_2$. The mechanical properties of banana fiber composites are improved by varying the number of layers and the orientation of the raw banana fibers. Fabrication of the composite is done with the hand layup method.

**Corresponding author*: nbt@nerist.ac.in; ORCID: https://orcid.org/0000-0001-6412-6049

R. Thanigaivelan, Pradeep Kumar Krishnan, Kamalakanta Muduli and Santosh Kumar Tamang (eds.) New Materials, Processing and Manufacturability: Fabrication and Processing of Advanced Materials, (351–368) 

***Keywords:*** Banana fiber composites, fiber orientation, hand layup, water absorption, composite fabrication

## 19.1 Introduction

The application of composite materials is gaining popularity in engineering disciplines due to their distinctive features and performance benefits. Composite materials are a blend of two or more constituents, either micro or macro, with diverse forms and chemical compositions that are not soluble in one another. These constituents consist of a matrix, which surrounds and binds the reinforcements. Composite materials offer a wide range of applications, including aerospace, automotive, construction, and sports industries, owing to their exceptional mechanical properties, such as high strength, stiffness, and fatigue resistance [1, 2].

Fiber-reinforced composites are a significant class of composite materials where fibers, such as glass, graphite, carbon, or natural fibers, are used as the reinforcement phase. The mechanical attributes of the composite can be improved with the integration of fibers, which leverages the fibers' high strength-to-weight ratio and directional properties. Furthermore, the mechanical properties of the materials can be enhanced by impregnating specific types of fillers like silicon carbide and sawdust [3, 4]. Apart from fiber type and fiber content, fiber orientation also plays a critical role in determining the final properties and behavior of the composite material [5, 6].

Natural fibers are increasingly popular in composite materials due to their renewable and eco-friendly qualities. Compared to synthetic fibers, they offer better environmental outcomes as they biodegrade quickly, can be recycled effortlessly, and leave smaller ecological footprints. Furthermore, they have notable mechanical strengths like firmness and toughness, qualifying them as proficient substitutions to usual support systems [7, 8]. Natural fiber composites are used in diverse industries, including automotive, construction, and packaging, offering performance benefits while promoting sustainability and resource conservation.

Banana fiber, derived from the pseudo-stem of the banana plant, is one of the extensively studied natural fibers for composite applications. Banana fiber exhibits excellent mechanical properties, including high tensile strength and modulus, making it an attractive reinforcement option. Banana fibers are biodegradable, readily available, and have a short growth cycle, making them sustainable and economically viable [9]. The utilization of banana fiber in composites offers the potential for lightweight and eco-friendly materials in various industries. Numerous studies have shown that the addition of various natural or synthetic fibers along with banana fibers can significantly improve the mechanical properties of composite materials, allowing for the customization of properties in accordance with particular application requirements [10].

Traditionally, manual techniques like hand layup have been used to create banana fiber–reinforced composites [11]. In this procedure, the mold is carefully filled with layers of banana fibers that are carefully positioned by hand. The matrix material, usually epoxy resin, is then applied to impregnate and encapsulate the fibers. The production of banana fiber composites can be done with remarkable flexibility and economy using the hand layup method [12]. However, cutting-edge production methods like vacuum infusion, compression molding, and filament winding have also been investigated in an effort to amplify the interaction between the fibers and matrix and thereby improve the composite's overall quality [13].

The layering of the fibers, their orientation, and the addition of interfacial modifiers all play crucial roles in improving the mechanical properties of composites made of banana fibers. This makes it possible to tailor strength, stiffness, impact resistance, and fatigue performance to meet the needs of particular applications [14]. The effects of fiber composition, alignment, manipulation, and interbreeding on the physical characteristics of banana fiber amalgamations have been the subject of several studies [15]. It is conceivable to increase the adhesion between the fibers and matrix and thereby improve the overall physical functionality by using coupling agents and surface modifications.

For composites made of banana fiber, testing to assess mechanical performance is of utmost importance. Tensile testing is used to calculate tensile strength and modulus and determine the material's resistance to applied forces. Flexural testing makes it possible to evaluate bending characteristics like flexural strength and stiffness. Impact testing measures the ability to absorb energy and resistance to sudden loads. Viscoelastic behavior and durability are better understood through fatigue testing and dynamic mechanical analysis (DMA). These various tests reveal the mechanical capabilities and limitations of composites made of banana fibers under various loading scenarios.

Testing for water absorption is an essential part of evaluating the durability and water resistance of composites made of banana fibers. Water has a propensity to interact with the fibers and infiltrate the matrix, potentially affecting the composite's long-term performance and mechanical properties. Researchers can develop strategies to improve the composite's water resistance by calculating the percentage of water absorption in order to gain a thorough understanding of the composite's capacity to withstand moisture exposure [16]. According to research, banana fiber composites' ability to absorb water can be significantly reduced through surface modifications like chemical treatments and coatings. This increases the composites' resistance to water-induced degradation. For the design and optimization of composite materials intended for applications in humid or wet environments, a thorough understanding of water absorption behavior is crucial [17].

Given these considerations, this study aims to examine the characteristics of banana fiber with a particular focus on enhancing its mechanical properties through the manipulation of layering and fiber orientation using the hand layup method while also conducting water absorption testing.

## 19.2 Materials

The fabrication is to be done on a composite material that is made of banana fibers as reinforcement in an epoxy matrix.

Banana, a member of the Musa family, is a large perennial herb characterized by its leaf sheath, which gives rise to leaves measuring up to 9 feet in length and 2 feet in width. The pseudo-stem of the banana plant serves as the source of banana fiber, a lignocellulose fiber known for its noteworthy mechanical properties. Notably, banana fiber exhibits superior spin ability, fineness, and tensile strength in comparison to natural bamboo fiber. Banana fiber is readily available in various regions of India, including Tamil Nadu, Maharashtra, Karnataka, Gujarat, Andhra Pradesh, Assam, and Madhya Pradesh. Its properties are as follows: Density ranges from 0.5 to 1.5 g/cm$^3$, elongation percentage falls within the range of 2.4 to 3.5, tensile strength ranges from 711 to 789 MPa, and Young's modulus is between 4.0 and 32.7 GPa. Additionally, the cellulose content of banana fibers ranges from 34 to 63%, the microfibril angle is approximately 11°, lignin content varies from 5 to 10%, lumen size measures 5 mm, tensile strength ranges from 529 to 914 MPa, and hemicellulose content falls within the range of 6 to 19% [18, 19]. Overall, these distinctive technical properties position banana fiber as a promising material for various applications in industries such as textiles, composites, and packaging.

The utilization of epoxy as a resin presents certain cost implications due to its relatively higher price compared to alternative options. However, the advantages offered by epoxy resins justify their expense. In terms of strength, epoxy resins exhibit remarkable performance, surpassing other types of resins by a factor of 3. Among the commercially available epoxy resins,

HS1100 stands out as a notable grade that can be effectively employed in conjunction with various fibrous reinforcing materials. Moreover, HS1100 demonstrates compatibility with a wide range of composite manufacturing processes, including vacuum-bag molding, filament winding, and hand layup techniques. This compatibility enhances the versatility of HS1100, making it a desirable choice for numerous applications in the field of composite materials. Considering these attributes, epoxy resins, despite their higher cost, present a convincing option for researchers and practitioners in need of superior mechanical properties and compatibility with diverse manufacturing processes.

## 19.3 Material Preparation

The initial stage involves a meticulous choosing and readying of the fundamental substances for the manufacturing procedure. For the construction of the compound in this analysis, the main components included epoxy resin and banana fiber. The quality of the composite material is greatly influenced by the ingredients employed, and their appropriate preparation assures the anticipated characteristics.

### 19.3.1 Banana Fiber

Banana fiber, a lignocellulosic fiber obtained from the pseudo-stem of the banana plant (Musa sapientum), was chosen as the reinforcement material due to its inherent mechanical properties and abundance. To acquire the raw brown banana fibers, indiamart.com [20] was identified and selected. The fibers were obtained in their natural state and required careful processing to remove impurities, such as waxes, oils, and other contaminants. This cleaning process involved manual separation and washing of the fibers to ensure high-quality reinforcement material for the composite fabrication.

### 19.3.2 Epoxy Resin

The matrix material used in this study was HSC-1100, an unmodified medium viscosity liquid epoxy resin derived from epichlorohydrin and bisphenol A. The epoxy resin was chosen for its compatibility with the banana fiber and its ability to form a strong matrix to bind the fibers together. The epoxy resin was acquired from Hindusthan Speciality Chemicals Limited [21], a reputable supplier known for providing high-quality epoxy resins. HSC-1100 possesses various desirable characteristics, such as decent chemical and thermal resistance, superior adhesion, and satisfactory mechanical and electrical properties. These aspects distinguish it as a fitting material for applications requiring high-performance composites.

### 19.3.3 Hardener

To initiate the curing process of the epoxy resin and promote the cross-linking of molecules, a suitable hardener is required. In this study, triethylenetetramine (TETA) was selected as the hardener. TETA is an organic compound with the formula $[CH_2NHCH_2CH_2NH_2]_2$. Amine hardeners, such as TETA, are known to cross-link epoxide resins either through catalytic mechanisms or by bridging across epoxy molecules. This cross-linking process enhances the mechanical strength and durability of the composite. TETA enables room-temperature curing, making it convenient for the fabrication process.

## 19.4 Fabrication Process

The fabrication of the composite involved a series of carefully executed steps to ensure proper fiber-matrix interaction and uniform distribution of the reinforcement material within the matrix. Additional materials and tools, such as brushes, disposable glasses, transparent plastic sheets, gloves, masks, wiper Waxpol polish, tiles, and weights, were utilized to facilitate the process. The fabrication process consisted of the ten stages listed below.

Step 1: Fiber cutting and preparation
The raw banana fibers were first cut into lengths of 200 mm to ensure uniformity in the reinforcement material as shown in Figure 19.1. Care was taken to avoid damaging or breaking the fibers during the cutting process. Uniform fiber length promotes consistent mechanical properties in the composite.

Step 2: Surface preparation
To ensure a clean and suitable surface for composite fabrication, the surfaces of the tiles were thoroughly cleaned. Any dirt, debris, or loose particles were removed by scraping the tile surfaces. After scraping, the tiles were cleaned with a cloth to achieve a smooth and clean surface.

**Figure 19.1** Cut banana fibers.

Step 3: Application of transparent plastic sheets
Two transparent plastic sheets of size 210 mm × 297 mm were laid and taped to the bottom tile. These plastic sheets served as a release agent to prevent the composite from sticking to the tiles during the curing process. The sheets were carefully aligned and secured to ensure proper coverage and prevent any leakage of the epoxy resin.

Step 4: Application of Waxpol polish
A layer of Waxpol polish, a silicon liquid polish, was applied on the transparent plastic sheets using a small cloth. The boundaries of the transparent plastic sheet were marked to ensure proper alignment during the fabrication process. Waxpol polish acts as a mold release agent, preventing the composite from adhering to the plastic sheets. The same procedure was repeated on the top tile, ensuring that the transparent plastic sheets were placed identically on both tiles (Figure 19.2).

Step 5: Preparation of epoxy resin mixture
The disposable cups were used to pour the HSC-1100 epoxy resin and TETA hardener separately, and the weights of both substances were measured with an analytical balance before they were blended in a 10:1 ratio, based on previous tests and optimizations. The mixture was thoroughly stirred using a suitable tool until the components were well-blended and the mixture became warm, indicating the initiation of the polymerization process.

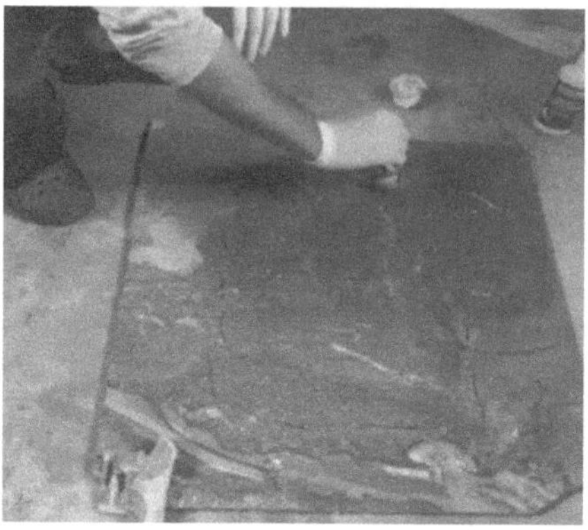

**Figure 19.2** Waxpol applied on plastic sheets.

Step 6: Application of the first matrix layer
A layer of the epoxy resin matrix was applied on the prepared transparent plastic sheet and evenly distributed using a brush as shown in Figure 19.3. The direction of the applied matrix was considered as 0° horizontal. Care was taken to ensure uniform coverage and the absence of any air bubbles or voids within the matrix layer. The brush strokes were carefully applied to ensure that the matrix layer was evenly spread and adhered well to the plastic sheet surface.

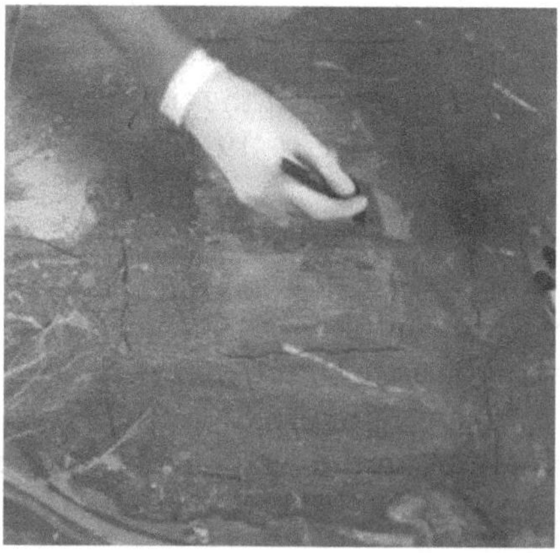

**Figure 19.3** Matrix applied to plastic sheets.

Step 7: Placement of banana fibers
Banana fibers were evenly placed on the applied matrix layer. The fibers were oriented at 90° with respect to the direction of the applied matrix, as shown in Figure 19.4. This orientation provided strength and improved mechanical properties to the composite structure. After placing the fibers, a wiper was utilized to apply an evenly spread layer of epoxy resin matrix over them, removing any excess resin and ensuring proper wetting of the fibers. Care was taken to prevent fiber displacement or entrapment of air bubbles during this process.

Step 8: Layering and matrix application
Step 7 was repeated for each subsequent layer of fiber, with each layer placed at a 90° angle to the previous layer. Two composite specimens were fabricated, one with three layers of fiber and another with four layers of fiber. The layering process involved alternating the fiber orientation between 0° and 90° to achieve a balanced and symmetric arrangement. The composition of the composite consisted of 50% banana fiber and 50% epoxy resin matrix by weight. This balanced composition aimed to achieve optimal mechanical properties and fiber-matrix interaction (Figure 19.5).

Step 9: Composite consolidation and excess matrix removal
After the final layer of the matrix was applied, the top tile was carefully placed on the composite stack. Weights were then placed on top of the tile to apply pressure and compress the composite as shown in Figure 19.6. This step served two purposes: first, it ensured proper consolidation of the composite layers, eliminating any air pockets or voids; second, it facilitated

**Figure 19.4** Fibers placed on a matrix layer.

**Figure 19.5** Matrix spread evenly on fibers.

**Figure 19.6** Weights placed on the tiles.

the removal of excess matrix material from the composite, enhancing its overall quality. A careful selection of weights was made to ensure that enough pressure was applied to the composite without causing any harm or distortion to its structure. Afterward, the composite was allowed to cure for a specified duration, usually about a day, to ensure that the fibers bond together completely through resin polymerization.

Step 10: Specimen cutting
After the specimens of the composite had cured, they were carefully cut into the required dimensions for testing their properties, as shown in Figure 19.7. Precise dimensions were crucial for conducting accurate mechanical, physical, and thermal tests on the composite material. Care was taken to ensure clean and straight cuts to avoid any inconsistencies or damage to the specimens.

The above fabrication process aimed to produce high-quality composite specimens with well-aligned fibers, uniform resin distribution, and minimal defects. To ensure consistent results and reproducibility in the fabrication of the composites, each step was carefully carried out. Figure 19.8(a and b) displays the specimen.

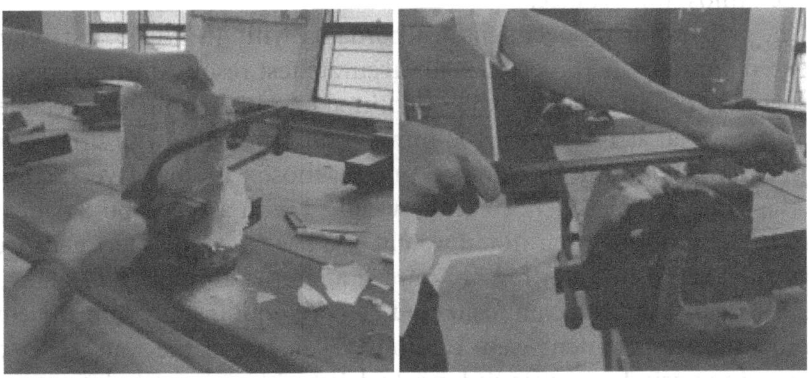

**Figure 19.7** Specimen preparation.

(a) (b)

**Figure 19.8** (a) 100 mm × 100 mm × 5 mm specimen, (b) 100 mm × 100 mm × 8 mm specimen.

## 19.5 Water Absorption Testing

For composites made of natural particles and fiber reinforcements, water absorption testing is an essential procedure for assessing performance and durability, especially in outdoor or humid environments. Evaluation of particle-matrix adhesion at the interface provides critical information about the composite's resistance to water infiltration and potential degradation. The water absorption test looks at how long the composite was exposed to moisture. The characteristics of water absorption are influenced by a number of variables, such as the type of plastic matrix, the presence of additives, temperature, and exposure time. In order to assess the composite's suitability for various applications and forecast its long-term performance, a comprehensive understanding of water absorption behavior is of the utmost importance.

The preparation of specimens in accordance with ASTM D570 standards ensured appropriate dimensions for accurate test results during the water absorption test. In order to achieve optimal representation, the specimens were measured at 100 mm × 100 mm x 5 mm for those with three layers of fiber and 100 mm × 100 mm × 8 mm for those with four layers of fiber.

The dried samples were carefully weighed with a precision balance to start the test, noting their initial weight as a baseline. The specimens were then submerged in water for 24 hours at a temperature of 26°C, allowing water to permeate the composite and achieve equilibrium with the environment. The specimens were carefully taken out of the water after the immersion period was over and gently dried with a clean cloth to remove any remaining surface moisture. To maintain the integrity of the test results, it was essential to avoid applying mechanical stress or causing harm. The specimens were weighed once more on the precision balance after drying to get their final weight. The difference between the final weight and the starting weight was used to determine the percentage of water absorption using the formula below:

(Final weight – Initialweight) / (Initial weight) × 100% = Water Absorption Percentage

This formula provides a numerical evaluation of the composite material's capacity to absorb water by expressing the weight change as a proportion of the initial weight. A higher water absorption percentage indicates that the composite will absorb more moisture.

Multiple purposes are served by the water absorption test when describing composite materials with natural particle and fiber reinforcements. It mainly helps in assessing how well the composite resists water infiltration because too much moisture absorption can cause swelling, dimensional instability, and possible structural damage. By quantifying the water absorption percentage, researchers and engineers can compare different composite formulations, optimize material compositions, and develop strategies to mitigate moisture-related issues.

Moreover, the water absorption test provides insights into the effectiveness of the bonding between the matrix and the reinforcing particles or fibers. A strong interfacial adhesion reduces the number of available sites for water absorption, resulting in lower water uptake. Therefore, this test indirectly assesses the composite's interfacial strength and the quality of the matrix-particle interface.

The results obtained from the water absorption test inform material selection, design optimization, and durability assessments of natural particle and fiber-reinforced composites. They guide engineers in identifying suitable applications, predicting the composite's behavior in different environmental conditions, and developing strategies to enhance its performance and longevity.

## 19.6 Results and Discussions

During the fabrication process of the composite, several observations were made that shed light on the characteristics and behavior of the materials used.

### 19.6.1 Observation on Fabrication of the Composite

Banana fiber, which is available in various forms such as woven or raw, was chosen in its raw form for this work due to its cost-effectiveness compared to other forms. This decision allowed for a more economical production process.

The tendency of raw banana fiber to exhibit increased moisture absorption when exposed to the atmosphere was observed. This distinctive quality is a result of the fiber's inherent hygroscopicity and porous structure. It is crucial to consider this moisture absorption property when designing composite structures to be used in moist or humid environments.

The raw banana fibers used in this study showed a less uniform arrangement and spacing than the woven banana fibers. As a result, more epoxy resin matrix was absorbed by the raw fibers during the fabrication process. The irregular fiber arrangement produced more interstices for the matrix to enter, increasing matrix absorption. This finding emphasizes the importance of handling and arranging raw fibers with care in order to achieve the best fiber-matrix interaction and exert control over matrix absorption.

Weights were applied to the composite during the fabrication process in order to compress it and enable proper layer consolidation. It was observed that this compression resulted in the extrusion of extra matrix material, which produced a significant amount of waste because the extra material could not be put to another use. This finding emphasizes how crucial it is to calculate the necessary amount of matrix material precisely in order to reduce waste and maximize the cost-effectiveness of the fabrication process.

Although not the most cutting-edge or sophisticated method, the hand layup process used in this study was chosen for its accessibility and simplicity. The composite was made using a method that did not require any specialized tools. Although there are other methods that can produce results that are more accurate and effective, the hand layup process used in this study was able to produce the desired outcome and proved to be a manageable and practical method of fabricating the composite.

### 19.6.2 Results of the Water Absorption Test

The composite's resistance to water penetration and its potential for outdoor applications were both assessed through the water absorption test. Table 19.1 lists the specimens' dimensions, initial dry weight, final weight after submersion, and percentage of water absorption as well as the results of the water absorption test.

**Table 19.1** Water absorption results.

| Dimensions | Dry weight | Final weight | Water absorption % |
|---|---|---|---|
| 100 mm × 100 mm × 5 mm | 56.15 g | 58.48 g | 4.149% |
| 100 mm × 100 mm × 8 mm | 101.34 g | 105.27 g | 3.878% |

The specimens had an initial dry weight of 56.15 g and measured 100 mm × 100 mm × 5 mm. After being submerged in water for 24 hours, the final weight increased to 58.48 g, giving a 4.149% water absorption percentage.

The specimens with the measurements of 100 mm × 100 mm × 8 mm had a dry weight of 101.34 g. The final weight after the water absorption test was recorded as 105.27 g giving the water absorption percentage a value of 3.878%.

These results conclusively show that the composite materials exhibited a detectable level of water absorption. It is noteworthy that the composite's raw banana fiber has a moisture regain value of 13%. The amount of water in a material as a percentage of its oven-dry weight is known as moisture regain. The properties of the banana fiber were strengthened through the incorporation of the epoxy resin matrix, leading to improved resistance to water absorption compared to the raw fiber alone.

The test results and observations offer valuable insights into how composite materials react when exposed to moisture, making it easier to optimize composite fabrication methods and performance under various environmental circumstances. In order to evaluate the water resistance and long-term durability of composites made of natural particles and fibers, water absorption testing is of utmost importance. This test provides vital information about the composite's moisture uptake behavior, interfacial adhesion, and potential long-term performance by quantifying the water absorption percentage. These results provide a direction for the choice of materials, design enhancement, and application-specific considerations, ensuring the creation of superior, moisture-resistant composites appropriate for a wide range of industries and outdoor applications.

## 19.7 Conclusion

The main goal of this investigation was to investigate the fabrication procedure and water absorption properties of a composite material made of epoxy resin and banana fiber. Raw banana fiber and HSC-1100 epoxy resin were among the raw materials carefully prepared and chosen for the fabrication process. To ensure the use of high-quality reinforcement material, the fibers underwent rigorous processing to remove impurities. The best option for the matrix material was epoxy resin, which is well known for its advantageous mechanical and adhesive properties.

The fabrication procedure used a methodical, step-by-step approach that included a number of tasks, including the careful cleaning and preparation of tiles, the cutting of banana fibers, the application of a mold release agent, and the methodical layering of fibers and matrix. For composite fabrication, the hand layup method was used because it provided a relatively simple and accessible method without the need for specialized equipment. While applying pressure during curing helped to consolidate the layers, it also caused some waste because extra matrix material had to be thrown away.

Testing the composite's water absorption behavior was an essential part of determining how well it performed in moist environments. The results showed that the raw banana fiber had a strong affinity for absorbing moisture because of its inherent properties. For composite specimens with dimensions of 100 mm × 100 mm × 5 mm and 100 mm × 100 mm × 8 mm, respectively, the water absorption percentages were 4.149 and 3.878%. This suggests that the composite had a low water absorption rate, which can be attributed to the banana fiber's improved characteristics as a result of being reinforced with a polymer matrix.

In addition, the observations made during the fabrication process provided insightful information. Cost factors led to the decision to use raw banana fiber rather than the woven version, but the raw fiber's unevenly distributed fibers resulted in higher matrix absorption. The hand layup method produced waste matrix material in excess, despite the fact that it was convenient.

In conclusion, this study successfully illustrated the viability of producing a composite material from epoxy resin and banana fiber. The composite demonstrated good water absorption resistance, making it appropriate for outdoor applications. The results confirmed that banana fiber's properties were strengthened by being reinforced with a polymer matrix, making it a competitive option for a variety of engineering and construction applications.

Future research projects might focus on streamlining the fabrication procedure to reduce matrix material waste and improve fiber distribution uniformity. Further investigation into the composite's mechanical and physical characteristics, such as its tensile strength and thermal stability, would provide a more thorough understanding of its performance traits.

Hence, banana fiber as a reinforcement material in composites shows promise and merits further research because of its potential in environmentally friendly and sustainable materials engineering.

## Declaration of Competing Interest

There are no known financial or personal interests of the authors that could have affected this paper.

## References


1. Yadav, S., Gangwar, S., Singh, S., Micro/nano reinforced filled metal alloy composites: A review over current development in aerospace and automobile applications. *Mater. Today Proc.*, *4*, 4, 5571–5582, 2017, https://doi.org/10.1016/j.matpr.2017.06.014.
2. Koli, D.K., Agnihotri, G., Purohit, R., Advanced aluminium matrix composites: the critical need of automotive and aerospace engineering fields. *Mater. Today Proc.*, *2*, 4-5, 3032–3041, 2015, https://doi.org/10.1016/j. matpr.2015.07.290.
3. Sarkar, D., Borah, P., Samanta, S., Teyi, N., Chetia, P., Borah, R., Tribo-performance of saw dust filled natural fiber–Glass fiber reinforced epoxy composites: Modelling and optimization. *Mater. Today Proc.*, *80*, 904–910, 2023, https://doi.org/10.1016/j.matpr.2022.11.326.
4. Sarkar, D., Chetry, D.R., Kumar, R., Borah, P., Samanta, S., Teyi, N., Study of Dry Sliding Wear Behaviour of Silicon Carbide Filled Glass-Jute Hybrid Epoxy Composite Using RSM and Taguchi Techniques, in: *Materials Science Forum*, vol. 1060, pp. 37–42, Trans Tech Publications Ltd, Switzerland, 2022, https://doi.org/10.4028/p-a12iqf.
5. Rana, N.K., Teyi, N., Borah, P., Stress analysis of laminated graphite fiber reinforced polymer Panel Stiffened joint with 450 fiber orientation using ANSYS. *Mater. Today Proc.*, *80*, 449–455, 2023, https://doi.org/10.1016/j.matpr.2022.10.196.
6. Teyi, N. and Tamang, S.K., Stress analysis of a flange stiffened FRP composite panel with varying stacking sequence, in: *Key Engineering Materials*, vol. 847, pp. 9–14, Trans Tech Publications Ltd, Switzerland, 2020, https://doi.org/10.4028/ www.scientific.net/KEM.847.9.
7. Thyavihalli Girijappa, Y.G., Mavinkere Rangappa, S., Parameswaranpillai, J., Siengchin, S., Natural fibers as sustainable and renewable resource for development of eco-friendly composites: a comprehensive review. *Front. Mater.*, *6*, 226, 2019, https://doi.org/10.3389/fmats.2019.00226.
8. Sanjay, M.R., Arpitha, G.R., Naik, L.L., Gopalakrishna, K., Yogesha, B.J.N.R., Applications of natural fibers and its composites: An overview. *Nat. Resour.*, *7*, 3, 108–114, 2016, http://dx.doi.org/10.4236/nr.2016.73011.
9. Sivaranjana, P. and Arumugaprabu, V. J. S. A. S., A brief review on mechanical and thermal properties of banana fiber based hybrid composites. *SN Appl. Sci.*, *3*, 1–8, 2021, https://doi.org/10.1007/s42452-021-04216-0.
10. Prabhakar, C.G., Babu, K.A., Kataraki, P.S., Reddy, S., A review on natural fibers and mechanical properties of banyan and banana fibers composites. *Mater. Today Proc.*, *54*, 348–358, 2022, https://doi.org/10.1016/j. matpr.2021.09.300.
11. Murugan, K., Venkatesh, S., Thirumalai, R., Nandhakumar, S., Fabrication and investigations of kenaf fiber and banana fiber reinforced composite material.

*Mater. Today Proc., 37*, 110–114, 2021, https://doi. org/10.1016/j.matpr. 2020.04.540.
12. Hossain, M.S., Mobarak, M.B., Rony, F.K., Sultana, S., Mahmud, M., Ahmed, S., Fabrication and characterization of banana fiber reinforced unsaturated polyester resin based composites. *Nano Hybrids Compos., 29*, 84–92, 2020, https://doi.org/10.4028/www.scientific.net/NHC.29.84.
13. Dinesh Kumar, S., Ponraj Sankar, L., Sathish, T., Vijayan, V., Parthiban, A., Kamalakannan, R., Rajkumar, S., Improving the mechanical properties of natural fiber composites of hemp fiber with ramie and banana fiber through compression molding method. *Adv. Mater. Sci. Eng., 2021*, 1–8, 2021, https://doi.org/10.1155/2021/7813634.
14. Venkateshwaran, N., ElayaPerumal, A., Jagatheeshwaran, M.S., Effect of fiber length and fiber content on mechanical properties of banana fiber/epoxy composite. *J. Reinf. Plast. Compos., 30*, 19, 1621–1627, 2011, https://doi.org/10.1177/0731684411426810.
15. Muralikrishna, M.V.V., Kumari, T.S., Gopi, R., Loganathan, G.B., Development of mechanical properties in banana fiber composite. *Mater. Today Proc., 22*, 541–545, 2020, https://doi.org/10.1016/j.matpr.2019.08.189.
16. Srinivasan, T., Suresh, G., Ramu, P., Ram, V.G., Giresh, M., Arjun, K., Effect of water absorption of the mechanical behavior of banana fiber reinforced IPN natural composites. *Mater. Today Proc., 45*, 1334–1337, 2021, https://doi.org/10.1016/j.matpr.2020.06.024.
17. Balaji, A., Kannan, S., Purushothaman, R., Mohanakannan, S., Maideen, A.H., Swaminathan, J., Premkumar, P., Banana fiber and particle-reinforced epoxy biocomposites: mechanical, water absorption, and thermal properties investigation. *Biomass Convers. Biorefin.*, 14, 7835–7845, 1–11, 2022, https://doi.org/10.1007/s13399-022-02829-y.
18. Rajasekaran, T. and Vigneshkumar, S., Experimental Analysis on the Wear Behavior of Natural Fiber Reinforced Polymer Composites. *J. Manuf. Eng., 10*, 3, 139–143, 2015, http://www.smenec.org/index. php/1/article/view/233.
19. Madhukiran, J., Rao, S.S., Madhusudan, S., Fabrication and testing of natural fiber reinforced hybrid composites banana/pineapple. *Int. J. Mod. Eng. Res.*, 3, 4, 2239–2240, 2013.
20. https://www.indiamart.com/proddetail/banana-fiber-21528685912.html.
21. www.hindusthanspeciality.com.

# 20

# Multi-Objective Optimization of Fusion Welding Parameters Using Non-Dominated Sorting Genetic Algorithm II

**Monoj Kanti Chakraborty, Arun Jyoti Kalita, Deepty Pandey, Nirsanametla Yadaiah, Md. S. Mujaheed Khan and Nabam Teyi***

*Department of Mechanical Engineering, North Eastern Regional Institute of Science and Technology (NERIST), Nirjuli, Arunachal Pradesh, India*

## *Abstract*

Laser beam welding is an efficient welding process known for its automated production of high-quality welds. Compared to other arc welding methods, laser welding offers advantages such as reduced passes, higher welding speeds, and minimized use of consumables. By optimizing welding procedures, the low heat input characteristic of laser welding can be fully utilized for various materials, resulting in welds with optimal penetration depth (PD), bead thickness (BT), and weld irregularity (WI). The selection of input parameters greatly influences the weld joint quality and the profile of the welding bead. This paper introduces a process parameter optimization approach utilizing the non-dominated sorting genetic algorithm II (NSGA-II) in laser beam welding. The key process parameters considered in this study are power (P), welding speed (WS), and focal position (FP). The application of NSGA-II enables the search for multi-objective Pareto optimal solutions, thereby achieving the desired outcomes.

***Keywords*:** Laser beam welding, penetration depth, bead thickness, weld irregularity, focal position, NSGA-II

**Corresponding author*: nbt@nerist.ac.in; ORCID: 0000-0001-6412-6049

R. Thanigaivelan, Pradeep Kumar Krishnan, Kamalakanta Muduli and Santosh Kumar Tamang (eds.) New Materials, Processing and Manufacturability: Fabrication and Processing of Advanced Materials, (369–388) © 2024 Scrivener Publishing LLC

## 20.1 Introduction

Fusion welding involves the utilization of melting to join materials that possess similar compositions and melting points. These welding procedures give rise to a heat-affected zone due to the occurrence of phase transitions at high temperatures. The term LASER is an acronym for light amplification by stimulated emission of radiation. It refers to a device that generates coherent and monochromatic light. This light can be efficiently amplified in a unidirectional manner with minimal divergence. Lasers have the ability to convert primary energy into a focused beam of electromagnetic radiation with a specific frequency [1, 2]. In laser beam welding, the welding equipment is initially positioned between the metal components that require joining. Once the setup is complete, the laser equipment is energized with high-voltage electricity, causing the machine's flash bulbs to emit photons. These photons are absorbed by atoms within the ruby crystal, subsequently activating electrons and inducing the emission of additional photons as they return to their ground state. This process leads to the generation of a concentrated laser beam. The laser beam is then directed toward the welding zone with the aid of lenses that focus the beam. The movement of the workpiece table and the control of the welding laser are managed by computer-aided manufacturing (CAM) systems. As the laser beam interacts with the metal pieces, it effectively burns them, resulting in their fusion within the cavity formed between them.

Laser beam welding is an incredibly advanced and versatile welding process that revolutionizes the automated production of welds with exceptional consistency and quality [3]. This notable technology surpasses traditional arc welding methods by enabling higher welding speeds, reducing the number of passes required, and minimizing the consumption of welding consumables [4]. The low heat input characteristic of laser welding empowers engineers and manufacturers to achieve optimal weld characteristics such as profound penetration depth (PD), ideal bead thickness (BT), and minimal weld irregularity (WI) through meticulous optimization of welding procedures.

PD is the depth to which fusion penetrates the base metal or previous pass from the molten surface during welding. A welding bead is produced when a filler material is deposited into a joint between two metal parts. Its thickness represents the BT. It increases as the current and voltage increase. WI occurs as a result of the expansion and contraction of the weld metal and nearby base metal throughout the welding process's heating and cooling cycles. WI is induced by the localized expansion and contraction of metal during the welding process. These three parameters are expressed in micrometers, millimeters, or meters.

The quality of a weld joint and the characteristics of the welding bead profile are greatly influenced by the input parameters meticulously selected for the laser welding process [5, 6]. There are many parameters that influence welding performance, but the three most critical in laser beam welding are power (P), welding speed (WS), and focal position (FP).

Power (P) is the amount of energy emitted by the laser beam during the welding process, expressed in watts or kilowatts. It controls the heat input into the workpiece, which influences the PD, weld width, and overall weld quality. Higher power results in more heat input, which results in deeper penetration and broader weld beads. Lower power, on the other hand, reduces heat input, resulting in shallower penetration and narrower weld beads. To obtain the desired weld qualities and avoid flaws such as under-penetration or an excessive heat-affected zone, power must be properly adjusted.

Welding speed (WS) is the rate at which the laser beams or workpiece moves during the welding process, expressed in millimeters per minute or meters per minute. Welding speed has a direct impact on the amount of heat transferred to the workpiece and the rate at which the weld pool cools. Higher welding rates cause less heat input and faster solidification, resulting in thinner weld beads and potentially lesser PD. Slower welding speeds, on the other hand, enhance heat input and allow for greater PD and broader weld beads. Optimizing welding speed is critical for balancing productivity and weld quality, as faster rates can increase production but compromise weld integrity, whereas slower speeds can produce greater weld quality at the expense of decreased productivity.

Focal position (FP) is the positioning of the laser focal point relative to the workpiece surface during welding, expressed in millimeters. The focal location is the concentration of laser energy at a certain point along the course of the laser beam. The focus spot size can be changed by modifying the focal position, which impacts the energy density and weld properties. A concentrated laser beam often has more energy density, resulting in deeper penetration and smaller weld beads. Defocused laser beams, on the other hand, distribute energy across a larger region, resulting in shallower penetration and bigger weld beads. The laser focus position is determined by the desired weld profile, joint geometry, and material parameters.

The selection of these appropriate process parameters represents a critical factor in achieving optimal welding results. Ascertaining the optimal combination of these parameters presents a complex challenge due to the intricate trade-offs inherent in the process.

Hence, this paper presents a method for optimizing process parameters utilizing the non-dominated sorting genetic algorithm II (NSGA-II) within the domain of laser beam welding. The NSGA-II algorithm amalgamates the principles of genetic algorithms with non-dominated sorting. This amalgamation effectively explores the parameter space to identify Pareto optimal solutions, offering superior trade-offs among the multiple objectives under consideration [7, 8].

The NSGA-II algorithm initiates with an initial population of potential solutions and employs genetic operators, including selection, crossover, and mutation, to generate novel candidate solutions. These solutions are subsequently ranked based on their dominance and non-dominance relationships, unveiling the non-dominated solutions that epitomize the Pareto optimal front. This front represents a realm wherein no alternative solution can enhance one objective without compromising another. Analyzing this front empowers decision-makers to make well-informed choices aligned with their specific requirements and preferences [9].

Summarily, for optimizing laser beam welding processes, the authors consider three pivotal process parameters: power (P), welding speed (WS), and focal position (FP). These parameters play a critical role in influencing weld quality and essential characteristics such as BT, PD, and WI. Using the NSGA-II algorithm, this paper attempts to negotiate the parameter space thoroughly, unveiling the Pareto optimal solutions that deliver the most favorable trade-offs among these parameters. This paper takes the experimental results from an established literature [10], comprising 32 sets of data for BT, PD, and WI corresponding to their respective P, WS, and FP. For this paper, the authors specifically utilize the data from 20 sets.

## 20.2 The Suggested Methodology

### 20.2.1 Genetic Algorithm

The genetic algorithm, inspired by natural selection and biological evolution, is a powerful approach to solving optimization problems, both constrained and unconstrained [11, 12]. The process by which the genetic algorithm works involves repeatedly changing a population of distinct solutions. The algorithm uses randomly selected parents from the current population in each iteration to create offspring for the following generation. As generations pass, the population gradually changes and converges on the best course of action. The genetic algorithm is particularly effective in addressing optimization problems that are not well-suited for traditional methods.

It can handle problems with discontinuous, non-differentiable, stochastic, or highly nonlinear objective functions. Additionally, it is adept at addressing mixed integer programming problems where certain components must be integer-valued. The algorithm incorporates three primary types of rules: selection rules for choosing parents, crossover rules for combining parents to create children, and mutation rules for introducing random changes to parents and generating offspring. Figure 20.1 shows this algorithm flow.

### 20.2.2 Non-Dominated Sorting Genetic Algorithm

Within the context of evolutionary computation, the NSGA is an example of a multi-objective optimization (MOO) algorithm. It is an expansion of the genetic algorithm designed specifically to handle optimization problems with multiple objective functions. The vector-evaluated genetic algorithm (VEGA), the strength Pareto evolutionary algorithm (SPEA), and the Pareto archived evolution strategy (PAES) are some of the other techniques that are included in the family of evolutionary MOO (EMOO) algorithms or multiple objective evolutionary algorithms (MOEA) [13, 14]. Over the years, the main criticisms of the NSGA-I approach have been as follows:

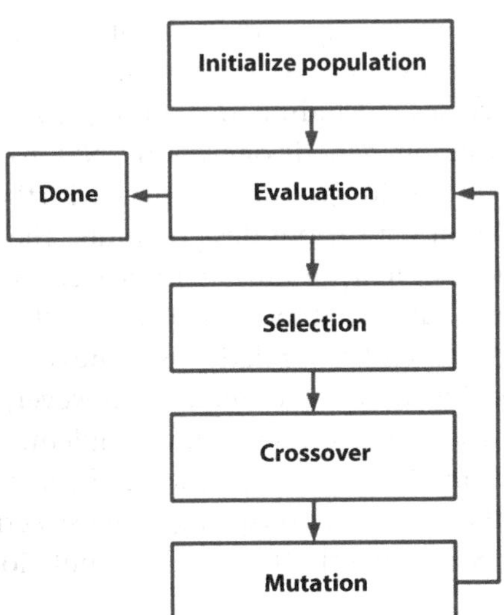


**Figure 20.1** Genetic algorithm flowchart.

1. The non-dominated sorting procedure in NSGA is associated with high computational complexity. The algorithm used for non-dominated sorting has a computational complexity of O(MN^3), where M refers to the number of objectives and N represents the population size. This computational burden becomes particularly noticeable for larger population sizes. The complexity arises from the intricate nature of the non-dominated sorting procedure performed in each generation.
2. Recent findings highlight the importance of incorporating elitism into the genetic algorithm to enhance its performance. Elitism enables the retention of promising solutions, preventing their loss once they are discovered, and can significantly expedite the optimization process.
3. NSGA necessitates the explicit specification of sharing parameters. These parameters are crucial for defining the extent of resource sharing among individuals within the population. Careful selection and calibration of these parameters are essential to achieve optimal performance and maintain diversity within the population.

### 20.2.3 A Fast and Elitist Multi-Objective Genetic Algorithm: NSGA-II

Due to its success in improving product quality, the NSGA-II has grown significantly in popularity and has been widely adopted in a variety of manufacturing processes, including machining, forming, and welding. Through the use of random genetic operations, NSGA-II uses a systematic approach to thoroughly explore the design space with the aim of identifying globally optimal solutions at various design points. The modified NSGA-II addresses these issues by incorporating an improved sorting algorithm and doing away with the requirement for sharing parameters, in contrast to the conventional NSGA, which has a relative weakness for its limited elitism and the need to specify sharing parameters. However, it still necessitates the specification of a sharing parameter in its implementation. Figure 20.2 provides an illustration of the technique, which begins with a population, $P_t$. The population then produces offspring using selection, crossover, and mutation operators. Subsequently, the current population and offspring are

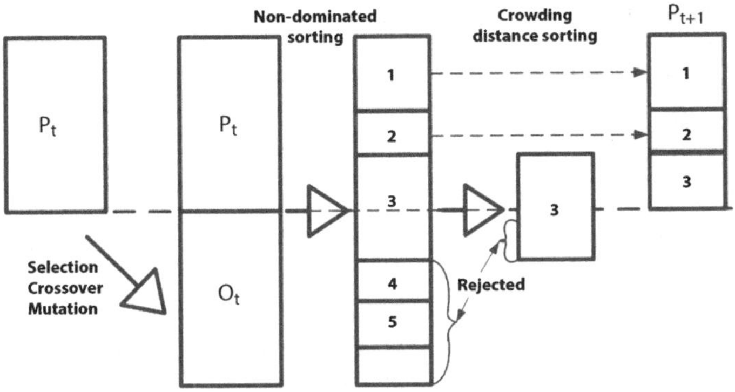


**Figure 20.2** NSGA II procedure.

merged into a single set, which is arranged on non-domination, as well as a certain crowding distance. Finally, the new set $P_{t+1}$ is obtained by selecting the finest N individuals from the set [10].

## 20.3 Results and Discussions

The results for the optimization of laser beam welding for process variables, namely, power (P), welding speed (WS), and focal position (FP), are validated with corresponding experimentally measured results from the literature [10] and discussed thoroughly.

### 20.3.1 Process Variable Considered for Optimization

Awareness and control of the welding variables are necessary to consistently produce defect-free welds with the desired attributes. The main variables that affect PD, BT, and WI are power, welding speed, focal position, welding current, arc voltage, travel speed, electrode extension, electrode orientation, electrode diameter, and shielding gas composition. However, in this paper, the considered process variables for laser beam welding are power, welding speed, and focal position. These variables are not completely independent, and changing one usually requires changing one or more of the others to produce the desired attributes in the weldment. Table 20.1 shows the algorithm setting.

**Table 20.1** Configuration of the algorithm.

| **Parameter** | **Value** |
|---|---|
| Size | 100 |
| Chromosome length | 14 |
| Iteration | 10,000 |
| Mutation probability | 0.10 |
| Selection type | Tournament selector |

### 20.3.2 Optimization of Penetration Depth, Bead Thickness, and Welding Irregularity by Regression Analysis

Using the data from the selected research paper [10] and using quadratic regression analysis, the relationship between the dependent and independent variables is obtained.

Three dependent variables are chosen. They are PD, BT, and WI. In addition, the three independent variables are power (P), welding speed (WS), and focal position (FP). The experimental results of 20 sets obtained from the literature [10] are shown in Table 20.2.

The obtained equations are the following:

1. Bead thickness (BT):
   BT = 4.56 – 0.22 × P – 1.34 × WS + 0.685 × FP + 0.029 × P2 + 0.148 × WS2 + 0.068 × FP2 + 0.045 × P × WS – 0.1399 × P × FP – 0.0502 × WS × FP
2. Penetration depth (PD):
   PD= 4.01 – 0.774 × P – 0.17 × WS + 0.156 × FP + 0.249 × P2 + 0.075 × WS2 – 0.0273 × FP2 – 0.145 × P × WS – 0.0432 × P × FP – 0.0116 × WS × FP
3. Welding irregularity (WI):
   WI = 1.85 + 0.195 × P – 1.250 × WS – 0.1555 × FP – 0.0691 × P2 + 0.1628 × WS2 – 0.00454 × FP2 + 0.1132 × P × *WS* + 0.0192 × P × FP + 0.0234 × FP × WS

The objective was to minimize BT; maximize PD; and minimize WI, subject to, 2.0 kW ≤ P ≤ 3.5 kW; 2.5 m/min ≤ WS ≤ 3.5 m/min; and –2.0 mm ≤ FP ≤ 0 mm.

**Table 20.2** Experimental result [10].

| P [kW] | WS [m/min] | FP [mm] | BT [m] | PD [m] | WI [m] |
|---|---|---|---|---|---|
| 2.87 | 3.37 | −2 | 0.00176 | 0.002621 | 0.000601 |
| 2.35 | 3.27 | 0 | 0.001667 | 0.002694 | 0.000441 |
| 2.16 | 3.21 | −2 | 0.001611 | 0.002553 | 0.000461 |
| 2.74 | 2.66 | −2 | 0.001905 | 0.002644 | 0.000575 |
| 2.90 | 3.11 | −1 | 0.001814 | 0.002734 | 0.000572 |
| 2.58 | 3.24 | −1 | 0.001611 | 0.002522 | 0.000519 |
| 2.97 | 2.89 | 0 | 0.001886 | 0.002739 | 0.000533 |
| 2.00 | 2.85 | −1 | 0.001632 | 0.002682 | 0.000396 |
| 2.52 | 3.34 | −2 | 0.00171 | 0.002575 | 0.000536 |
| 2.06 | 3.08 | 0 | 0.00189 | 0.002739 | 0.000379 |
| 2.03 | 2.95 | −1 | 0.001619 | 0.002679 | 0.000402 |
| 2.94 | 3.31 | 0 | 0.001833 | 0.002731 | 0.000548 |
| 2.26 | 3.18 | −1 | 0.001605 | 0.002700 | 0.000453 |
| 2.71 | 3.15 | 0 | 0.001793 | 0.002731 | 0.000517 |
| 2.13 | 3.44 | −1 | 0.001527 | 0.002675 | 0.000453 |
| 2.65 | 2.69 | 0 | 0.001957 | 0.002739 | 0.000499 |
| 2.68 | 2.79 | −1 | 0.001885 | 0.002739 | 0.000537 |
| 2.55 | 3.4 | −1 | 0.001716 | 0.002689 | 0.000513 |
| 2.61 | 3.05 | −1 | 0.001716 | 0.002689 | 0.000513 |
| 2.32 | 2.82 | −1 | 0.001786 | 0.002693 | 0.000524 |

### 20.3.3 Validation of Results with Experimental Result

To evaluate the precision and dependability of the optimization outcomes, three solutions were chosen from the Pareto frontier and compared against experimental results obtained from relevant literature [10]. Table 20.3

**Table 20.3** Optimal results.

| P [kW] | WS [m/min] | FP [mm] | | Experimental | Optimized | Error percentage |
|---|---|---|---|---|---|---|
| 2.74 | 2.66 | −22 | BT [μm] | 1905 | 1882 | −1.20 |
| | | | PD [μm] | 2664 | 2719 | 2.81 |
| | | | WI [μm] | 575 | 578 | 0.52 |
| | | | | | | |
| 2.00 | 3.00 | −1 | BT [μm] | 1632 | 1601 | −1.89 |
| | | | PD [μm] | 2682 | 2693 | 0.41 |
| | | | WI [μm] | 396 | 400 | 1.01 |
| | | | | | | |
| 2.92 | 2.50 | 0 | BT [μm] | 1935 | 1886 | −2.53 |
| | | | PD [μm] | 2723 | 2739 | 0.58 |
| | | | WI [μm] | 659 | 533 | −19.11 |

presents the laser welding process parameters along with their values for BT, PD, and WI. This analysis aimed to assess the consistency and correctness of the optimization data through a direct comparison with experimental data.

The data in Table 20.3 show that the optimum results and the related experimental verification display some variation. For instance, for PD at 1, the optimum is 2719 μm, showing an error of 2.81% when related to the experimental data. Similarly, for BT at 1, the optimum is 1882 μm, with an error of −1.20%. The WI at 1 shows an optimal result of 578 μm, while the experimental result is 575 μm, resulting in a relative error of 0.52%. It is worth noting that these discrepancies can be attributed to factors such as measurement errors, material in-homogeneity, and inherent variations in laser welding. Furthermore, at validation points 2 and 3, the errors for BT, PD, and WI exhibit different magnitudes, further emphasizing the need for careful consideration and analysis in future studies.

### 20.3.4 Pareto Frontier Chart

Figure 20.3 demonstrates the obtained results of laser process parameters when only BT and WI are taken into consideration for minimization purposes. Using the plotted graph, it is found that the most optimal point having rank 1 has a BT of 1.593 mm and WI of 0.406 mm.

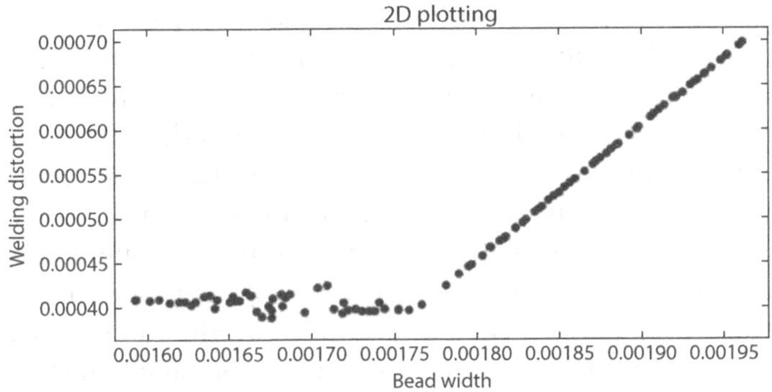


**Figure 20.3** Variation of the bead with and welding irregularity with process variables.

The provided figure illustrates the Pareto frontier obtained for optimizing process parameters, considering three variables: power, focal position, and welding speed. Every point here represents a specific blend of process parameters that satisfy the preferred potentials. Engineers can select an optimal combination based on the specific welding requirements.

Figure 20.4 reveals that no solution is inferior to others on the Pareto frontier. This is due to the trade-off among the BT, PD, and WI. For instance, achieving a maximum PD of approximately 2664 μm results in significant deviations in both BT and WI from the desired values. Conversely, obtaining a minimum BT of about 1591.83 μm does not align well with the preferred PD and WI. In summary, optimizing one objective function comes at the cost of the two others.

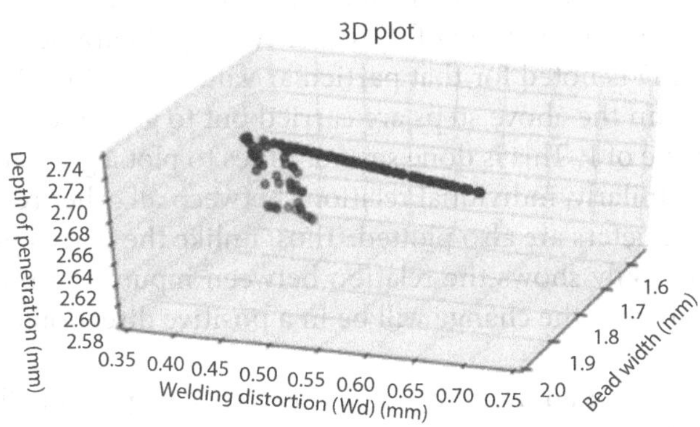


**Figure 20.4** Variation of bead thickness, welding irregularity, and penetration depth with process variables.

Now, the first step is to identify the impact of particular input parameters on particular output parameters and also the combined effects of input parameters on particular output parameters. For this purpose, effect Pareto graphs were plotted. A Pareto graph shows the effects of different inputs on the output, thus giving an idea of the weightage of different input parameters on the output. A Pareto graph shows how much the value of a particular output changes with a unit change in the value of a particular input. Thus, it determines which parameters are significant for a particular output and which parameters are insignificant. The individual effects of power, welding speed, and focal position; the combined effects of power with welding speed, welding speed, and focal position; and the effect of squares of power, welding speed, and focal position were plotted separately for PD, BT, and WI.

Separate Pareto charts are given in Figure 20.5(a to c) below.

From the above charts, it was concluded that power had the most significant effect in determining the WI, followed by focal position. Similarly, welding speed and laser focal point contribute heavily to the determination of BT and PD, respectively. The significance of combined effects, two input parameters, or the effects of squares of input parameters can also be seen in the Pareto charts shown above. The Pareto chart tells us how much output is expected to change with a unit change in a particular input, but it does not tell whether the change will be positive or negative. So now, to see how exactly the output changes with the change in input parameters, factorial plots and contour plots were plotted. A factorial plot or main effects plot shows the relationship between one response and one individual input parameter at a time.

For example, to plot a PD vs. P factorial plot, P is kept constant, and PD is noted at every combination of the other two input parameters, and then the mean of PD is noted for that particular value of P. Now P is increased a little and again the above steps are carried out to get a mean PD for that particular value of P. This is done several times to plot a graph of the mean of PD vs. P. Similarly, individual relations between all other responses and the input parameters are also plotted. Thus, unlike the Pareto plot, the factorial plot not only shows the relation between inputs and responses but also shows whether the change will be in a positive direction or a negative direction.

Factorial plots of all individual responses with respect to all individual input parameters are shown in Figure 20.6(a to c) below.

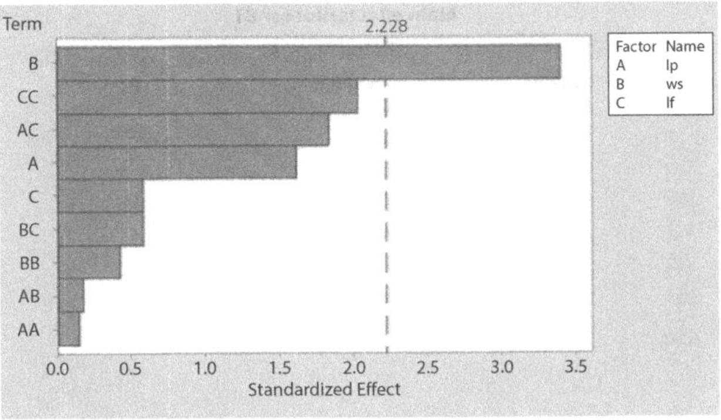


(**a**) Pareto chart for effects on bead thickness

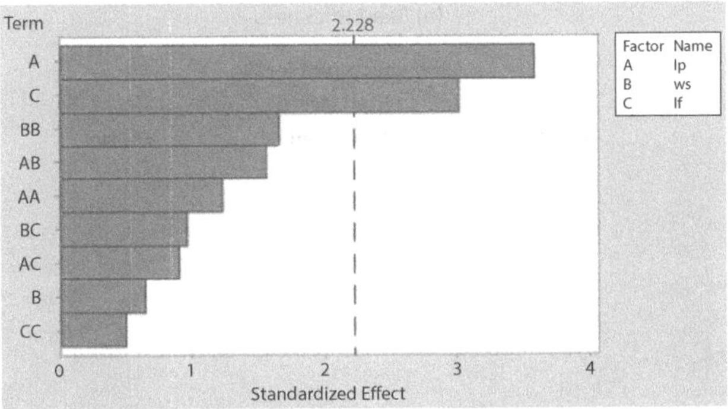


(**b**) Pareto chart for effects on welding irregularity

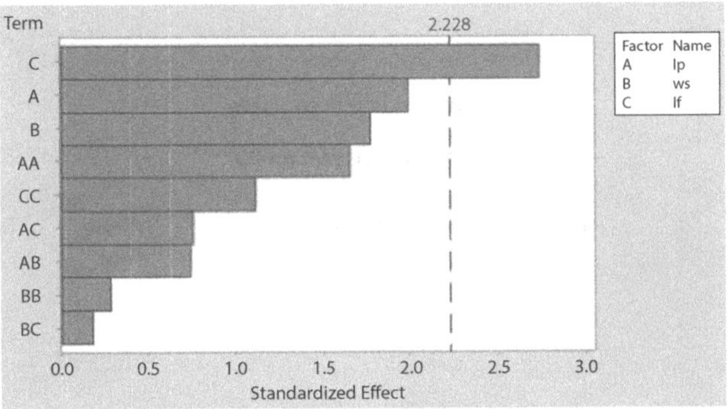


(**c**) Pareto chart for effects on penetration depth

**Figure 20.5** (a) Pareto chart for effects on bead thickness, (b) Pareto chart for effects on welding irregularity, (c) Pareto chart for effects on penetration depth.

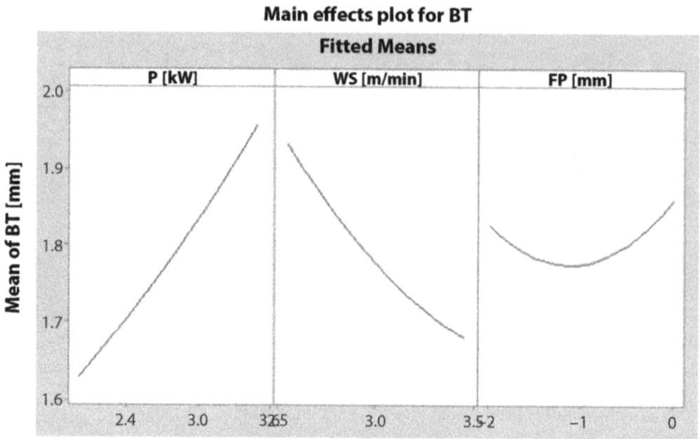


(**a**) Bead thickness

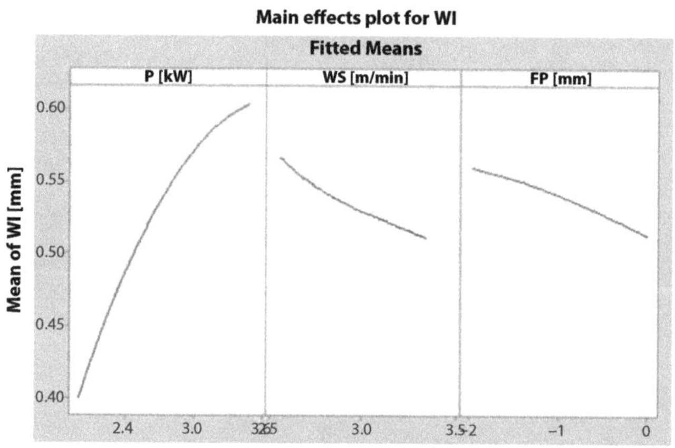


(**b**) Welding irregularity

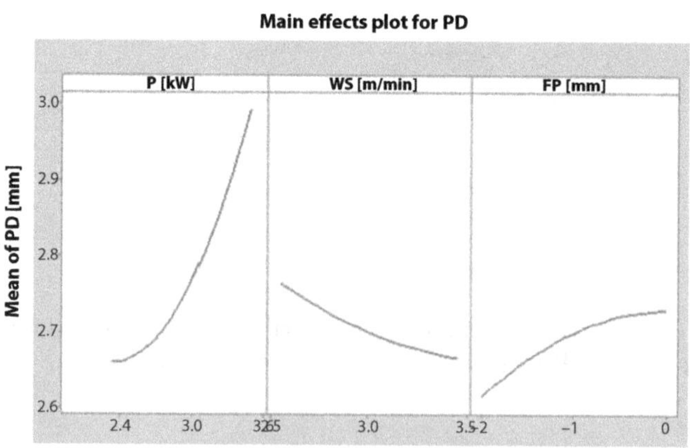


(**c**) Penetration depth

**Figure 20.6** (a) Bead thickness, (b) Welding irregularity, (c) Penetration depth.

After the Pareto and factorial plots, next come the response surfaces and surface plots. A typical graph shows the relationship between only two parameters (one input and one output). However, a surface plot gives a three-dimensional plot of the relation between three parameters, thus forming a 3D surface. Different surface plots were plotted for PD, BT, and WI, showing the relationship between two different input parameters while keeping the third input parameter fixed at certain values.

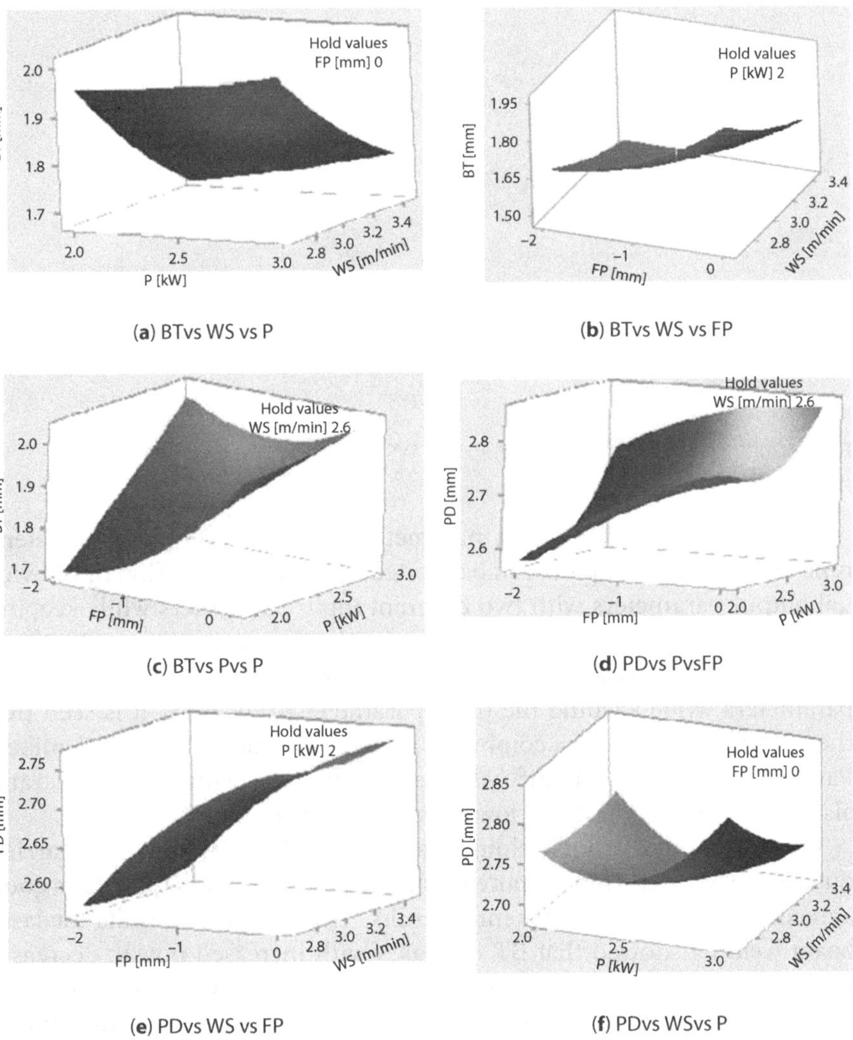


**Figure 20.7** (a) BT vs. WS vs. P, (b) BT vs. WS vs. FP, (c) BT vs. P vs. P, (d) PD vs. P vs. FP, (e) PD vs. WS vs. FP, (f) PD vs. WS vs. P.

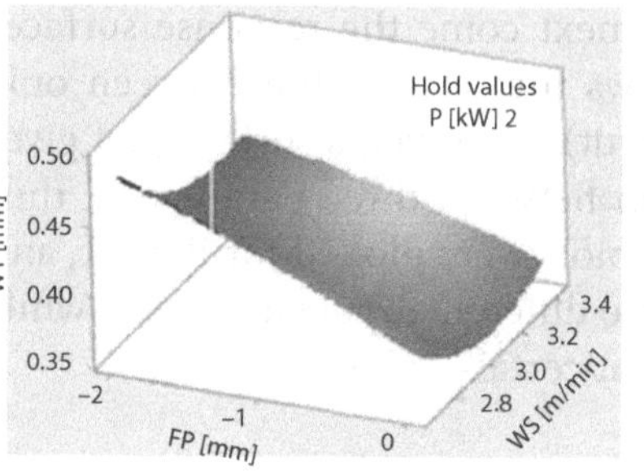


(**g**) WIvs WSvsFP

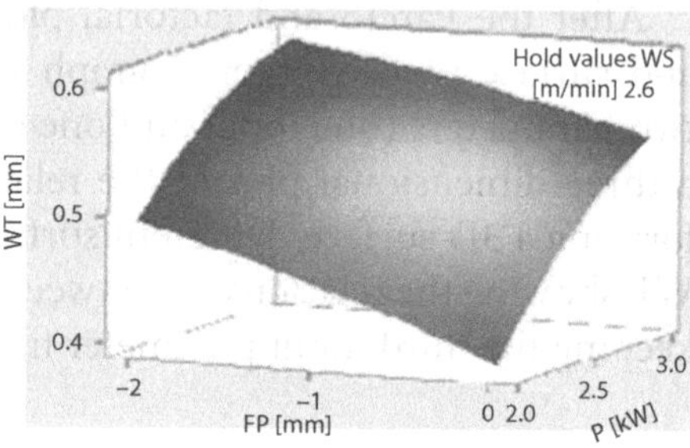


(**h**) PDvs WS vs P

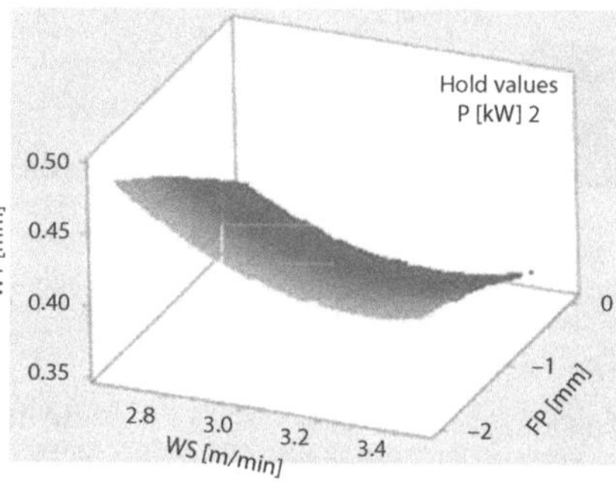


(**i**) WIvs Pvs FP

**Figure 20.7** (g) WI vs. WS vs. FP, (h) PD vs. WS vs. P, (i) WI vs. P vs. FP.

Since there are three input parameters and three output parameters, nine different surface plots can be plotted, showing the relation of individual output parameters with two different input parameters while keeping the third parameter static at different levels separately [Figure 20.7(a to i)]. Three output parameters show relationship with the other two input parameters while keeping the third parameter fixed. Here, it is seen that the output is affected by a combined change in input parameters. Unlike a Pareto chart, one may identify whether the effect is positive or negative and also identify the highest and lowest values of the outputs.

Optimization of fusion welding, namely, laser beam welding was carried out using NSGA-II. The computed results for PD, BT, and WI showed a good agreement with the experimental results. The computed results for laser beam welding showed that BT increases with increased power, decreases with welding speed, and, though minutely, increases with focal position. For WI, trends show an increase, decrease, and decrease for gradual increments of power, welding speed, and focal position, respectively. For PD, trends show an increase, decrease, and increase for gradual increments of power, welding speed, and focal position, respectively. Table 20.3 presents the optimization results for the PD, BT, and WI at various points, along with their relative errors compared to the experimental results. At point 1, the optimum PD is 2719 μm, with an error of 2.81%. The optimal BT is 1882 μm, with a relative error of −1.20%. The optimal WI is 578 μm, with a relative error of 0.52%. These errors can be attributed to factors such as measurement errors, material in-homogeneity, and process variations. At validation points 2 and 3, the relative errors for BT, PD, and WI are −1.89, 0.41, and 1.01% and −2.53, 0.58, and −19.11%, respectively. The results from NSGA-II demonstrate a trade-off between the process variables, with no inferior solutions. In conclusion, the preferred optimum for one objective function comes at the cost of the two others. The results obtained from NSGA-II are confirmed with the corresponding experimental measurements from the selected independent literature [10].

## 20.4 Conclusion

During the optimization process, the main effects of various process parameters on welding quality were examined, and the method's accuracy was confirmed through experiments. The method successfully generated the anticipated welding results and offered credible laser beam welding guidance.

The following conclusions can be drawn from the above investigation:

- The Pareto frontier analysis of laser beam welding revealed that no solution is deemed inferior to another. This is due to a trade-off existing between the parameters of BT, PD, and weld width (WI). The other two objectives must be compromised in order to reach the optimum for one objective function.
- The Pareto chart for effects on objective variables showed that power had the most significant effect in determining the WI, followed by laser focal position. Similarly, welding speed and focal position contributed heavily to the determination of BT and PD, respectively.
- BT increases with an increase in power, decreases with welding speed, and increases minutely with focal position.
- For WI, objective variable trends show an increase, decrease, and decrease for the gradual increment of power, welding speed, and focal position, respectively.

- PD increases for a gradual increment of power, welding speed, and focal position, respectively. But it decreases with an increase in welding speed.

Future research in laser beam welding may focus on optimizing other process parameters, such as beam diameter and material properties. To minimize heat-affected zones and residual stresses, MOO can be investigated. To accurately simulate laser-material interaction, advanced modeling techniques should be refined. Material-specific investigations can be used to examine the behavior of dissimilar materials and devise techniques for producing high-quality welds. In-process monitoring and control, coupled with application-specific studies, can improve the quality and productivity of welded joints. These efforts will advance laser beam welding, allowing for enhanced welds in a variety of industries while taking into account specific requirements and problems.

### Declaration of Competing Interest

The authors of this paper do not have any known financial or personal interests that could have influenced its content.

## References

1. Hecht, J., A short history of laser development. *Appl. Opt.*, *49*, 25, F99–F122, 2010, https://opg.optica.org/ao/abstract.cfm?URI=ao-49-25-F99.
2. Herrmann, T.R., Liatsikos, E.N., Nagele, U., Traxer, O., Merseburger, A.S., EAU Guidelines Panel on Lasers, T., EAU guidelines on laser technologies. *Eur. Urol.*, *61*, 4, 783–795, 2012, https://doi.org/10.1016/j. eururo.2012.01.010.
3. Martukanitz, R.P., A critical review of laser beam welding, in: *Critical Review: Industrial Lasers and Applications*, vol. *5706*, pp. 11–24, 2005, https://doi.org/10.1117/12.601655.
4. Sokolov, M. and Salminen, A., Improving laser beam welding efficiency. *Engineering*, *6*, 09, 559, 2014, http://dx.doi.org/10.4236/eng.2014.69057.
5. Saravanan, S., Raghukandan, K., Sivagurumanikandan, N., Pulsed Nd: YAG laser welding and subsequent post-weld heat treatment on super duplex stainless steel. *J. Manuf. Processes*, *25*, 284–289, 2017, https://doi.org/10.1016/j.jmapro.2016.12.015.
6. Wang, W., Xiao, Y., Wu, X., Zhang, J., Optimization of laser-assisted glass frit bonding process by response surface methodology. *Opt. Laser Technol.*, *77*, 111–115, 2016, https://doi.org/10.1016/j.optlastec.2015.09.007.

7. Deb, K., Agrawal, S., Pratap, A., Meyarivan, T., A fast elitist non-dominated sorting genetic algorithm for multi-objective optimization: NSGA-II, in: *Parallel Problem Solving from Nature PPSN VI: 6th International Conference Paris, France, September 18–20, 2000 Proceedings 6*, Springer, Berlin Heidelberg, pp. 849–858, 2000, https://doi.org/10.1109/4235.996017.
8. Verma, S., Pant, M., Snasel, V., A comprehensive review on NSGA-II for multi-objective combinatorial optimization problems. *IEEE Access*, *9*, 57757–57791, 2021, https://doi.org/10.1109/ACCESS.2021.3070634.
9. Jiang, P., Wang, C., Zhou, Q., Shao, X., Shu, L., Li, X., Optimization of laser welding process parameters of stainless steel 316L using FEM, Kriging and NSGA-II. *Adv. Eng. Softw.*, *99*, 147–160, 2016, https://doi. org/10.1016/j. advengsoft.2016.06.006.
10. Yang, Y., Gao, Z., Cao, L., Identifying optimal process parameters in deep penetration laser welding by adopting Hierarchical-Kriging model. *Infrared Phys. Technol.*, *92*, 443–453, 2018, https://doi.org/10.1016/j. infrared.2018.07.006.
11. Katoch, S., Chauhan, S.S., Kumar, V., A review on genetic algorithm: past, present, and future. *Multimedia Tools Appl.*, *80*, 8091–8126, 2021, https://doi. org/10.1007/s11042-020-10139-6.
12. Mirjalili, S. and Mirjalili, S., Genetic algorithm, in: *Evolutionary Algorithms and Neural Networks: Theory and Applications*, pp. 43–55, 2019, https://doi. org/10.1007/978-3-319-93025-1_4.
13. Mendoza, F., Bernal-Agustin, J.L., Domínguez-Navarro, J.A., NSGA and SPEA applied to multiobjective design of power distribution systems. *IEEE Trans. Power Syst.*, *21*, 4, 1938–1945, 2006, https://doi. org/10.1109/ TPWRS.2006.882469.
14. Liu, H.L., Chen, L., Deb, K., Goodman, E.D., Investigating the effect of imbalance between convergence and diversity in evolutionary multiobjective algorithms. *IEEE Trans. Evol. Comput.*, *21*, 3, 408–425, 2016, https://doi. org/10.1109/TEVC.2016.2606577.

# Index

Printed in the USA/Agawam, MA
September 18, 2026

910661.122